Fine Pitch Surface Mount Technology

Quality, Design, and Manufacturing Techniques

Phil P. Marcoux

PPM Associates, Inc.
Sunnyvale, CA

VNR VAN NOSTRAND REINHOLD
New York

Library of Congress Catalog Card Number 92-3241
ISBN 0-422-00862-7

Printed in the United States of America.

Van Nostrand Reinhold
115 Fifth Avenue
New York, New York 10003

Chapman and Hall
2-6 Boundary Row
London, SE1 8HN, England

Thomas Nelson Australia
102 Dodds Street
South Melbourne 3205
Victoria, Australia

Nelson Canada
1120 Birchmount Road
Scarborough, Ontario MIK 5G4, Canada

16 15 14 13 12 11 10 9 8 7 6 5 4 3 2 1

Library of Congress Cataloging-in-Publication Data

Marcoux, Phil P.
Fine pitch surface mount technology : quality, design, and manufacturing techniques Phil P. Marcoux.
p. cm. (Electrical engineering series)
ISBN 0-442-00862-7
1. Integrated circuits—Design and construction. I. Title.
II. Series.
TK7874.M259 1992
621.381'5—dc20 92-3241
CIP

Dedication

This text is dedicated to my father, Phil J., and my mother, Cecelia, for all of their sacrifices that enabled me to acquire the education and incentive to accomplish this text, the founding of my three companies, and so much more.

This text is also dedicated to my wife, Ann for her support and understanding. And to my children Danielle and Paul, so they can build from here.

And finally this text is dedicated to each reader in the hope that it will provide a few seeds of wisdom from which high quality electronic assemblies will grow.

Contents

Preface

Fine pitch high lead count integrated circuit packages represent a dramatic change from the conventional methods of assembling electronic components to a printed interconnect circuit board. To some, these FPT packages appear to be an extension of the assembly technology called surface mount or SMT. Many of us who have spent a significant amount of time developing the process and design techniques for these fine pitch packages have concluded that these techniques go beyond those commonly useed for SMT. In 1987 the present author, convinced of the uniqueness of the assembly and design demands of these packages, chaired a joint committee where the members agreed to use fine pitch technology (FPT) as the defining term for these demands. The committee was unique in several ways, one being that it was the first time three U.S. standards organizations, the IPC (Lincolnwood, IL), the EIA (Washington, D.C.), and the ASTM (Philadelphia), came together to create standards before a technology was in high demand.

The term fine pitch technology and its acronym FPT have since become widely accepted in the electronics industry. The knowledge of the terms and demands of FPT currently exceed the usage of FPT packaged components, but this is changing rapidly because of the size, performance, and cost savings of FPT.

I have resisted several past invitations to write other technical texts. However, I feel there are important advantages and significant difficulties to be encountered with FPT. My motivation in writing this text is to explain the advantages and obstacles, and to offer some of the possible solutions to users of FPT so they can enjoy the benefits sooner.

In this text, I present two strategies for the assembly of the FPT packages to the substrate. One strategy is to incorporate the assembly directly in with the SMT process. The other is to treat FPT as a secondary operation using a simultaneous placement and reflow operation. I also present a strategy for developing design guidelines to optimize for what I call "design for ability" which incorporates manufacturability, quality, reliability, and repairability into the design of the product. These strategies will be useful to all users regardless of whether they are in low or high volume electronic assembly.

FPT offers many of the benefits sought, but not achieved, by users of SMT. With SMT, product cost savings and reliability improvements have been elusive. With proper planning, design, and hard work, many cost and reliability improvements are achieved with FPT. I hope this text also helps the reader achieve these and other benefits from FPT.

Phil P. Marcoux

Acknowledgments

This text is only possible due to the support, experience and intellect of several people. My appreciation and thanks is extended to the following individuals for their technical input or review:

Joe Fjelstad, Elf Technologies
Harvey Miller, Kirk-Miller Associates
Dieter Bergman, IPC
Norbert Socolowski, Alpha Metals
Richard Vaughn, Sony Corp.
Hideo Yokoyama, Sony Corp.
Hiroshi Takamizama, Sony Corp.
Toshihumi Nakamura, Sony Corp.
Douglas Peck, SRT
Harold Hyman, SRT
Phil Zarrow, Vitronics
Rob Rowland, Teradyne Inc.
Jim Bell, EPE.
Takayoshi Komatsu, Hitachi AIC
Hiroyoshi Yokoyama, Hitachi AIC
Hakan Sodersved, Expira
Werner Engelmaier, Engelmaier Associates
Gail Stout, Lake Publishing
Dr. Robert Parden, University of Santa Clara
Dr. Charles Wosjlaw, San Jose State University
Foster Gray, Texas Instruments
Bill Coleman, Photo-Stencil, Inc.
Bob Jillings, Stencil Manufacturing Technologies, Inc.

and especially to Danielle Blayzor of PPM Associates for composing the drawings, typing the text, and trying to keep the author focused and organized.

1

Introduction to Fine Pitch Technology (FPT)

1.0 INTRODUCTION

The family of fine pitch packages and the accompanying assembly technology evolved to satisfy the ongoing trends and demands of the consumers of electronic products. These trends and demands can be summarized as follows:

1. As silicon complexity grows, the number of leads required for input and output grow (Rent's Rule).
2. The operating speeds and frequencies are increasing with functional capability, thus limiting the allowable parasitics in a package.
3. Consumers want more functions in the same or smaller size as a product's predecessor.

Fine pitch technology provides another resting point in the packaging and assembly evolution of electronic products. The prior and future resting points are shown in Figure 1.1; these include through-hole, surface mount (SMT), FPT, and direct chip mounting technology (CMT).

From the 1960s until the mid-1980s through-hole assembly was the prevailing method. Then surface mount (SMT) became the dominant package and assembly methodology, offering significant size savings but mixed cost savings. Fine pitch is becoming the major method for the 1990s, as it provides size and cost savings for many users. Fine pitch packages are an essential link in the evolutionary chain. However, the revolutionary proliferation of different fine pitch package types that has occurred will not hasten their acceptance.

Beyond FPT, direct chip mounting (CMT) techniques will emerge as the dominant technology. Early CMT assemblies are taking form as multichip

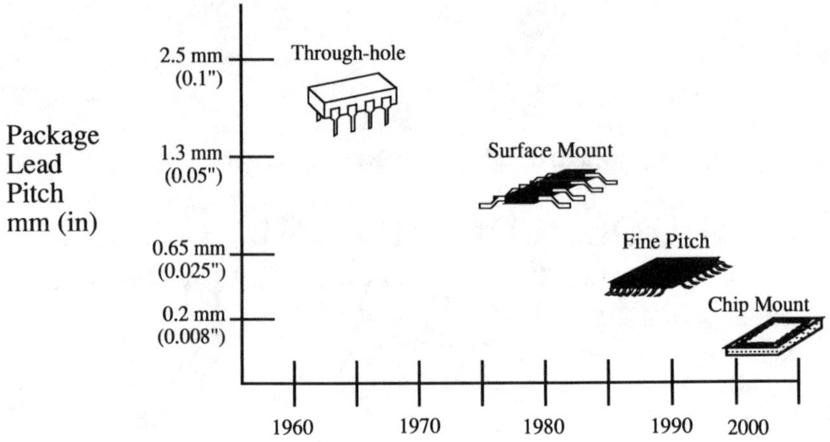

FIGURE 1.1. Electronic packaging and assembly continues to evolve into denser and potentially less costly alternatives.

modules and simple chip on board (COB) subassemblies. CMT will replace FPT when reliable bare integrated circuits are available and a CMT infrastructure develops.

The increase in the number of leads for a given increase in integrated circuit functionality is predicted by an algorithm known as Rent's rule. Rent's rule states that the number of package leads needed is a function of the number of logic circuits in the package. It can be expressed as:

$$\overline{\text{Circuits}} = \left(\frac{N}{K}\right)^{n}$$

where

$\overline{\text{circuits}}$ = the average number of circuits supported by N
N = number of signal terminals (lead count)
K = a constant, the value of which depends on the ability to share signal lines; a typical value for high-performance applications is $K = 2.5$.
n = a constant, in the range of 1.5 to 3.0.

Transposing this general expression shows:

$$N = K^{n}\,\overline{\text{circuits}}$$

As shown in Figure 1.2 the microprocessor and gate array functions closely follow Rent's rule. This figure also shows the trend in memory devices, which also

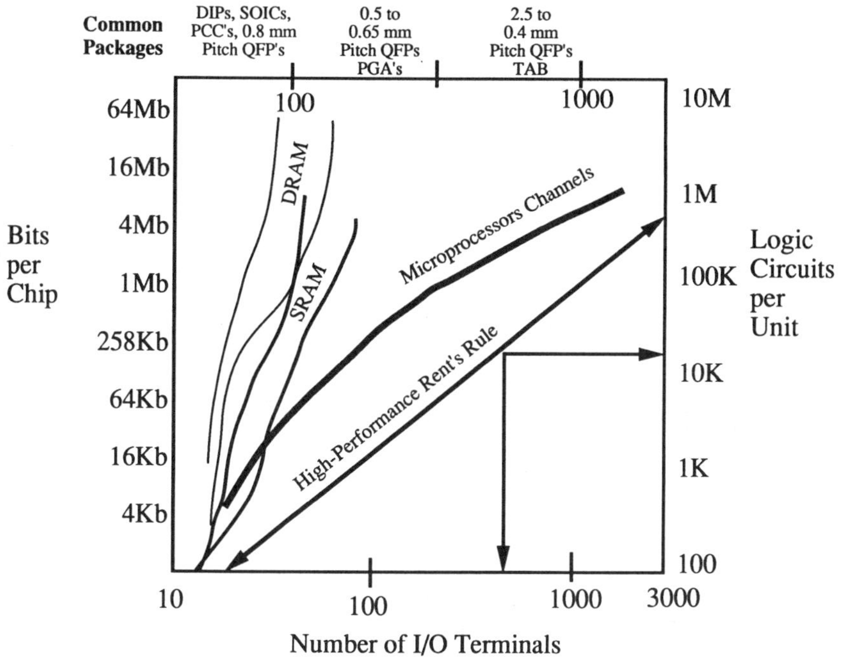

FIGURE 1.2. As the number of logic functions in a circuit increase so does the number of I/O leads required to interconnect these functions.

generally follow this rule but require fewer leads because memory devices can share signal lines.

The ability to put more transistors on a tiny piece on silicon has grown tremendously in the past two decades. This growth can be appreciated by examining Figure 1.3, which shows the number of transistors contained in the various commercial microprocessors. Just as the number of transistors has increased with each microprocessor generation, so have operating speed and clock frequency. The speed increases are customer driven demands. The faster the operating speed of the microprocessor, the more operations it can accomplish in a given period of time. For integer operating microprocessors this is expressed in millions of instructions per second (MIPs) and for floating point processors it is expressed in floating operations per second (FLOPs). The limiting factors on a product's operating speed are on-chip gate delays, package delays, and printed circuit board delays. Package delays are of concern in this section. Printed circuit board delays will be covered in Chapter 4. On-chip delays are beyond the scope of this text.

All three delay categories (on-chip, on-package, and on-board) are affected directly by the interconnection distance. On a chip this is the aluminum or silicon nitride distance. In a package it is the wire bond length or the TAB lead length from

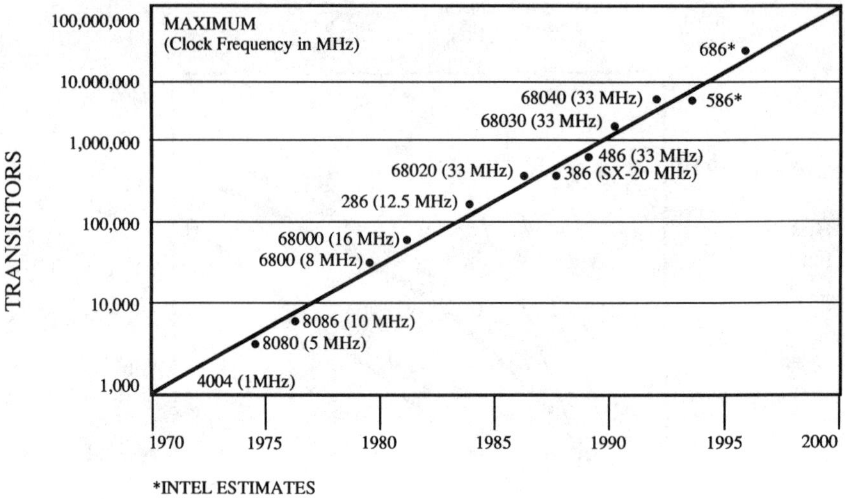

FIGURE 1.3. The number of transistors that can be fabricated onto a small area of silicon has grown tremendously in the past two decades. Correspondingly, so has the clock frequencies. (Source: Intel Corporation)

the chip to the solder joint. On the printed circuit board it is the copper laminate trace distance. The delay as a function of these interconnection distances is shown in Figure 1.4. This figure plots signal delay as a percentage of the total delay versus the interconnection distance. For example, if the total interconnection distance is 25 mm, then from Figure 1.4 the chip accounts for approximately 20% of the total signal delay, the package contributes 30%, and the board trace adds 50%.

The personal computer provides an excellent example of the trend toward more functions in the same or smaller size. The main selling points of a laptop personal computer are weight, size, storage capacity, and functional capability. Figure 1.5 shows the evolution in the personal computer toward fewer and fewer components with more and more memory storage and functional capability. By 1996, for example, it is estimated that a laptop computer will contain all the necessary microprocessor and necessary logic in one chip and house eight megabytes of memory in two chips.

1.1 WHAT IS FPT?

Fine pitch technology (FPT) is the term used to describe the printed circuit board assembly technology for a family of integrated circuit packages distinguished by the fine spacing between their small leads. Fine pitch packages are defined as those with lead spacings of 0.65 mm (0.026 inches) or less. The finest defined pitch is 0.15 mm center to center. Since the main role of these packages is to house large

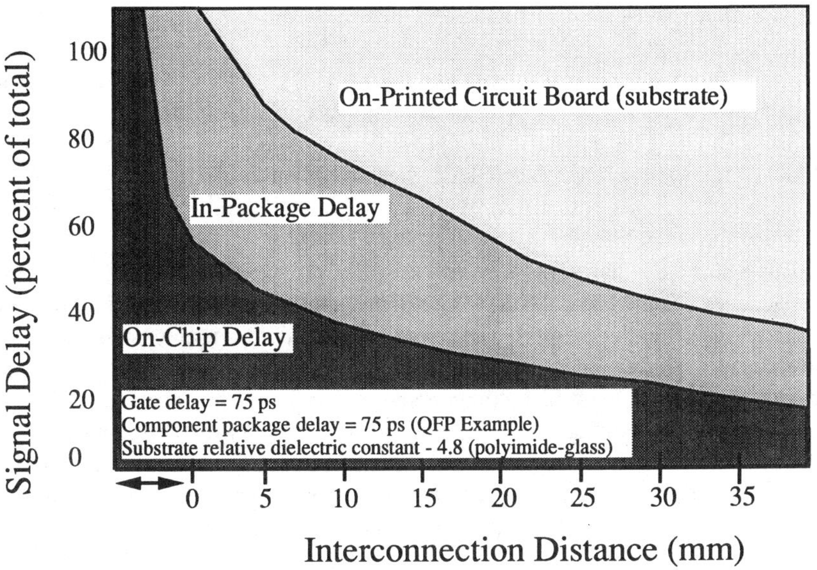

FIGURE 1.4. Signal delays are a function of the chip, package and board size. Shorter interconnects on the board and in the package result is most of the delays to occur on the chip. On chip delays, depending on the IC technology are very small. (Source: Electronic Packaging and Production Magazine, August, 1990)

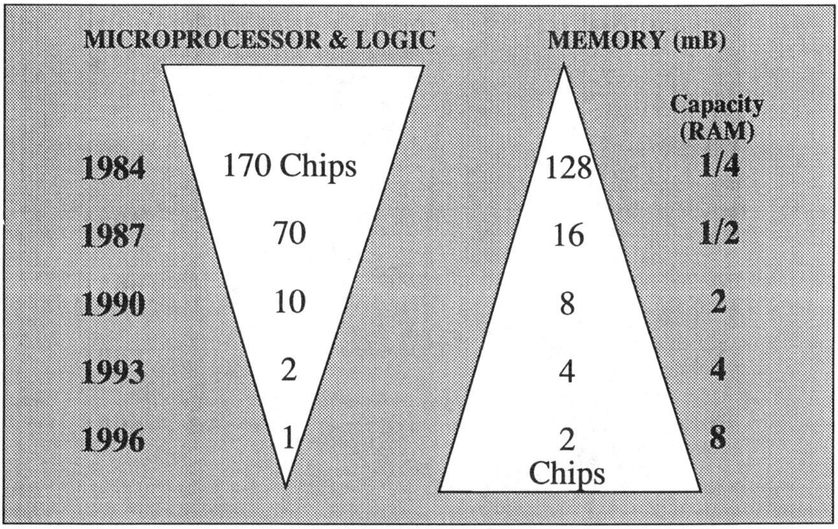

FIGURE 1.5. As the ability to put more transistors and functions on a single silicon chip has expanded, certain functions, such as personal computers, require fewer different components to make a complete product. As illustrated here, a laptop computer contains a fraction of the number of components required in prior years. (Source: Intel Corporation)

complex integrated circuits they generally have high lead counts. Lead counts range from 40 to 1256 leads, with the average usage between 100 and 200 leads. Included in the FPT package family are encapsulated tape automated bonded (TAB) packages, unencapsulated TAB, very small outline (VSOP) and thin small outline (TSOP) packages, and quad flat packages (QFP and PQFP). Examples of these are shown in Figure 1.6.

These packages represent the fastest growing and most challenging category of IC packages available. They offer significant component cost and interconnect area savings over their through-hole and surface mount counterparts, provided they can be reliably interconnected with other packages. Fine pitch packages provide as much as a 25-50% component cost savings over older types of packages, such as dual-in-line (DIP) and pin grid array (PGA). FPT packages usually house the most expensive and largest components in the circuit, so product savings can be very significant. In some cases, the FPT components represent as much as 50% of an entire product's cost (not including chassis and hardware).

FIGURE 1.6. Fine pitch packages have lead pitches of 0.65 mm or less. They are available in a variety of formats. (Photo: Courtesy of Universal Instruments)

FPT packages are generally square or rectangular in shape. The connection of the integrated circuit to the external package lead may use a small wire bonded to the IC and the lead, or the lead may be attached directly to the IC. The IC and lead assembly is usually encapsulated in a plastic resin or ceramic to protect it from the environment. However, many package styles are much thinner than the older packages, and TAB packages completely omit the encapsulation.

The assembly of some of the fine pitch packages onto the interconnecting substrate, commonly called the printed circuit board, uses many of the same techniques as the assembly method called surface mount technology, or SMT. The printed circuit board contains the sites, called *lands*, where the package leads are soldered and interconnect with other components to form a functioning circuit. Sufficient volume of a metal (called solder) that has a low melting temperature and is electrically conductive is preapplied to this site so the package lead can be soldered to the printed circuit board land. The package leads are placed on this alloy and joined using either soldering or wielding. Soldering is the most common method using various heat sources, such as convection, laser, or thermode. Welding is not viable for most FPT applications because of the potential damage from high heat or electrical energy, but may be viable for some of the tabbed packages.

The assembly and design of these packages demand an attention to detail beyond what is required in the older electronic assembly methods; this is especially true when FPT is mixed with the older methods, surface mount and through-hole. Such a mixed approach is a common result of limited component availability. Greater placement accuracy, greater solder volume control, tighter design tolerances, and militant process controls are required to maintain a capable high yielding sole or mixed FPT process. A typical printed circuit board assembly is shown in Figure 1.7, the large rectangular component is an example of a quad flat pak style 0.65 mm lead spaced fine pitch package.

1.2 THE BENEFITS OF FINE PITCH TECHNOLOGY

Size Advantages

The most obvious benefit of the fine pitch packages is the smaller package size, the high number of leads, and the resulting increased circuit density. A product using fine pitch packages either is smaller than or holds more circuit functions in the same size as its earlier product generation. FPT packages allow roughly a fourfold increase in circuit density over SMT packages. FPT packages offer a twelvefold savings over the conventional through-hole (TH) devices, with the exception of the pin grid array (PGA) package where it offers a two to fivefold savings. The pin grid array offers the same package area to lead count ratio as an

FIGURE 1.7. Most printed circuit assemblies today use a mixture of component package types. This board shows Through-hole, Surface mount and Fine Pitch package types. These different package types necessitate a complex, carefully controlled assembly process. (Photo courtesy of Ampro, Inc.)

FPT package for lead counts up to 250. The relative area comparison is illustrated in Figure 1.8. FPT package lead capacities currently range from 40 to 1256 leads on lead centers ranging from 0.65 mm (0.026″) down to 0.15 mm (0.006″). Beyond this, the only packaging options denser than FPT are the direct chip mounting techniques of wire-bonded and flip-chip.

Smaller package area resulting in denser circuits is only one size savings benefit. Fine pitch packaging also offers size advantages to the IC chip itself. For many commodity integrated circuits reducing the size of the IC chip translates into a competitive advantage. Smaller chips generate more die per silicon wafer and higher sales dollars per wafer. To provide enough space for the conventional 0.05 mm (0.002″) diameter wire bonded leads, the IC's bonding pads must be at least 0.15 mm (0.006″) square on 0.1 mm (0.004″) center to center spacings. Fine pitch inner lead interconnect options, such as TAB lead attachment, can utilize 0.03 mm (0.001″) wire which enables 0.15 mm (0.006″) pads on 0.10 mm

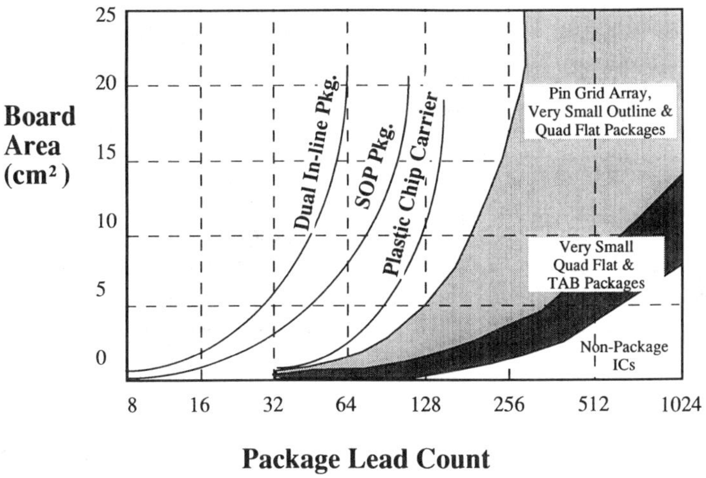

FIGURE 1.8. There are many different package options for integrated circuits. Fine pitch and "non-packages" integrated circuits offer the highest number of potential leads in the smallest area. (Source: Task Technologies)

(0.004") spacings. This savings is illustrated in Figure 1.9. The ability to use the thinner wire is, however, dependent on the thermal requirements of the IC (see Section 2.7).

Table 1.1 shows the impact smaller IC chip size has on the number of good chips and sales dollars per wafer. In this example, the processed wafer cost and test yield are assumed to be the same regardless of the chip size.

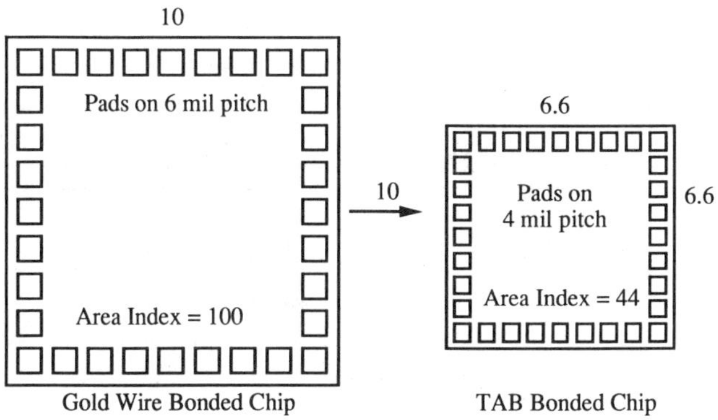

FIGURE 1.9. TAB and thin wire bonding techniques allow the bonding pads and the chip to be smaller, which results in cost savings. (Source: Olin Interconnect)

TABLE 1.1 Smaller Size Integrated Circuits Result in Cost Savings.

Pad Size/ Pad Pitch	Chip Size	Good Chips per Wafer	Wafer Cost	Chip Cost	% Savings
.01"/.006"	.23x.23"	509	$500	$0.98	--
.008"/.005"	.18x.18"	725	$500	$0.69	30%
.006"/.004"	.15x.15"	1104	$500	$0.45	54%

Cost Advantages

Cost is usually of prime importance in all products. To be cost viable, fine pitch packages must offer cost reductions on a per package, or per component, or some other basis of interest to the user. Table 1.2 shows the relative packaging costs including the cost of the package, the IC, and the assembly labor. The significant cost advantages of these fine pitch packages gives FPT a clearer decision opportunity for designers and purchasing people looking for high lead count solutions. This is different from the SMT cost situation, where users find that most SMT packaged IC components lack sufficient cost savings since most of the low lead count ICs are the same or a higher price than the equivalent through-hole packaged component.

Package Cost Advantages

Reducing the chip size was shown to be one potential cost savings as was shown in Table 1.1. A comparison of packaging costs (Table 1.2) was developed by a TAB

TABLE 1.2 This Table Shows the Packaging Costs which Includes the Cost of the Package, the IC, and the Assembly Labor. (Source: Olin Interconnect)

132 Lead Package	Packaging Cost*
Ceramic PGA	$15.00
Plastic PGA	$10.50
Plastic QFP	$4.50
Tape Auto Bonded	$3.50

*Includes Chip, Package, and Assembly Costs

The TAB version provides the greatest potential cost savings, 77% over the ceramic PGA and 67% over the plastic PGA.

tape supplier for two through-hole and two FPT 132 lead count packages. The through-hole packages are a ceramic and a plastic pin grid array (PGA) package with 132 leads. The fine pitch packages are a 0.65 mm pitch 132 lead quad flat package (QFP) and a 132 lead unencapsulated TAB package. As shown in Table 1.2, the FPT packages are 70-77% less expensive than the PGA packages ($3.50 and $4.50 vs. $10.50 to $15). These package costs include the cost of a typical CMOS gate array integrated circuit plus the package and the package assembly costs.

An example of a PGA package is shown in Figure 1.10. PGAs have discrete pins seated in either a ceramic or plastic base that contains the package leads and a base support for the chip. The pins connect to the chip pad with a wire bond. The QFP and TAB packages use a chemically etched leadframe. The QFP leads may attach directly to the chip or may require a separate wire bond; the TAB lead attaches to the chip.

The TAB package assembly (see Section 2.5) requires the additional step of preapplying a metal bump (gold or tin/lead) to the surface of the IC bonding pads. This is necessary to assure a reliable joint between the IC and TAB leads. Unfortunately, this can be a relatively costly and risky process to perform because of the limited number of sources available. The alternative method of using prebumped tape will, in time, eliminate this problem and possibly reduce the cost of TAB package assembly further. Today, tape automatic bonding is considered a custom option by many IC suppliers. As such, the customer is charged for the initial design and tooling of the tape design. When this is amortized over the number of

FIGURE 1.10. The pin grid array, or PGA, is a through-hole package that provides high pin counts. However, the package area and cost are usually high.

integrated circuits, it can add a significant cost, especially if the purchased volume of ICs is small.

Component Cost Savings

Reduced packaging and chip costs translate into lower component cost to the user. Table 1.3 is a comparative list of retail prices of various surface mount, through-hole, and fine pitch packaged components. These prices assume an annual purchasing volume of 1000 units per year. In the list of FPT prices, the 80386SX is the only component available in fine pitch, in a plastic quad flat (PQFP) style package. In the through-hole column the 80386SX device is available only in a PGA package. The lower price of the 80386 in the fine pitch package (48% lower!) can be understood by examining the packaging cost discussion in the previous section.

Product Cost Savings

Component prices comprise the majority of the typical electronic product's total cost. In most applications the components can account for 60-80% of the product cost. Therefore, any component price savings can provide a significant reduction in product cost, as much as 25-50%. The circuit board containing the components listed in tables 1.3 and 1.4 provides an excellent example to illustrate the savings that can be achieved by changing the packages and mixing the assembly technologies. The SMT/FPT version of this product appears similar to that shown in Figure 1.11.

TABLE 1.3 This Table Shows the Relative Cost of Various Components Such as Those that Might be Used in a Small Computer. As Shown Most of the Surface Mount Parts are Compatibly Priced as the Through-Hole Equivalents. Significant Cost Savings are Experienced with the High Lead Microprocessor—The 80386sx which is in a Fine Pitch Package.

Component	SMT Price $	FPT Price $	Through-Hole Price $
5% Resistor	0.017	---	0.011
1% Resistor	0.027	---	0 .015
.1 µf Capacitor	0.061	---	0 .063
1N4148 Diode	0.04 (MELF)	---	0.04
74LS00	0.10	---	0 .09
74S74	0.20	---	0 .20
27C512-250	4.45	---	4.10
2764	4.25	---	3.00
27128	4.50	---	3.50
2N3904	0.069	---	0.13
80386SX	---	103.00 (PQFP)	198.00 (PGA)

FIGURE 1.11. Photo of a predominately SMT and FPT computer-printed circuit assembly.

In Table 1.4, the first column lists the major components on a possible circuit board. The second column lists the prices and the resultant board area (last line) of the packages used on the SMT/FPT version of the circuit board. This version of the product may also include a small number of through-hole components which are unavailable or unreliable in SMT or FPT packages. The mixed SMT/FPT/TH version is considered an SMT/FPT version when the majority of parts are surface mounted. This mixed mode is very common given the lack of availability and questionable process compatibility of a small number of the components and the need for through-hole mounted connectors. The consequence of this is the need to solder the components using two or possibly three different soldering methods, the options being reflow, thermode, and wave soldering.

The third column lists the prices and board area for the completely through-hole version.

The total component cost in table 1.4 is calculated by multiplying the unit prices of the components times the total quantity of each component type. Then the extended component prices are summed to give the total component cost of $126.88 for the SMT/FPT version of the assembly and $219.04 for the through-hole version. Omitting the 80386SX device shows the interesting comparison of "SMT only" components costing $23.88 vs the through-hole components costing $21.04, a 13% increase. Without the large difference between the FPT and through-

TABLE 1.4 This Table Shows the Total Estimated Product Cost for Manufacturing a Small Computer Board.

Qty	Elements	Total SMT/FPT Cost	Total Through-Hole Cost
1	80386SX	103.00	198.00
2	27128	9.00	7.00
2	27C512-250	8.90	8.20
9	74S74	1.80	1.80
10	74LS00	1.00	0 .90
10	1N4148	0.40	0.40
40	.1µf CAPS	2.44	2.52
20	1% RES.	0.34	0 .22
Total Component/Cost		$126.88	$219.04
Difference		-$92.16 (-42%)	
Product Cost Savings - Case Study - An "Almost" Controller Application			
Board Cost		$9.60	$11.52
Difference		-$1.92	-16.1%
Mfg./test (incl. equip)		$17.40	$15.20
Difference		$2.20	+14.50%
Total Product Cost		$153.88	$245.76
SMT/FPT Savings		$91.88 (-42%)	
Size Savings		19 vs 36 sq. in. (- 47%)	

hole prices of the 80386, the user would have a difficult decision if price savings is the only justifiable reason to choose between surface mount and through hole assembly.

Board area savings further enhance the savings of the SMT/FPT version in this example. Board area is 19 sq. in. for the SMT/FPT and 36 sq. in. for the through-hole. This represents a 47% area savings and a 16% board cost savings. It should be noted that the cost per unit area of the SMT/FPT board ($0.51 per sq. inch) is higher than that of the through-hole board ($0.32 per sq. inch). This is because the precision required for FPT adds additional handling and care in the board fabrication, resulting in higher unit area costs.

The higher assembly cost to assemble the SMT/FPT version ($17.40 vs. $15.20) is due primarily to the extra equipment and care needed to attach the 80386SX in the PQFP package.

Performance Advantages

The shorter lengths and smaller peripheral area of the fine pitch package leads significantly reduce the parasitic "gremlins" that create slow, dirty electrical signals. Reduction of these parasitics by switching to fine pitch packages does much to increase the speed and to maintain the signal integrity compared to the older, larger packaging methods. Figure 1.12 demonstrates the impact on a signal as it traverses the diagonal length of a package. In this example, a wire is bonded to the two farthest leads of the packages shown in Figure 1.13, then a step function voltage signal is applied to one lead and the output monitored on the other end. The output wave form and the time required to settle to within 99% of the final amplitude is recorded.

As shown in Figure 1.12, for a typical 64 lead dual-in-line (DIP) package, the output response can take up to 260 picoseconds to settle. This is due to the high lead to lead capacitance and the wire inductance that the signal encounters as it propagates through the package and wire. The inductance in the wire and the capacitive coupling causes a ringing of the output, with the amplitude dependent

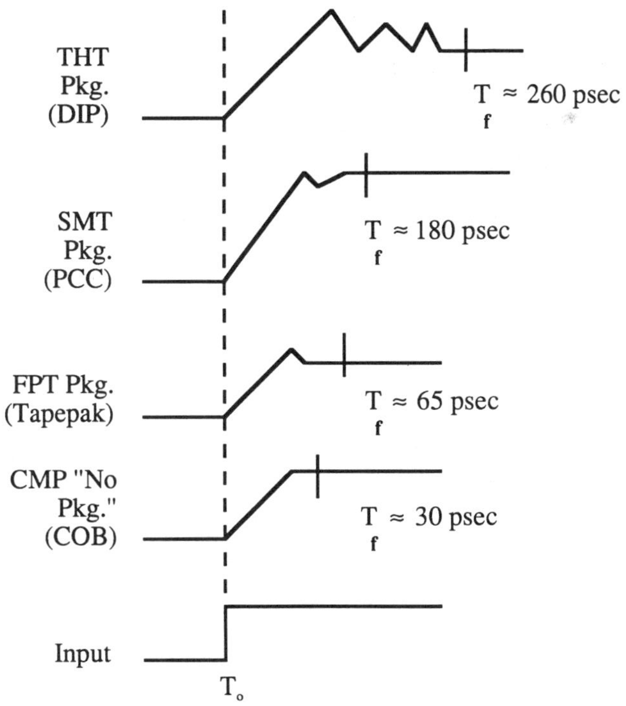

FIGURE 1.12. The shorter the interconnect lengths and the lower the lead peripheral area, and the cleaner the signal propagation. (Courtesy of National Semiconductor)

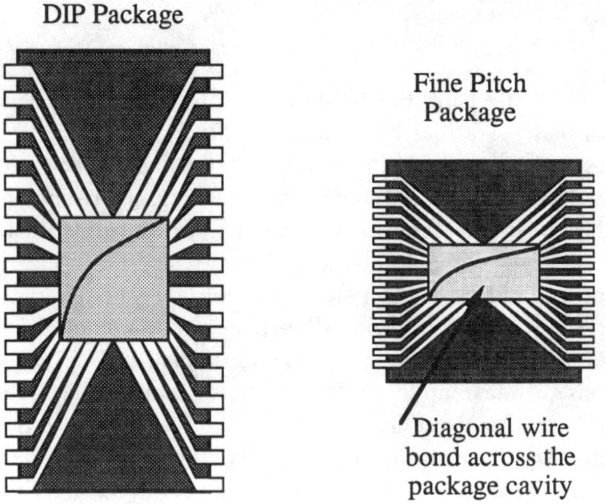

FIGURE 1.13. This figure shows the density of the inner portion of the leads for a typical dual-in-line (though-hole) package and a SOIC (surface mount) package.

on the amount of inductance and parasitic capacitance. The measured values of the parasitics for the DIP and the other packages tested are shown in Table 1.5.

The surface mount plastic chip carrier (PCC) package improved the speed and reduced the peak ringing amplitude of the output signal compared to the DIP. This is because the PCC has less parasitics than the DIP, due to shorter distance between the farthest spaced leads (35 mm vs. 85 mm for the DIP). This reduces the effective inductance from 22 nanoHenries to 4.9 nanoHenries. The lead frame mass used in the PCC is about the same as in the DIP, therefore, the parasitic capacitance (lead to lead) of the PCC lead is about the same as a DIP lead in its shortest length (0.12 pF). This is shown in Table 1.5.

The fine pitch guard-ring quad flat package (a GQFP TapePak©, TapePak© is a registered trademark of National Semiconductor) that was tested contained 0.08 mm (0.003″) thick copper leads on 0.5 mm (0.02″) lead centers. This package yielded a 65 picosecond response time with minimal ringing.

The longer response time of the GQFP compared to the direct mounted IC chip on the board is due to the extra lead length on the GQFP to allow for encapsulation of the chip.

1.3 THE MOTIVATION FOR FINE PITCH TECHNOLOGY

The fundamental motivation for fine pitch IC packages and the FPT assembly technology is to meet the customer demands listed in Section 1.0. To meet these

TABLE 1.5 The Measured Values of the Parasitics for Various Ic Package Configurations. The Smaller, Denser Leads in the Fine Pitch Packages Result in Lower Parasitics and Cleaner Electrical Performance. (Table courtesy of National Semiconductor)

COMPARISION OF FPT, SMT AND TH						
	TH DIP 40L		SMT PCC 44L		FPT TapePak 40L	
Lead Thickness (Mils)	10.0		10.0		2.8	
Lead C-C (Mils)	100		50		20	
Pkg. Length (Mils)	2050		650		350	
Pkg. Width (Mils)	600		650		350	
Pkg. Thickness (Mils)	175		190		72	
Volume Ratio	24.4		9.1		1	
	Long	Short	Long	Short	Long	Short
Lead Length (Inches)	0.99	0.13	0.20	0.15	0.12	0.06
Resistance (mOhm)	125	123	98	98	3.6	2.2
Inductance (nH)	22	3.9	4.6	3.3	2.1	0.8
Capacitance (pF)	0.68	0.12	0.12	0.16	0.04	0.02

demands electronic producers need to find more economical, manufacturable, and testable packages for high lead count integrated circuits such as microprocessors and custom application specific integrated circuits (ASICs) that are being designed into the newest electronic products.

A questionnaire was given to 125 attendees at advanced electronic design and assembly seminars conducted by the author. Their responses to the question, "What is the main motivation for your company's investment in FPT?" were as follows: 60% said they wanted to implement FPT "to achieve size savings," 30% said "to improve product performance (higher speed, lower noise, etc.)," and only 10% responded "to reduce manufacturing cost."

To the question "What is the second most important motivation?" the respondents answered: 50% "for performance improvements," 30% "for size savings," and again a minority or 20% "for manufacturing cost savings"

To explain why manufacturing cost seemed to be of least importance, the respondents indicated that they assumed there would be a product cost savings and that most of that would result from lower cost components in FPT packages as opposed to assembly cost savings. This implies that the respondents aren't overly concerned that fine pitch technology may produce higher assembly cost.

The respondents to this questionnaire represented U.S.-based companies consisting of the following product categories:

20%	Computer peripherals
10%	Small computers
10%	Large high performance computers
25%	Instrumentation
30%	Military
5%	Other

The family of FPT packages follow some easily predictable trends. These include the need for higher lead count, lower cost, smaller size, and higher reliability. As silicon complexity grows, the number of leads to interconnect all of the increased inputs and outputs of the circuit increases. For example, the eight bit microprocessors of the past required less than 64 leads to interconnect all of the processor's functions. A microcomputer offering 32 bit address capability requires at least 132 leads and a 64 bit unit requires approximately 200 leads.

FPT packages fill what appears to be the last package frontier. Beyond FPT, users enter the era of the "no package." This means direct chip mounting of the silicon components using various methods such as solder bumps (as with flip-chip chip attachment) or wire bonds (as with wire bonded IC chips). This direct chip mounting technology is designated as chip mounting technology (CMT) by the IPC standards organization.

Several major obstacles prevent the IC suppliers from leaving the package off the IC for most applications. This situation is expected to continue for a number of years. The most commonly mentioned reason for an IC package is to permit testability. Most IC chips can only be partially tested without some protection and leads to allow temperature, environmental, and dynamic testing. Seldom mentioned, but probably a larger obstacle to users, is the lack of a warranty when purchasing un-packaged integrated circuits. Virtually all IC suppliers today will sell only un-packaged IC chips "as is" and without a warranty or, at best, with a limited warranty. This is because the suppliers cannot test and guarantee all of the parameters on their specification sheets if the IC is not packaged. This applies particularly to dynamic operation and testing at hot and cold temperature extremes. As a result many suppliers refuse to specify all but the most basic component and readily guaranteed parameters for unpackaged components.

Studies show that the expected yield of an unpackaged ASIC IC averages 97% compared to the fully packaged and tested IC yield average of 99.5%. The low yield of the unpackaged IC (average defect rate of 30,000 ppm) means that a user can expect an 100% probability of needing to rework when several IC chips are attached to a substrate using CMT. The impact of lower chip yields on product yield is shown in Figure 1.14.

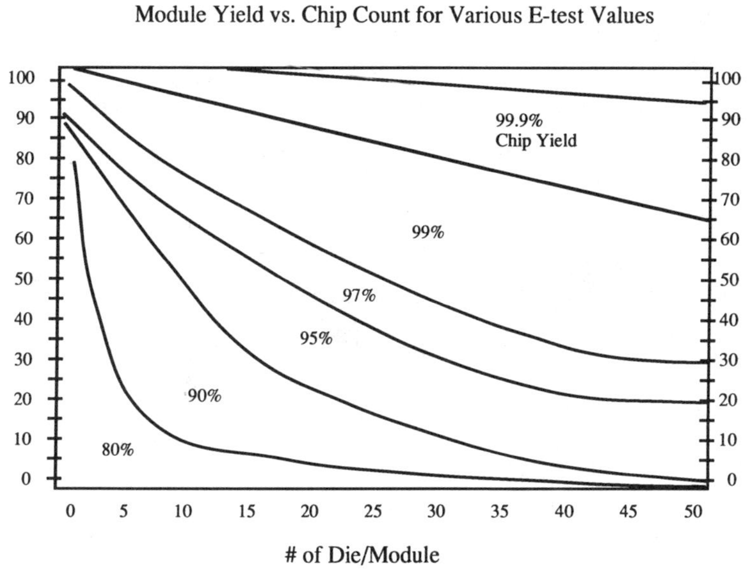

FIGURE 1.14. Packaging an IC prior to board assembly allows the IC to be fully tested. This results in a higher yield after assembly which minimizes rework, inspection and sometimes, cost. (Adapted from "Initiates for High Value Electronics" Workshop, MCC 1991)

1.4 COMPARISON OF FPT WITH THROUGH-HOLE AND SURFACE MOUNT TECHNOLOGIES

Through-hole assembly technology generally uses packages with leads on 2.5 mm (0.1 inch) centers. These leads are inserted and soldered into plated holes that have been drilled in the printed circuit board. The soldering is done by transporting the board and package leads over a pot containing molten solder. Through-hole technology was the prevailing assembly technology from the 1960s until the mid-1980s.

Starting in the late 1970s surface mounting of some components such as passive components and low lead-count ICs along with several through-hole components became popular. The low availability of SMT components and their higher cost kept SMT from overtaking through-hole until the mid-1980s. Surface mount IC packages have leads on centers ranging from 0.8 mm (0.03 inch) to 1.3 mm (0.05 inch). Comparison of a through-hole, SMT, and FPT package is shown in Figure 1.15.

Surface mount assembly technology consists of applying a solder-bearing paste to the board, placing the package leads into the paste, and forming the solder joints by reflowing the solder in an oven.

FIGURE 1.15. Comparison photo of a through-hole (DIP), and SMT (SOIC) and Fine Pitch (TSOP) packages.

FPT is becoming the prevailing assembly technology in the mid-1990s largely because of the cost savings and apparent similarity of the FPT assembly process to the SMT assembly process.

The package attachment techniques for fine pitch packages are more demanding than those for through-hole and surface mount packages. The prime reasons for this are the large number and close spacings of the leads. The ability to get all the leads onto the solder sites is important, as is the ability to form a reliable junction between the bottom of the lead and the top of the land. The issue of joint integrity is of particular concern, given that the combined dimensional variations of the lead and land give rise, particularly in the finer lead pitches of <0.4 mm, to a situation where the land width is smaller than the lead width. This eliminates the formation of solder fillets that wet and cover the top of the lead during the reflow of the solder. In through-hole and surface mount assembly, it is customary to inspect visually for the existence and appearance of the solder fillets as the prime means of assuring that the solder joint is properly formed and that the process is under control. In contrast, x-ray and laser vision systems are required to perform quality assurance and process feedback for FPT soldered leads. These tools are the most viable nondestructive means to test the bonding of the bond area of the fine pitch lap joint. The lap joint is the joint between the bottom of the lead and the top of the land. This is the area where most or in certain FPT package cases, all of the mechanical bonding occurs.

The basic assembly steps for medium and high volume attachment of FPT packages are similar to those used in SMT: preapply solder to the board, place the components, mass reflow solder, and, if required, clean. This is illustrated in Figure 1.16.

For lower volumes and some finer pitch (<0.3 mm) applications the assembly flow shown Figure 1.16 will need to be changed to overcome the inadequacies of the mass attachment process. Alternative processes that replace solder paste, use simultaneous place and reflow, replace the existing reflow heat source, and use more patient operators, permit very cost effective low volume and very fine pitch

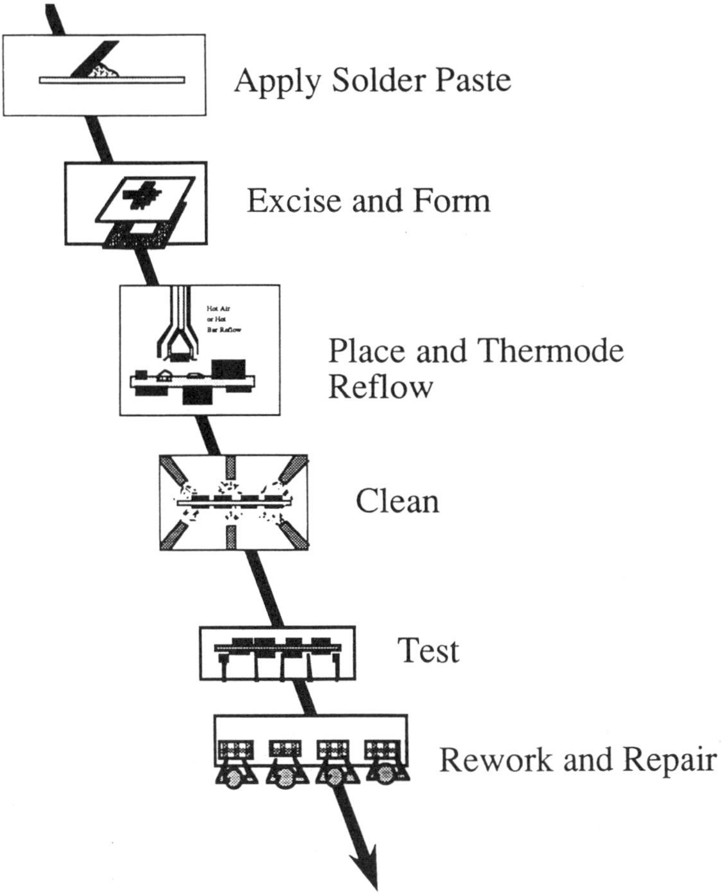

FIGURE 1.16. The assembly diagram showing the steps to solder a fine pitch package to a board at the same time the surface mount packages are soldered.

(<0.3 mm) package to board assembly. For example, solder paste may be replaced with pre-reflowed solder. For most SMT applications and 0.5-0.65 mm lead spacing packages, solder paste is still the most economical means to add additional solder to the board.

When volumes are low or when mass reflow methods become highly defective, the best means of reflow soldering fine pitch packages are hot bar (thermode) machines or fine tip soldering irons. Most current studies in mass reflow techniques are favoring forced gas convection heating from an embedded heat source as opposed to infrared or vapor phase sources. Regardless of the heat source, the heat rampup and cooldown phases must be carefully controlled in order to achieve repeatable results.

Rework and repair are more difficult for FPT than for through-hole or SMT. As a result most users will invest in equipment and operators they feel will provide almost microsurgeon skills.

All in all, implementation of FPT design and assembly capability may require new capital equipment and training expenditures even if a company has recently invested in SMT capability.

The major equipment areas affected are the following:

1. Stencil printers with vision
2. Placement machines with vision assist
3. Convection heating for solder reflow
4. Cleaning
5. Rework and repair
6. Direct CAD data linkage to the printed circuit board fabricator and placement equipment
7. X-ray or laser automated optical inspection equipment
8. Excise and lead form equipment.

Below is a brief discussion of the FPT attributes that may justify new equipment and training.

Stencils and Stencil Print Machines

The stencils used for applying solder paste must be fabricated with precision, improved registration, and smoother etched openings. What is in SMT a marginally acceptable paste misalignment of 25% stencil to land registration, translates in FPT to an unacceptable 0.08 mm (0.003″) offset for a 0.3 mm (0.012″) wide land.

As the stencil openings trend smaller and smaller the sidewalls need to be smoother to release the paste so it adheres to the board and not the stencil. Using lasers instead of chemicals to etch the openings is one solution, a nickel plate over the stainless steel stencil is another, and micropolishing the stencil openings is an additional option.

Printer considerations include vision assist, automatic stencil cleaning and more squeegee pressure to help assure uniform printing, without smearing, for openings above 0.25 mm (0.01″). Current paste, stencil, and printer technologies make printing paste in openings of less than 0.25 mm (0.01″) impractical.

Placement Machines

Placement accuracy is demanding for the large fine pitch packages, particularly for rotational precision. A 25.4 mm (1″) square 0.65 mm pitch 100 lead package will require a theta resolution of better than ± 0.07° (see Chapter 6). This accuracy is much tighter than many of the commercially available "FPT" class placement systems are capable of holding in production. The rotational resolutions range from 0.05° to 0.2°, the typical data sheet specification being 0.1°, with the area and accuracy unspecified.

A new class of placement system has emerged to address the low volume and high rotational accuracy FPT assembly needs. Since the number of fine pitch packages on a typical product is generally small in number, many users assemble them as a secondary operation, after the SMT and through-hole assembly, rather than *en masse* with the SMT packages. To do this, they use manual and semi-automated systems that combine the placement with a video-camera assist and a method of localized reflow. The reflow is either done by hot bar (thermode) or hot air. Focusable IR reflow is another option that may be introduced onto placement machines. The economics of these approaches show that the semi-automated assembly of selected FPT packages onto the board can cost substantially less than full automation using many of the automated machines available today.

Solder Reflow

Mass solder reflow of FPT packages, when desired, will enjoy the same benefits from forced gas convection type reflow systems as with SMT packages. In recent years, a slower indirect heating method, known as convection heating, has steadily replaced the faster heating vapor phase and radiant infrared heating as the preferred heat source for solder reflow systems.

There are several reasons for this, the first being economics. Convection is the least expensive heating method to operate and control. Second, convection heating has a slower heat transfer rate so there's less chance of solder wicking up the lead and more time for the flux to clean the metal surfaces for the solder. Also there is less chance of excess heat absorption by the package, which has caused package cracking and chip failure in some SMT packages and components. Package cracking is the result of too rapid an expansion of trapped gas, package defects, or moisture in the package.

Cleaning

Cleaning, if required, presents more of a challenge with FPT packages than with SMT packages. These packages are larger and closer to the board surface (0.05 mm vs. 0.1 mm) and the leads are closer together, which prevents easy access for cleaning agents. Another reason is environmental concerns. Many users have transitioned from groundwater polluting and ozone-layer reducing solvent cleaners to alternative processes. These include high pressure water sprays and more benign cleaning agent combinations. Some packages can't be cleaned in these newer systems because of the type of package encapsulant, the encapsulation technique, and the potential for lead damage by high pressure sprays.

Rework and Repair

The fragile leads and lands on the board will not tolerate much heat if they need to be repositioned or replaced. The rework equipment will resemble something used for microsurgery and will require highly skilled operators.

Printed Circuit Board Design

Design of printed circuit boards using FPT is more demanding than with through-hole or SMT because of tighter spacings, non-grid-conforming leads, and higher precision. Direct CAD data linkage will need to be installed with the printed circuit board fabricator and placement equipment in order to reduce tolerance levels to within acceptable limits. Additionally, key "ability" features must be designed in. These include testability, reliability, repairability, and manufacturability.

Automated Optical Inspection

In some cases, such as lead spacings below 0.4 mm, the land area is narrower than the lead width. This is due to the variation in land width and lead width, which are controlled by the fabricator and component packager, respectively. As a result, the solder joint is difficult to inspect without the aid of an automated or semi-automated approach, such as an x-ray or laser scanning system.

Excise and Form Equipment

One new FPT process step that adds to the complexity of package to board assembly is the need to excise and form the package leads prior to placement on the board. This is required on several of the package styles, such as GQFP and TAB. The reason the suppliers offer the package leads tied together is to prevent lead damage during shipping and handling. Lead damage has been a major problem for the plastic chip carrier (PCC) and quad flat (QFP) style packages. Routine handling and testing of the PCC and QFP packages may cause leads to be nonplanar by more than 0.1 mm (0.004″). The stencil and paste limitations in fine pitch demand that lead planarities be within 0.05 mm (0.002″) to avoid solder opens.

Excise and form is not without problems. The first is tooling, which is expensive because of the precision and number of tools required. A different tool size or fixture is needed for each package size, lead count, and lead shape style. Lead shapes may vary, since different lead shapes result in different levels of long term reliability. Additionally, the excise and form time lag before soldering is critical. When the leads are excised, the base metal is exposed on the end or "toe" of the lead. If it is left exposed to the atmosphere for too long, the toe of the lead may not solder.

1.5 FINE PITCH TECHNOLOGY OBSTACLES AND ISSUES

Most users view fine pitch technology as a result of the increasing use of ASIC devices in fine pitch packages. Some feel forced into the technology, others welcome it, striving to achieve its benefits. They also see the trend continuing toward more complex ICs and even higher lead counts as ASIC users increase their IC densities.

Those who feel forced into the technology will create obstacles to slow the use of the FPT packages. Many of these obstacles will be in the form of the following complaints:

1. Fine pitch is too demanding
2. FPT is capital intensive
3. Package sources are limited
4. FPT lacks standards
5. Knowledgeable resources are hard to find

Those who welcome the technology will easily overcome most of the obstacles but will encounter challenging roadblocks.

First, let us address some of the "obstacles" above. The statement that fine pitch technology is more demanding than through-hole and SMT is true; how much more demanding is very relative. It is a function of present user capabilities, expectations, and resources. This obstacle is easily overcome by a user's mental resolve. The first successful implementers of fine pitch in the U.S. were small contract assembly factories, who by the nature of their business have very limited capital and personnel resources.

High volume fine pitch technology is capital intensive if a user is converting from an existing through-hole capability or starting from no present electronic assembly capability. If the conversion is being made from an existing surface mount capability then the cost can be very modest. In some small and medium volume applications, implementation may cost less than $50,000.

The issue of limited package sources is valid and is discussed below. However, the issue of "lacking" standards is not valid. Users can create their own standards, usually in the form of procedures and purchasing agreements that spell out the acceptable techniques, dimensions and controls that allow a process to work and foster a viable user-supplier relationship. These may not guarantee perfection or zero defects, but in most cases they come close, and they are more practical than waiting for an industry-wide standard.

If a company wants a "jump start" into a new technology such as fine pitch, it is sometimes helpful to use an outside knowledge source. For a very new technology, such outside sources are difficult to find. However, for FPT today, they do exist in the form of consultants, contract assemblers, suppliers, and educators, however few they may be.

The truly significant issues facing all FPT users can be summarized as follows, in approximately descending order of significance.

Package Availability

No one integrated circuit supplier will carry each type of available FPT package. This is due to the high tooling and support cost of each package, which can be as much as $300,000 to $500,000 per package lead style. As a result some users will find themselves buying a sole-source IC in a sole-source package. This problem plagued the adaptation of SMT in its early years also. Over time this issue will resolve itself, since users need multiple sources for the packages. The packages of choice will be those that users find to fit all of their "ability" needs. Lack of availability and component shortages can spell economic ruin for any user company.

Package availability also includes the correct electrical performance and meeting the specified mechanical tolerances allowed by the process.

Core Assembly Techniques

The present core assembly process consists of applying a solder paste, placing the package, reflowing the solder, and, if necessary, cleaning and removing the flux.

The issue with applying the solder is that neither the package nor the board supplier are able to create enough or a suitability uniform solder volume on the package or the land to assure a reliable solder joint. This is due to the limitations in today's solder plating and the dip and level processes. Therefore, users are required to add solder volume using a solder paste that is stenciled or syringed onto the land areas before the package is placed. Fine pitch also means fine stencil openings, which can easily trap the paste and prevent uniform application. This results in solder opens and shorts after solder reflow. An alternative to screened or stenciled solder paste is needed to eliminate the high occurrence of solder opens and shorts.

Currently most installed placement equipment is not adequate to meet the needs of FPT. Some machines are capable of accurately placing the fine pitch packages but they sometimes lack other key features. As was discussed earlier, more accurate rotational control is needed for placement.

Substrate Compatability

Printed circuit board fabricators need direct access to CAD data to improve registration, automated optical inspection, and plating/solder coating capabilities. Blind and buried vias on the printed circuit boards are needed to complement the area savings of FPT and to enhance the process.

There is also a potential reliability problem due to the coefficient of thermal expansion mismatch between the substrate material and the package. The probability of the problem is primarily a function of the flexture of the leads and the substrate material.

Testing

The traditional approach of in-circuit testing (ICT) each component on a board after soldering will be costly and error prone with the complex FPT devices. This is due to the small lead sites available to be probed by the larger ICT probes. Better on-chip testing and supplier quality is needed to free the user from ICT dependency.

Elimination of Rework and Repair

FPT packages represent more than a tough rework problem, they also represent up to 50% of the product's cost. If they are reworked incorrectly, the cost consequences are significant. Users need to be free of the need to rework, reposition, or replace these packages. This can be accomplished only by purchasing electrically and mechanically correct components and by properly assembling the packages to the printed circuit board using capable processes.

Process Controls

Users need to incorporate better control procedures from the start. A process must be established that is capable from the initial design. The capability limits of the process must be measured and supplied to designers and vendors. Manufacturers and suppliers need to work together to maintain and improve process capability. Designers need to have quantitative access to process capability (process capability, Cp = maintained limits / specified limits) so they can constantly design within those constraints. This is known as design for manufacturability.

People Resources

One aspect requiring significant attention in implementing any new, complex technology, such as fine pitch, is the people aspect. Several teams of different disciplines must work together to assure success. These include the circuit designers, the printed circuit board designers, the materials purchasers, the manufacturing operators, and the process developers. Managing these groups and encouraging them to readily accept change is possibly the most significant obstacle.

The economic rewards for overcoming these obstacles will be worthwhile. FPT packages, as discussed, cost 25-50% less than the same components housed in an alternative SMT or through-hole package. Even with increased material and assembly costs, the product savings, can amount to 10-45% when the FPT packages are used.

References

1. Ginsberg, G. 1990. The Impact of New Technology on PCB Processing. In *Electronic Packaging and Production.* August, 1990.
2. Marcoux, P. 1990 *Fine Pitch Technology—A Market Study.* Saratoga, CA: Electronic Trend Publications.
3. Marcoux, P. 1989. *SMT: Design for Manufacturability.* Sunnyvale, CA: PPM Associates and Dearborn, MI: Society of Manufacturing Engineers.
4. Messner, G. 1987. Cost-Density Analysis of Interconnections. IEEE Transactions on Components, Hybrids, and Manufacturing Technology CHMT-10(2): 143–151.
5. Turmmala, R., and Tymaszewski, E. 1989. *Microelectronics Packaging Handbook.* New York: Van Nostrand Reinhold.
6. —1991. Initiates for High Value Electronics Packaging/Interconnect Program. 1991 Membership Workshop, Sept. 1991, MCC, Austin, TX.

2

The Family of FPT Packages

2.0 THE EVOLUTION OF FPT PACKAGES

Fine pitch packages are an essential link in the evolutionary chain. However, the revolutionary proliferation of package types will not hasten their acceptance.

2.1 INTRODUCTION AND CONSTRUCTION OF FINE PITCH PACKAGES

The variety of different surface attachable FPT packages currently available is staggering. There are 12 fine pitch packages that can be generally classified into three groups for general description purposes. The general classifications are the quad flat (QFP), the guard-ring, and the small outline (SOP) packages. These 12 fine pitch packages are in contrast to the four surface mount packages (SOIC, PCC, LCC, and QFP).

In Table 2.1 the 12 current packages are listed by trade name. The original package innovator or supplier is listed where known, otherwise the country of origin is given. If the package's mechanical outline is registered or standardized, the registering organization is listed. The registering organizations include JEDEC (Joint Electronic Device Engineering Council), a subcommittee of the Electronic Industries Association (EIA, Washington D.C.), the Electronic Industries Association of Japan (EIAJ, Tokyo), and the American Society for Testing and Materials (ASTM, Philadelphia).

Internal Package Construction

The internal construction of the QFP and most of the fine pitch packages is the same as that of the conventional plastic chip carrier or dip assembly (see Figure

TABLE 2.1 The List of Some of the Various Fine Pitch IC Package Options by Trade Name.

Package Trade Name	Originator	Standards Registration
Quad Flat Packages:		
1.Plastic Quad Flat Pack - PQFP	ATT	(JEDEC)
2.Plastic Quad Flat Pack - PQFP (metric-bumperless)	-	(JEDEC)
3.Small Quad Flat Pack - SQFP	Japan	(EIAJ)
4.Very Small Quad Flat Pack - VQFP	Japan	(EIAJ)
5.Thin Quad Flat Pack - TQFP	Japan	(EIAJ)
6.Ceramic Quad Flat Pack - CQFP	ATT	(JEDEC)
7.Axial Quad Flat Pack - CQUAD	NTI	(JEDEC)
Small Outline Packages:		
8.Very Small Outline Pkg. - VSOP	Philips	(EIAJ, JEDEC, DIN)
9.Thin Small Outline Pkg - TSOP	Japan	(EIAJ)
Guard-ring Packages		
10. Guard-ring Quad Flat Pack - GQFP	NSC	(EIAJ)
11. TapePak		NSC (JEDEC)
12. Tape Automated Bonding - TAB	GE	(EIA,EIAJ, ASTM)

2.1). The exception is the tape automated bonding construction, which is shown in Figure 2.2 and described in Section 2.5.

Conventional wire-bond IC assembly starts with the attachment of the IC chip to a metal pad which is part of the lead frame; this pad is commonly called the *die flag*. The chip bonding pads are connected to the leads of the lead frame via the wire bonds (shown in Figure 2.1). The lead frame and chip assembly are encased in molded epoxy for protection and testability. Then the lead frame is solder plated or dipped after the shorting bars and epoxy flash are removed. The solder plating or solder dip is critical to maintaining solderability during storage and handling. The commonly used copper lead frames would readily oxidize and become unsolderable if not solder coated.

2.2 QUAD FLAT PACKAGES

The quad flat packages (QFPs) have leads on all four sides. The leads are usually precut and formed into the popular gull-wing lead shape. The molded plastic body is available in various thicknesses and in either a rectangular or square shape.

The quad flat packages permit a large number of leads by extending the leads out all four sides of the package. The QFPs are also the most testable and most available packages. Higher lead counts may be available from future packages that combine the QFP gull-wing leads with surface mountable pin grid arrays. Therefore, it should be no surprise that the QFPs are expected to evolve as the industry

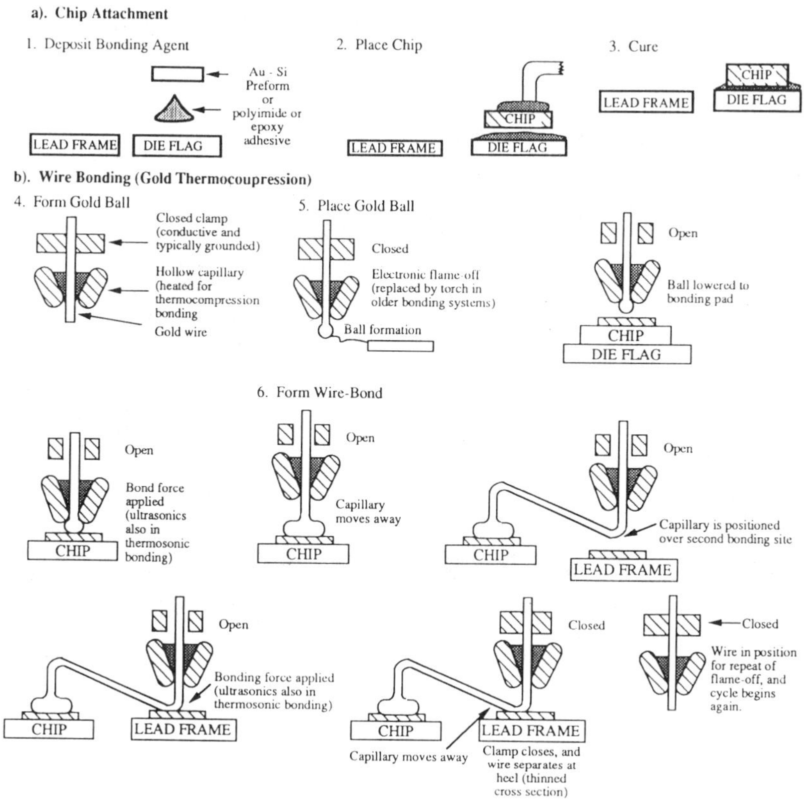

FIGURE 2.1. The steps needed to assemble, mount, and wire bond an IC into a package.

favorite with the largest volume usage of all the FPT packages. It is expected that up to 70% of all the FPT packages above 80 leads will be of the QFP style. Which of the quad lead spacings will be most used will depend on user success with soldering the closer lead spacings. As user skills will take time to evolve, even with the help of this text, it is anticipated that the wider spaced QFPs will dominate for the next several years. The wider spaced QFPs in our context are those with 0.5 mm (0.02″) and 0.65 mm (0.026″) lead spacings. These package spacings provide up to 376 and 240 leads, respectively, an example is shown in Figure 2.4. Figure 2.5 lists the range of lead counts (20 to 1256 leads) available by varying the QFP body size (10 × 10 mm to 48 × 48 mm) and lead pitch (0.15 mm to 1.0 mm).

The other quad styles consist of the small and thin designations. The small quad flat packages, or SQFPs offer up to 576 leads on spacings of 0.4 mm (0.016″), 0.3 mm (0.012″), and 0.25 mm (0.01″). The thin quad flat package (TQFP) is much thinner than the SQFP, with a body thickness of 1.2 mm (0.05″) vs. 2.7 mm (0.11″)

FIGURE 2.2a. A Tape Automated Bonded, or TAB, package lead attachment. (Photo courtesy of International Micro Industries)

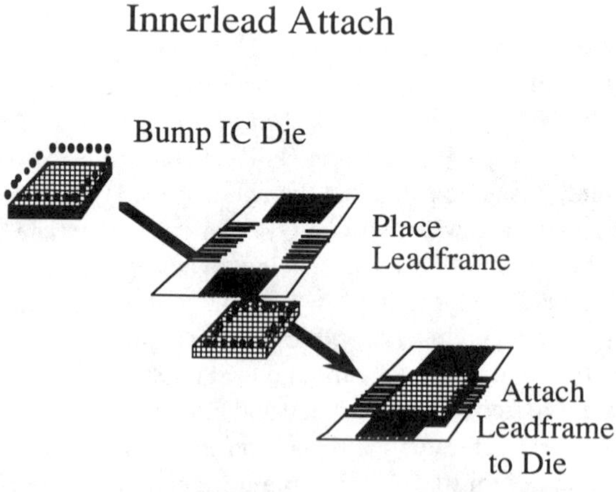

FIGURE 2.2b. The TAB assembly steps.

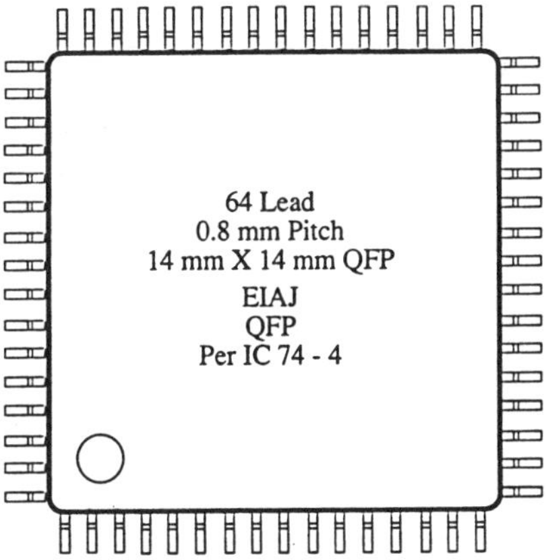

FIGURE 2.3. A 0.8mm lead pitch quad flat package example.

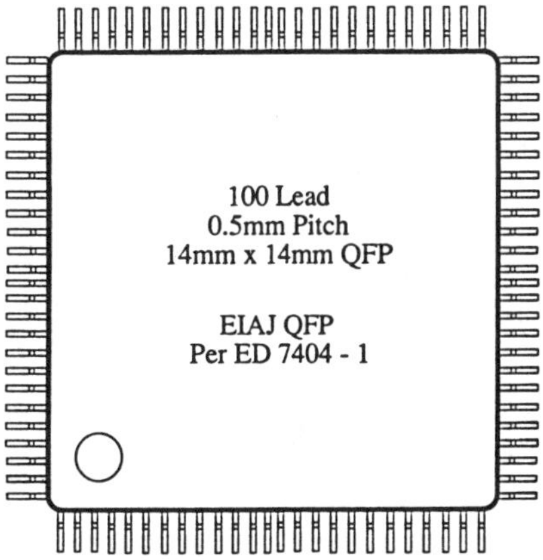

FIGURE 2.4. A 0.5mm lead pitch quad flat package example.

thickness for the QFP and SQFP. The TQFP is intended for low profile consumer applications. As will be discussed in Section 2.8, the thin bodies of the TQFP and the TSOP (thin small outline package) present a reliability problem: if excess moisture and too rapid a reflow heat transfer are combined, the result is package cracking.

The Original Flat Packages

The original flat package was developed in the mid-1960s. It was based on 2.5 mm (0.1″) and 1.3 mm (0.05″) lead frames sandwiched between a ceramic base and lid.

LEAD PITCH IN MM	MAXIMUM LEAD COUNT								
Body Size in MM	1.00	0.08	0.65	0.5	0.4	0.3	0.25	0.2 (Future Availability)	0.15 (Future Availability)
10 x 10mm	20	32	40	72	88	120	144	180	236
12 x 12mm	36	48	56	88	108	144	176	224	296
14 x 14mm	52	64	80	108	128	176	208	264	344
20 x 20mm	76	96	120	152	192	256	304	384	504
24 x 24mm	108	136	168	184	232	304	368	464	616
28 x 28mm	156	192	216	240	272	360	432	544	720
32 x 32mm				248	312	416	496	624	824
36 x 36mm				280	352	464	560	704	936
40 x 40mm				312	392	520	624	784	1040
44 x 44mm				344	432	576	688	864	(1144)
48 x 48mm				376	472	624	752	944	(1256)

FIGURE 2.5. This figure shows the available number of leads for the various QFP lead pitches. These numbers reflect the standards established by the EIA and EIAJ. (Source: Motorola)

It is still used in some critical military and hermetic applications. The flat packages are thin, rectangular packages with bodies constructed of alumina or beryllia, with glass-to-metal seals. The long leads extend out from the package body on two or four sides and are formed in a gull-wing style lead or soldered straight onto the board. Lead counts generally range from 12 to 28. Figure 2.6 shows a picture of the original ceramic quad flat package.

The Surface Mount QFP (1.0 and 0.8 mm Spacing)

As commercial development of surface mount became prevalent in the early 1980s, the Japanese, coordinated by the Electronics Industry Association of Japan (EIAJ) standards organization, began to develop their own plastic versions of the quad flat package. These packages use a standard package body size and vary the lead pitch, thus increasing lead count density without increasing the molding and handling costs.

The lead pitches of 1.0 mm (0.039″), and 0.8 mm (0.032″) form the standards that define the QFP family of packages from 80 to 100 leads. An example of this package is shown in Figure 2.3.

The QFP uses a gull-wing lead form, which complicates the automatic handling of the packages since they cannot be supplied in plastic tubes like the dual-in-line (DIP) package. However, each package is housed in its own protective compartment and shipped in trays known as "waffle packs" or matrix trays. The mechanical design of these was approved, in 1990, as registered standards by a joint EIA-US (JEDEC) and EIAJ standards committee.

Table 2.2 shows the advantages and disadvantages of the surface mount pitch (1.0 and 0.8 mm) QFP.

The QFP offers a much lower profile than the surface mount plastic chip carrier (PCC) (2.8 mm vs.) 4.2 mm). This is shown in Figure 2.7. The QFP's standoff height of 0.1 to 0.4 mm makes it more difficult to assure cleanliness under the package than the PCC or PQFP which have a standoff of 0.4 mm to 0.5 mm.

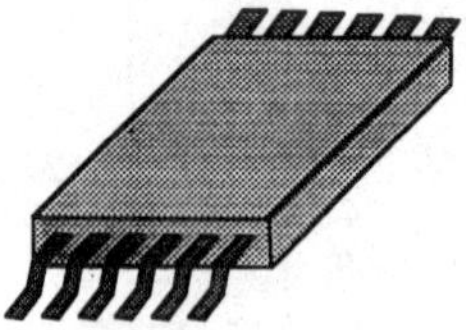

FIGURE 2.6. A diagram of the original flat package. The leads of the flat pak were either surface mounted or soldered into holes.

TABLE 2.2 A Summary of the Advantages and Disadvantages of the QFP Packages.

Advantages and Disadvantages of 1.0 and 0.8mm Pitch QFPs
QFP Advantages • Finer pitch (1.0 to 0.8mm) than other SMT packages (1.3mm) • Small body sizes than PCC • Lower profile (2.7mm) than a DIP or PCC (5.7mm) • Low printed circuit board assembly cost • Worldwide acceptance • Registered outlines with EIAJ and Jedec (EIA-US) QFP Disadvantages • No lead protection (no bumpers) • Can't be shipped in tubes • Excess lead coplanarity • Limited thermal dissipation using Alloy-42 metal leads • High printed circuit board assembly cost if automated • Few U.S. based suppliers

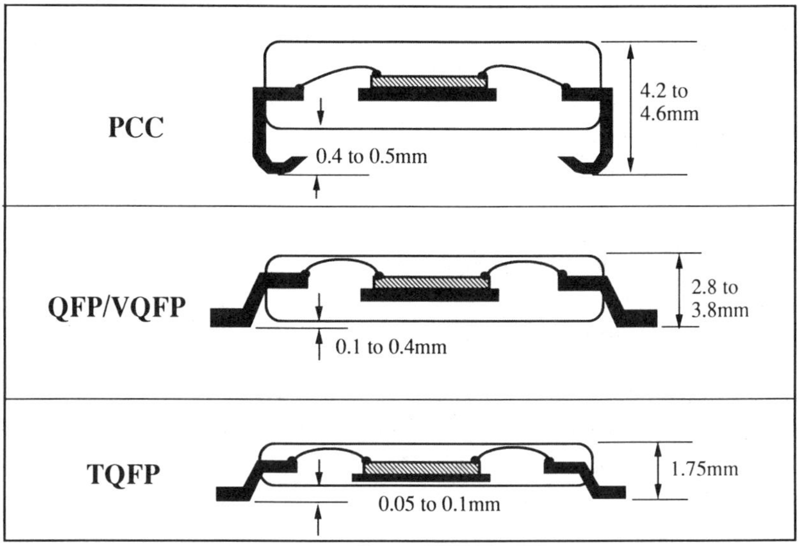

FIGURE 2.7. A comparison of the standoff heights of the PCC, QFP, and TQFP packages.

The Fine Pitch QFPs (0.15 to 0.65 mm Pitch)

With the consumer market favoring smaller and lower cost electronic gadgets and the ASIC market needing yet higher lead count packages, Japanese-based semiconductor suppliers have developed an extension to the QFP package family. These include finer pitch packages down to 0.25 mm, and possibly down to 0.15 mm pitch in the future. Body sizes range beyond the surface mount QFP size of 40 × 40 mm up to a size of 48 × 48 mm. This permits a maximum lead count of 752 for the 0.25 mm, 48 × 48 mm package and 1256 leads (when available) with the 0.15 mm, 48 × 48 mm package.

The fine pitch QFP packages are labeled the SQFP, short for the small quad flat packages (EIAJ) which have 0.5 to 0.65 mm pitch; the VQFP, short for the very small quad flat packages (EIAJ), with pitches ranging from 0.15 to 0.4 mm, and the TQFP for thin quad flat packages (EIAJ) with pitches from 0.15 to 0.65 mm.

The TQFP (EIAJ) is thinner than the VQFP and the QFP: 1.75 mm thick, compared to 3.0 mm for the QFP as shown in Figure 2.7.

2.3 PLASTIC QUAD FLAT PACKAGES

In the early 1980s several U.S. manufacturers agreed that placing leads on all four sides of a package increased density and offered attractive features. But they assumed that bending the leads underneath the package in the form of a J instead of forming them in the gull-wing shape would increase density even further. And this package style would also be compatible with a ceramic leadless chip carrier (LCC) footprint. Thus the J-bend, plastic leaded chip carrier (PCC) was developed, with lead counts ranging from 18 to 84 leads on 1.3 mm (0.05″) center lead spacing (see Figure 2.8). A ceramic version of the PCC, designated as the LDCC, or leaded ceramic chip carrier, was also developed for military and high reliability applications. It is footprint compatible with the PCC.

The PCC, however, being on 1.3 mm (0.05″) spacing, does not address the demand of large application specific ICs, or ASIC products, such as gate arrays. ASICs need higher lead counts than the PCC's maximum 84 lead capacity. So several U.S. suppliers, lead by ATT, developed the plastic quad flat package, (PQFP) to satisfy this requirement.

The PQFP (which is shown in Figure 2.9) uses approximately the same plastic body sizes as the PCC, but with leads on 0.635 mm (0.025″) centers, and a molded bumper protruding from each corner for lead protection during handling. Lead counts for this package family range from 44 to 244 leads. The gull-wing lead shape is used for these packages since the PCC style J leads on lead spacings closer than 1.3 mm (0.05″) are impractical. The PQFPs can be shipped in tubes or trays, thus providing more flexibility in automated handling.

One compatibility problem between the QFP and PQFP packages is the differ-

FIGURE 2.8. An example of a plastic chip carrier, or PCC, package. (Photo courtesy of the IPC)

FIGURE 2.9. An example of a plastic quad flat pak, or PQFP package. (Photo courtesy of the IPC)

TABLE 2.3 A Summary of the Advantages and Disadvantages of the PQFP Packages.

PQFP Advantages

- Lead Protection - molded bumpers
- Standardized body sizes - Jedec
- Copper lead frame for higher thermal dissipation
- Shippable in tubes
- Higher standoff for easier cleaning

PQFP Disadvantages

- No vertical lead protection (planarity)
- Manual soldering slowed by bumpers
- Higher package profile
- Few Non-US suppliers
- Non-compatible lead spacing

ence in lead pitch. The QFP lead pitch closest to the PQFP is the version on 0.65 mm spacings. When converted to English units, this becomes 0.026 inch. The PQFP is 0.025″ (0.635 mm). Additionally, the QFP lead is nominally 0.31 mm wide by 0.8 mm long versus the PQFP lead, which is 0.31 mm wide by 0.635 mm long.

When an attempt is made to place a QFP package on a PQFP site, the mismatch of 0.015 mm (0.0006″) in lead pitch will result in a lead being greater than 50% off the land after 10 leads. And a lead will be completely off the land after 20 leads. The 100 lead QFP and PQFP have 25 leads per side, resulting in five completely offset leads per side if the above situation occurs.

Ceramic Quad Flat Pack (CQFP)

The ceramic quad flat pack is based upon JEDEC registration outlines. Its footprint is compatible with the PQFP, but is intended for high reliability and military applications. The CQFP looks physically the same as the square or rectangular QFP but is made of ceramic with leads side brazed or co-fired onto the package. A CQFP is shown in Figure 2.10. The features and benefits of the ceramic quad flat package are shown in Table 2.4.

2.4 SMALL OUTLINE PACKAGES

The small outline packages (SOPs) are smaller versions of the surface mount small outline packages, commonly called SOICs. The SOPs incorporate finer lead pitches and thinner bodies to provide increased lead capacity and volume savings.

FIGURE 2.10. Photo of two ceramic quad flat paks, or CQFP packages soldered to a board.

Leads on the SOPs are limited to two sides of the rectangular shaped package. The fine pitch SOPs use the gull-wing lead shape universally, unlike the SOIC which uses either the gull-wing or the J lead. The J-lead shape is used for most of the high capacity (1 Meg, 4 Meg, 16 Meg, etc.) DRAM memory devices.

TABLE 2.4 A Summary of the Features of a CQFP Package.

<u>Advantages</u> • Gull wing lead form • Common pitch at 0.0635mm • Compliant leads - form reliable joint • Solder joint inspection • Easily assembled in small lots • Non-conductive tie bar • Cavity up or cavity down options to improve thermal management
<u>Disadvantages</u> • Excess coplanarity • Larger land area than "J"lead form • Non-compatible lead pitch to metric equivalents

The SOP and SOIC were originally developed by a joint effort, known as Fascalac of Philips Corporation and three other semiconductor companies in the early 1970s. The package development was intended to find low cost solutions to making electronic watches and other consumer items. The original Fascalac developments used a fine pitch style, but the pitch was increased to 1.3 mm when it was recognized that the broader user base was not yet prepared for fine pitch. Examples of the SOIC, the SOP, and the thin SOP (TSOP) are shown in Figure 2.11.

The SOIC is available in 8 to 28 leads in a body area of 7.8 × 18 mm, and body thickness of 2.5 mm. The SOP is available with 24, 28, 32, and 40 leads. There is also a limited availability of a very small outline package, or VSOP providing 48, 52, and 56 leads. The SOP occupies areas ranging from 5 × 4 mm up to 20 × 14 mm.

Users will find the SOJ, a small outline package with J shaped vs. gull-wing leads, and the 0.8 mm pitch SOP (or VSOP from a few suppliers) not unlike using the 1.3 mm pitch SOIC package. These packages have proven easy to assemble onto printed circuit boards and very reliable in field service.

The SOP of greatest concern, however, is the thin SOP or TSOP. The TSOP is shown in Figure 2.12. A comparison between the SOJ and SOP is shown in Table 2.5. From Figure 2.12, one can see that the TSOP has a large body to lead foot area ratio. This creates a solder profiling problem, as will be discussed in Section 7, (Reflow Soldering) when using certain reflow systems such as lamp and focused infrared reflow.

FIGURE 2.11. Examples of the small outline, or SOP package types.

TABLE 2.5 Comparison of the SOJ and SOP Packages.

ITEMS		TSOP	SOJ
Package size	Area (mm^2)	16.0 x 6.0 (96.0)	17.15 x 8.45 (144.9)
	Height (mm)	1.2	3.45
Lead pitch (mm)		0.5	1.27
Standard Memory Module Size 1M x 9 (mm^3)		88.9 x 25.4 x 2.54 (5735)	88.9 x 20.3 x 5.08 (9167)
Reliability		Requires Care (Thermal Stress)	Very Good
Weight (g)		0.22	0.75
Production cost (Ratio)		1.1	1.0

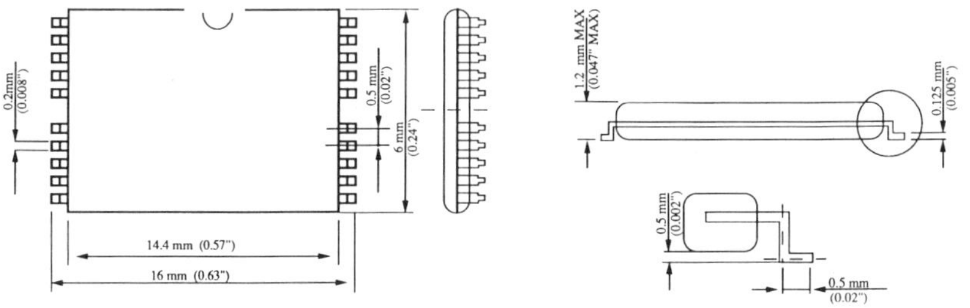

TSOP OPTIONS

Pin Count	Pkg. Width	PACKAGE LENGTH 13.4mm	14mm	16mm	20mm
24 Pin	6mm		256K V-RAM	1M D-RAM	
28 Pin	8mm	256K S-RAM			
32 Pin	8mm				1M S-RAM
40 Pin	10mm		1M OTP		

FIGURE 2.12. Diagrams the TSOP style packages and its options. (Source Mitsubishi Electric)

The TSOP has a thin body (1.2 mm (0.05″)) and a low printed circuit board standoff (0.05 mm (0.002″)). The thin body increases the probability of package cracking. The low standoff height aggravates the problem of cleaning under the package. The TSOP, as shown in the table in Figure 2.12, is available in a variety of body sizes, and with lead counts ranging from 24 to 40 leads.

2.5 TAPE AUTOMATED BONDED AND GUARD RING PACKAGES

The tape automated bonded (TAB), the Tapepak, and the GQFP offer a new twist to packaging. The outer leads are secured together with a guard ring to prevent lead damage during shipping and handling. The guard ring is either molded plastic or mylar tape. It is the user's task to excise the guard ring and to form the leads for reliable attachment to the printed circuit board. This saves potential yield losses for the supplier and the user, due to planarity problems, but imposes an increased cost on the user because of the added excise and form assembly operation.

The guard ring protected packages may be encapsulated or unencapsulated in a protective polymer, plastic or ceramic covering. Many of these packages use the tape automated bonded attachment method. TAB forms an interconnect between the pad on the IC chip and the outside world by connecting thin copper lead frames directly to the chip instead of the conventional wire-bond to a lead frame, the opposite side of the copper frame forms the external lead. The leads may be secured on the outer perimeter by mylar tape as shown in Figure 2.13. They can also be secured with plastic guard rings as in the case of the GQFP shown in Figure 2.14.

TAB packaged ICs consist of a bumped tape that is bonded to a bumped IC chip. The IC chip is processed with additional pad metalization to permit reliable bonding of copper leads to the IC. Conventional wire-bonded IC packaging assembly uses aluminum or gold wires that bond well to aluminum.

The tape is usually a single layer of two ounce etched copper foil that is used as the lead frame, bonding, and test medium. The foil is 0.07 mm thick rolled copper glued to a mylar or equivalent tape base for mechanical support. Rolled copper provides a very smooth surface with a pronounced horizontal grain structure that results in improved bondability. This type of copper provides a dense, ductile, pinhole free metal. With adequate process control this tape assembly permits maximum control in the imaging and etching processes and yields a uniform electrical resistivity for any given thickness.

The lead pattern that serves as the inner lead bond to the chip and the outer land bond to the board is formed by chemically etching the unused copper off of the tape. The chemical etch process is the same as that used to etch copper laminate off of the fiberglass-epoxy material for printed circuit boards. Either before or after

FIGURE 2.13. A photo of a tape automated bonded package. (Photo courtesy of International Micro Industries)

FIGURE 2.14. A photo of a guard-ring QFP. (Photo courtesy of Rogers Corp.)

etching, depending on the tape supplier, the sprocket holes are punched into the mylar tape. These holes are used to feed and coarsely align the part for placement and bonding.

The copper lead pattern is then thermo-compression bonded to the chip in either a single bond per pad step or a simultaneous bond of all the leads to all the accompanying pads. The combined process steps are illustrated in Figure 2.15. The thermo-compression bonder is a conduction heated metal assembly in the shape of the innerlead bond outline, if the pads are simultaneous bonded, or a flat point for single point bonding applications. As shown in Figure 2.16, the chip can be mounted onto the top or bottom of the tape. The assembly can then be encapsulated to assure safe handling or left unencapsulated until after assembly to the printed circuit board.

TapePak (JEDEC)

The TapePak, developed by National Semiconductor in the late 1980s was the original molded guard ring package and it utilizes TAB. In the TapePak, a chip is tabbed or wire bonded onto a strip of etched or stamped metal tape. This strip of chip-on-tape is then molded simultaneously in plastic to produce the package and protective plastic rings. The dam bars, which also support the tape leads during the chip bonding and molding operation, are then removed. This is followed by a deflash operation and solder plating. The devices are singulated and packed into tubes, ready for test.

The chip-on-tape combination is molded in plastic in such a manner that an outside ring is formed apart from the inside encapsulated chip. This outside ring

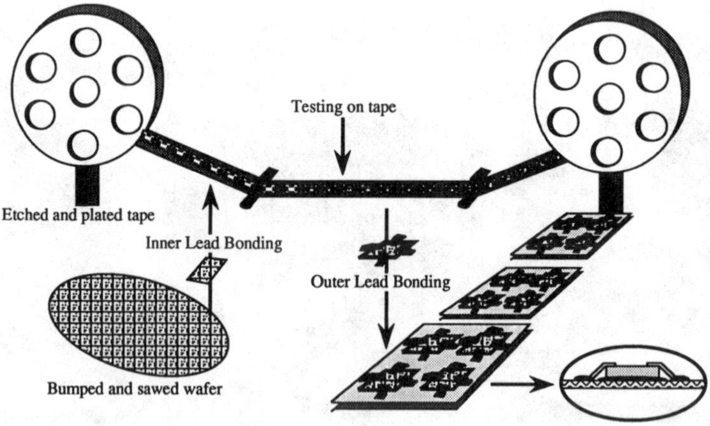

FIGURE 2.15. The process steps for TAB package assembly.

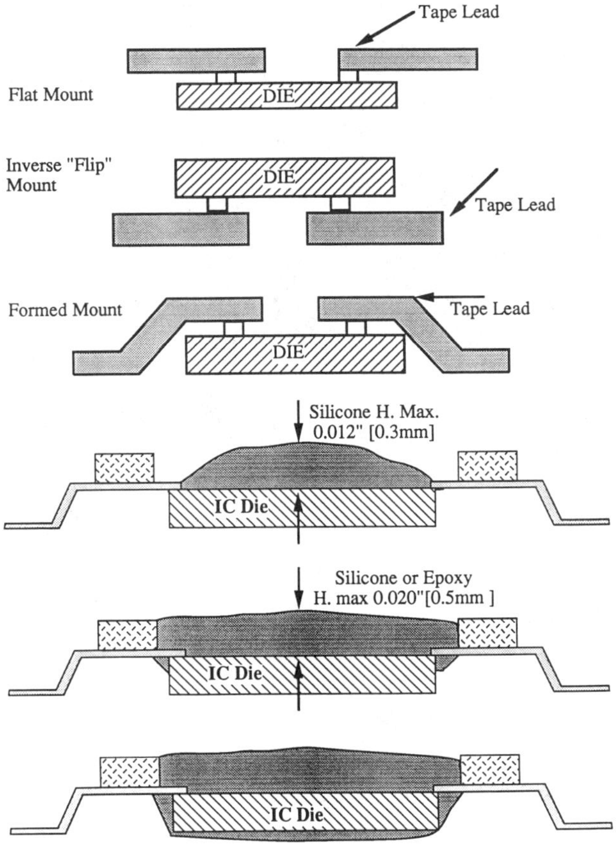

FIGURE 2.16. The various options for mounting an IC chip to TAB tape.

provides for lead protection and test capabilities. The package body is excised from the carrier ring by the pick-and-place machine and is subsequently attached to the printed circuit board. Like the Japanese quad flat pack, the TapePak family uses standard body sizes with lead counts from 40 to over 460 leads on 20, 15, and 10 mil pitch. This package is shown in Figure 2.17.

Outer lead bonding to the printed circuit board is done as a modification to existing surface mount pick-and-place and attachment processes. A presentation system consisting of a feeder and forming unit adapts to the pick-and-place machine. This unit removes the TapePak package from a tube and excises the final package from the outside plastic ring. The leads are formed into gull wing leads and presented to the placement head. The device is then placed on the board for subsequent soldering.

FIGURE 2.17. A photo of a guard-ring QFP that is sold under the trade name, Tapepak, a trademark of National Semiconductor.

2.6 PACKAGE HANDLING AND SHIPPING

One of the most common problems encountered in assembling fine pitch packages onto printed circuit boards is open and short solder joints. Several factors including the board design, the paste application, package placement, and reflow method contribute to these defects. However, package lead damage appears to overshadow these as the main defect cause, unless precautions are followed.

After assembly and test, the non-guard-ringed packages are placed in matrix trays or tubes. The guard ring protected packages may be stored on tape reels or singulated and put into tubes.

Regardless of whether the packages are placed in trays, tubes, or reels, the shipping carrier must be antistatic. This means that the carrier is conductive so any electrical voltage potential buildup during shipping and handling can be safely discharged to ground, avoiding possible electrostatic discharge (ESD) damage to the sensitive IC in the package. This precaution extends to the handling of the final assembly also. ESD-safe methods that have been incorporated for through-hole and surface mount assembly are required and adequate for the handling of the fine pitch packages. The damage created by ESD is to the silicon IC chip and is not a function of the package, or lack of a package, that surrounds the chip.

Matrix Trays

Matrix trays are flat, stackable trays designed to safely hold the packaged ICs during shipping and handling. The typical carrier tray is molded from high temperature plastic. Added into or sprayed on the plastic is a conductive material to provide for ESD protection. A tray will hold 10, 30, or 66 IC packages, depending on the package type, and several trays may be stacked. Figure 2.18 shows a photo of a JEDEC registered tray for the quad flat packages. The tray shown in this photo is known as the CO-012 which designates it as the carrier outline for the metric quad flat packages (MQFPs). The other carrier tray outlines designated by Jedec, are the CO-010 for pin grid and land grid arrays and the CO-011 for the ceramic quad flat packages (CQFP).

The carrier tray must be capable of holding the IC package without introducing lead planarity damage during shipping and routine handling. Early tray designs failed at this task, due in part to the poor design of the package support.

The cross section of the present matrix tray is shown in Figure 2.19. The tolerances and sizes of the package holding posts in the earlier tray allowed the packages to move around, resulting in lead damage. The present tray standards are of a tighter tolerance thus minimizing this problem.

To enable shipping of several packages and trays, the carriers are designed to be stackable. The carriers have grooves, chamfers, and locating pins to assure

FIGURE 2.18. A photo of a common matrix tray used for transporting fine pitch packages in shipping and also used for support during placement.

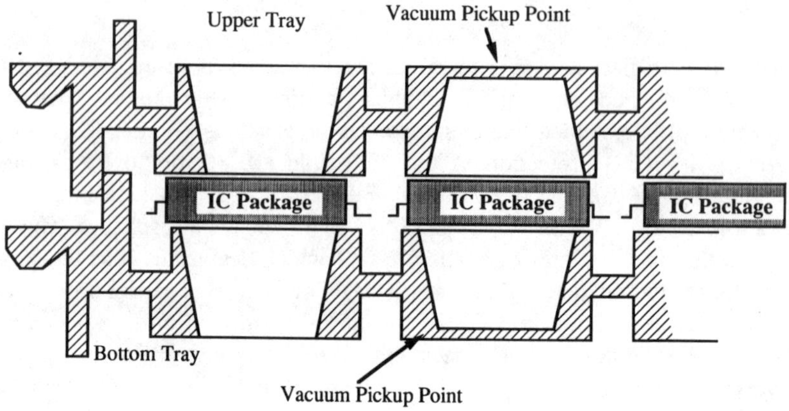

FIGURE 2.19. A cross sectional view of the improved matrix tray. Earlier versions allowed many packages to get damaged during shipment due to excess tolerances.

precise positioning of the stacked carriers. Within the center area of each tray are two filled areas large enough to allow vacuum pickup and placement. These are shown as the vacuum pickup points in Figure 2.19. This allows automated stacking and unstacking by the supplier and user. This automation is either an external or internal machine to the placement machine that picks and places the IC packages.

Another role of the matrix trays is to allow an easy way to hold the parts during the final bake at the IC suppliers facility. This bake is essential for the larger and thinner packages to purge most of the moisture that may have been trapped in the package. If unpurged, this moisture may cause package cracking. Broader discussion of the package cracking issue is covered in Section 2.7. To minimize the problem and the need for any further baking by the user, the supplier seals the baked packages and trays in a marked pouch. The label on the pouch (example shown in Figure 2.20) warns the user of the potential moisture problem and lists precautions and handling instructions.

Tubes

Carrier tubes have been used reliably for the past two decades as the conventional method of shipping and handling IC packages. The tubes allow easy gravity transfer from one tube into a process, such as IC test, and then into another tube. They also are easy to load and feed into placement machines. This is shown in Fig. 2.21. Tubes are generally made of extruded conductive plastic or metal.

Tube dimensions are standardized under the carrier outline, or CO, section of the JEDEC outlines. Copies of the tube and other carrier dimensional standards

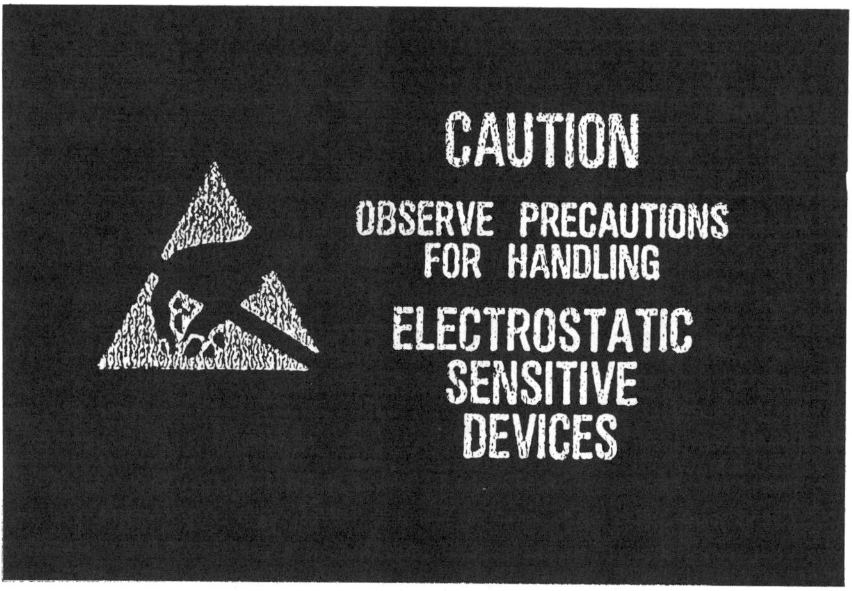

FIGURE 2.20. Since many of the fine pitch packages are prone to absorbing moisture, precautions are required to prevent moisture-induced cracking during reflow soldering. (Courtesy of Signetics-Philips)

are available from the Electronic Industries Association (See Appendix A for Address) as part of their publication 95 "JEDEC Registered Outlines for Semiconductor Devices." The tubes are available in different widths and lengths depending on the package style. The number of packages per tube varies with the package and the tube length. Figure 2.22 shows a tube containing PQFP packages.

Tape and Reel

For very high volumes, certain placement machine feeders, and the unencapsulated TAB (or GQFP) packages, the tape and reel carrier may be required. The tape and reel carrier is the carrier of choice for the SMT chip resistors, capacitors, diodes, and transistors. Particularly if they are placed using a "chip shooter" placement system, which is the common designator for high speed (>10,000 parts per hour) systems. Tape and reels are standardized and defined in the EIA-481 specification. For IC packages, embossed conductive plastic tape material is preferred over paper or nonconductive plastic options. If the IC packages on the tape and reel are susceptible to the moisture cracking, then the reels should be baked and enclosed

FIGURE 2.21. Photo showing tube stored ICs ready for placement. (Photo courtesy of Dynapert)

FIGURE 2.22. A photo of PQFP packages in shipping tubes.

in sealed pouches by the supplier. An example of the tape and reel carrier is shown in Figure 2.23.

Bulk Shipments

Bulk shipments of electronic components means the components are dropped into a plastic bag and handed to the user. This practice is common when parts are kitted for prototyping or when small quantities are purchased from distributors. Never allow handling of fine pitch IC packages in bulk, severe lead damage will occur. It is possible to handle small quantities of the molded and encapsulated guard-ring packages, the GQFPs, in bulk, but again this practice is best avoided.

Table 2.6 summarizes the carrier options for the various fine pitch IC packages.

2.7 PACKAGE RELIABILITY AND QUALITY

Reliability and quality of fine pitch packages is a complex set of issues, as the packages may show degradation in quality and reliability as a result of both external and internal forces.

To treat these issues fairly, they are separated here into internal and external issues. Internal issues include package cracking, bond or lead breakage, chip

FIGURE 2.23. A photo of tape and reel mounted components.

TABLE 2.6 The Shipping and Handling Methods Available for the Various Fine Pitch Packages.

FPT Package Carrier Options				
	Carrier Option			
Package	Bulk	Tape and Reel	Matrix Tray	Tube
QFPs	Note 1	Possible	Yes (Note 3)	No
PQFPs	Note 1	Possible	Yes	Yes (Note 3)
TSOPs	Note 1	Yes	Yes	Yes (Note 3)
VSOPs	Note 1	Yes	Yes	Yes (Note 3)
GQFPs Encapsulated	Note 2	Yes	Possible	Yes (Note 3)
GQFPs Un-Encapsulated	Note 1	Yes (Note 3)	Possible	No

NOTE:
1. Don't even think about this as an option!
2. May be a viable option but not advised
3. The common preferred carrier.

cracking and thermal failures. External issues are solder joint cracking, deviations from lead planarity, lead to package separation, and package crazing. The internal package issues will be discussed in this section and the external will be discussed later in Chapter 10.

TABLE 2.7 A summary of the reliability and quality attributes for the various fine pitch packages.

Internal Issues	External Issues
• Package Cracking • Bond or Lead Breakage • Chip Cracking • Thermal Failures	• Solder Joint Cracking • Lead Planarity • Lead to Package Separation • Package Crazing

Package Cracking

Package cracking of fine pitch packages generally shows as cracks emanating from the area under the IC chip and metal base, called the die flag. An example is shown in Figure 2.24. The probability of package cracking can be roughly approximated from the graph shown in Figure 2.25.

The cause of this package cracking seems to be due to the combination of large IC chip size, very thin body thickness under the die flag, the amount of trapped moisture in the package, and the thermal ramp rate during solder reflow. If the reflow thermal ramp is too fast, say of the order of 10°C/sec near 100°C, then the moisture that has been absorbed by a plastic package will become steam and expand. If the plastic is weak or thin, especially under the die flag, then the rapidly expanding moisture will form a crack. This crack will continue until it exits the package, allowing the moisture to escape.

Referring to the graph in Figure 2.25, it can be assumed that an IC die size of 19 mils^2 and a package thickness below the die flag of 1.3 mm would have a 15 to

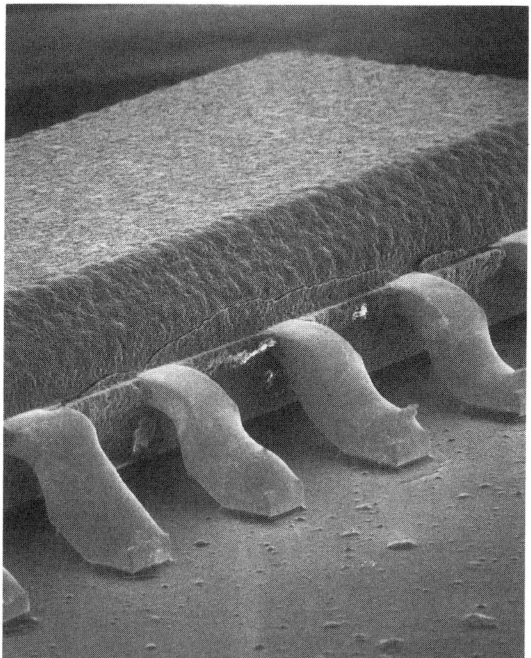

FIGURE 2.24. An extreme close-up photo of a crack in a plastic IC package. (Photo courtesy of Signetics-Philips)

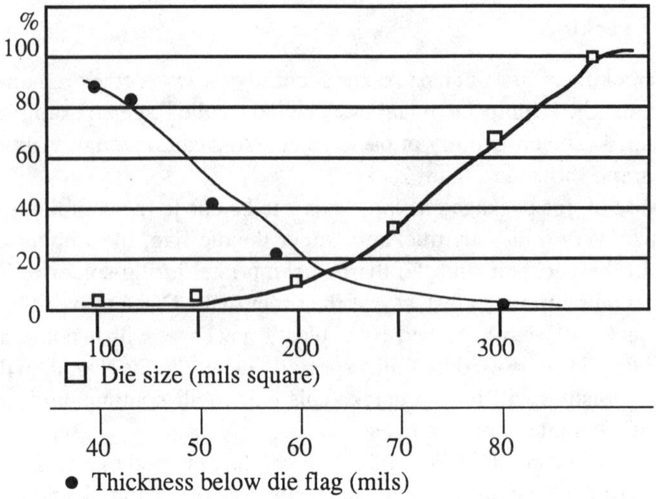

FIGURE 2.25. A graph of the probability of package cracking. As the chip die size increases and the package thickness under the IC bonding tab (known as the die flag) becomes thinner, the probability for package cracking increases. (Source: Signetics-Philips)

25% probability of package cracking when exposed to certain conditions, such as rapid heating.

The relative humidity in the package is dependent on where the package is assembled and how much of a prebake it receives prior to solder reflow. The relative humidity ranges for different geographical points of assembly is shown in Table 2.8. In this table the worst conditions occur in Bangkok, Thailand which is

TABLE 2.8 The Range of Humidity for Different Parts of the World. (Source: Signetics-Philips)

	TEMPERATURE			HUMIDITY		
Time Period	Minimum	Maximum	Average	Minimum	Maximum	Average
14 Apr to 21 Apr	28.°C	32.9°C	31.2°C	63%	74%	69%
22 Apr to 29 Apr	29.5°C	32.8°C	31.6°C	58%	75%	67%
06 May to Jun	27.2°C	32.1°C	29.7°C	61∞	85%	71%
14 Jul to 17 Aug	26.3°C	28.4°C	28.4°C	63%	87%	72%
18 Aug to 13 Sep	26.5°C	29.2°C	28.0°C	69%	86%	81%
14 Sept to 18 Oct	26.0°C	29.5°C	28.0°C	71%	89%	81%
Median Values			29.0°C			74%

the site of several of the largest IC package assembly factories. Using Bangkok as its test site, Philips-Signetics measured the package percentage weight gain for various surface mount packages. The results are shown in Figure 2.26.

The mold compound may contribute some variability to the moisture weight gain. The graph shown in Figure 2.27 showing the results of moisture weight gain over time as a function of the type of mold compound. The general conclusion from the various studies is that there is no appreciable difference in the commonly used molding compounds.

The typical plastic molding compound for the fine pitch IC packages is an epoxy resin polymer. The epoxy resin is mixed with a filler and a hardener prior to transfer molding. The transfer molding system is illustrated in a cross sectional view of a transfer mold as shown in Figure 2.28.

Transfer molding is the process of molding a thermoset plastic. Thermoset plastics are polymers that are soft and viscous at low temperatures. When heated the polymers form complex crosslinked hydrocarbon chains that no longer display

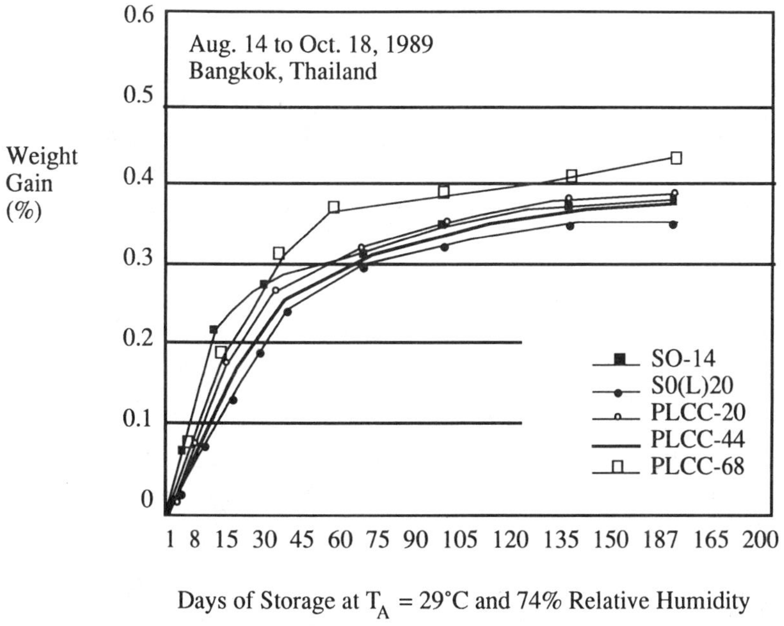

Figure 2.26. The % weight gain due to humidity during the package transfer cracking tests conducted by Signetics—Philips. (Source: Signetics-Philips)

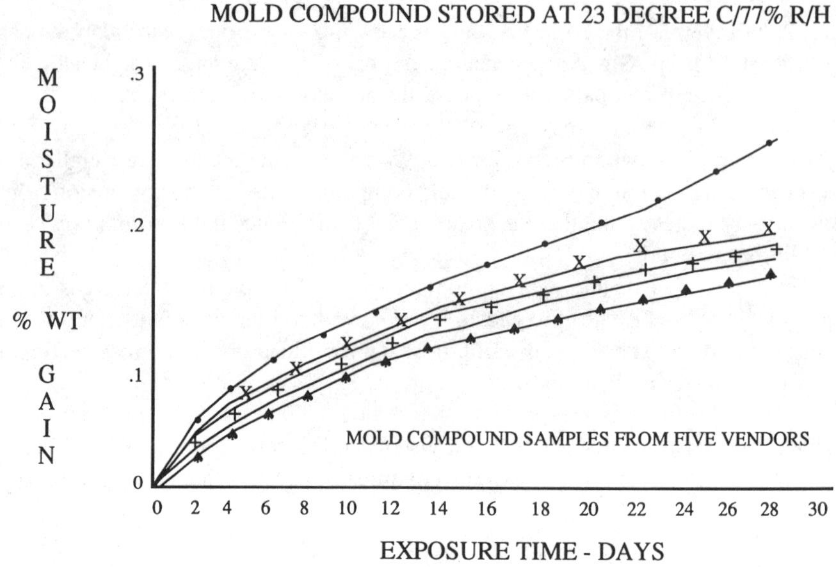

FIGURE 2.27. The amount of moisture absorbed by several common IC package molding compounds. Source: SMTA Journal of SMT, Oct. 1989, pp 24.

a low temperature *melting*, or plastic, point. The most common epoxy resins are diglycidylether of bisphenol A (DGEBA), and Novolac epoxy. The chemical structure of these are shown in Figure 2.29. These epoxies are made of a high purity and must be free of hydrolyzable chlorine, sodium and phosphorus. If not, then damaging chemicals such as hydrochloric acid (HC1) may form when the epoxy is exposed to moisture.

The common hardeners are amino acids or acid anhydrides. These hardeners bond to the epoxy when exposed to heat, and cause the chemical reaction that hardens the plastic. When the hardener and heat are introduced to the epoxy the resultant chemical structure is as shown in Figure 2.30. Often a filler material is mixed with the epoxy. The purpose of the filler is to add strength and to match the epoxy coefficient of thermal expansion (CTE) with the lead metal. The typical filler is fused silica, which reduces the epoxy CTE from 40 to 50 ppm/°C down to 20 to 25 ppm/°C. This assures a more consistent bond of the copper lead (CTE of 15 to 20 ppm/°C) to the plastic and assures a reliable bond over time and temperature changes.

Review of several sources indicate that the package thickness under the die flag, more than the mold compound and the die size causes the cracking phenomenon. This is best shown by the Philips-Signetics tests that showed no package cracking with IC packages that have a molded "dome" under the

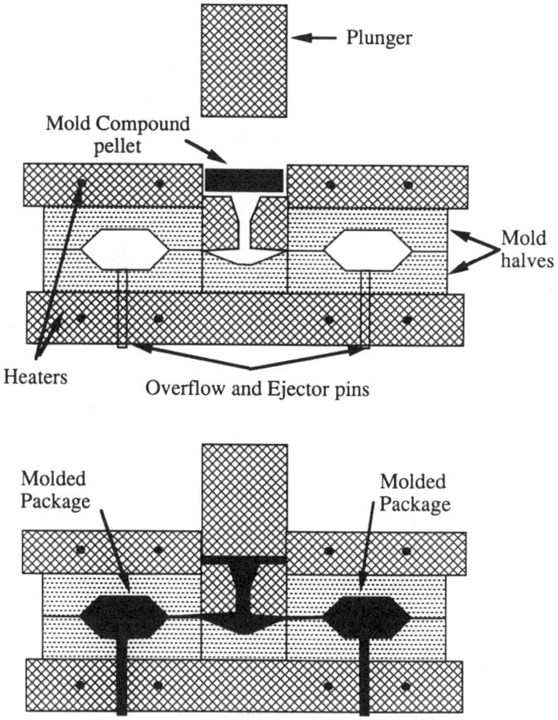

FIGURE 2.28. A cross sectional view of an IC package molding system.

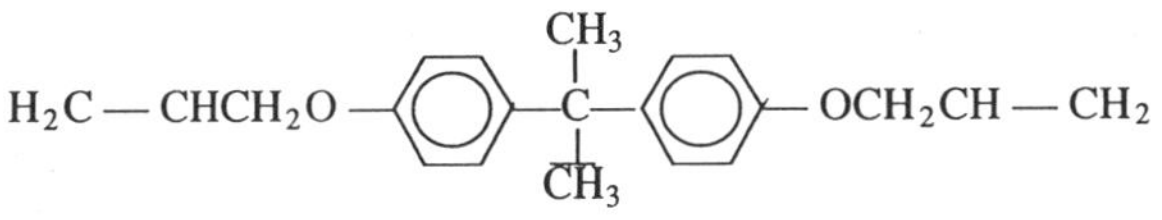

Diglycidylether of Bisphenol A (DGEBA)

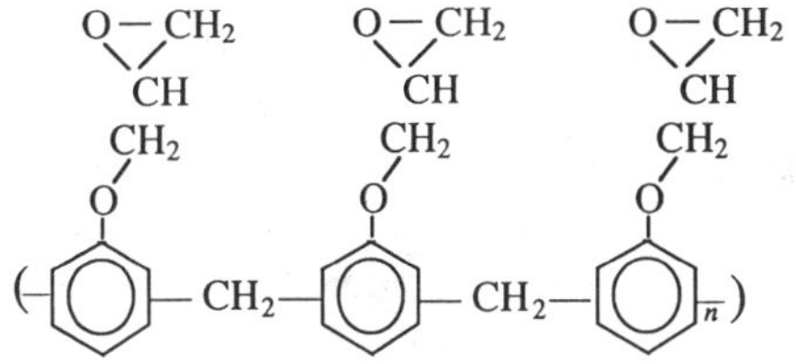

Novolac Epoxy

FIGURE 2.29. The chemical diagrams for the most commonly used epoxy molding compounds for IC packages.

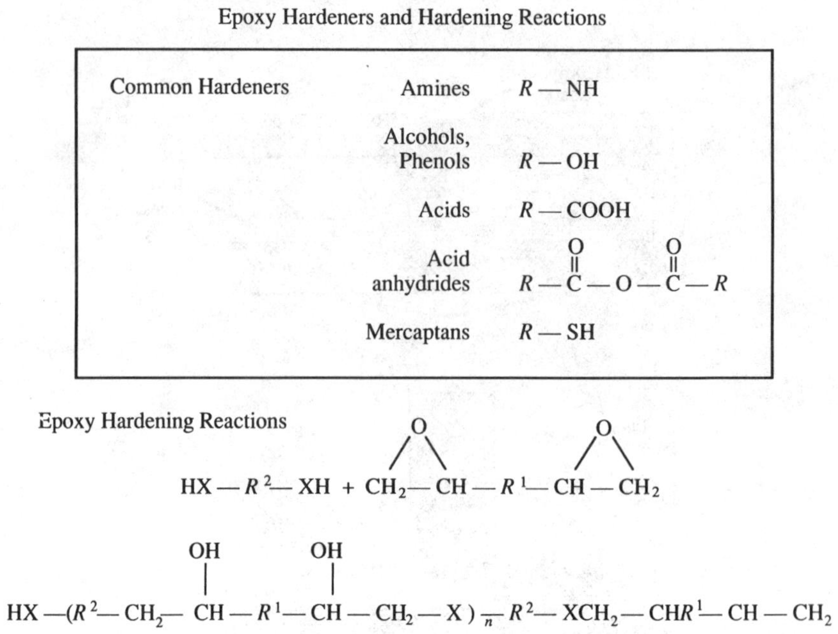

FIGURE 2.30. The chemical structure for hardened molding compound after heat curing.

package. The dome mold, shown in Figure 2.31, effectively increases the package thickness under the die flag, thus reducing the probability of cracking as shown in Figure 2.25.

The IC Suppliers Provide a Solution

For those packages molded with a layer of plastic under the flag thin enough to have an unacceptability high probability of cracking. A solution is detailed in the IPC-SM-786 report. Basically, all susceptible IC packages are baked, pouched, and sealed by the supplier. The baking dries out the absorbed moisture that when rapidly heated, expands against the thin layer of plastic causing it to crack.

One issue that arises with supplier baked packages is what to do with any unassembled parts after opening a pouch containing several tubes or trays of packages. The suppliers' recommended procedure is to store the parts in a low humidity (RH<50%) container until they are needed for assembly. If the packages have been exposed to high humidity (RH>60%) for several hours (>6 hours) then they should be baked to remove the moisture prior to storage and assembly to the printed circuit board. IPC-SM-786 recommends a bake at 125°C for approximately 24 hours.

For those packages with sufficiently thick molding under the flag, Philips-

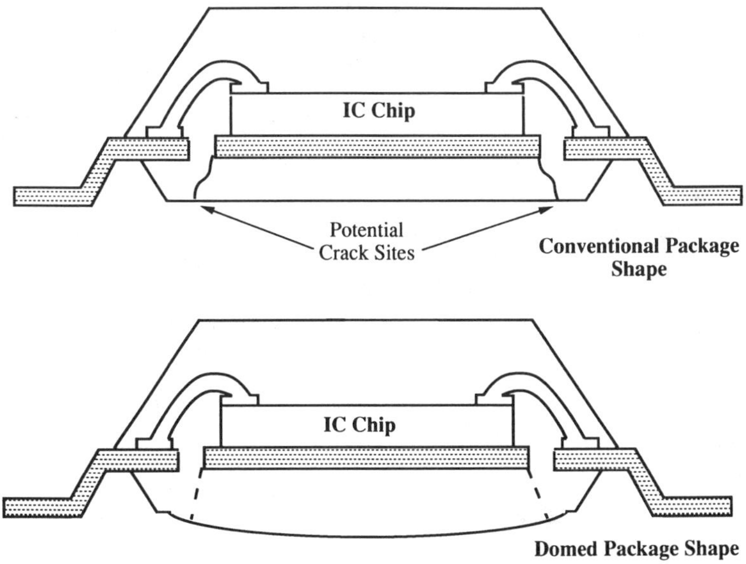

FIGURE 2.31. The addition of a dome under the package may increase the package's resistance to cracking. (Courtesy of Signetics—Philips)

Signetics concludes that no prebaking is necessary. They conclude that the additional plastic thickness provides enough strength to withstand the moisture expansion and release.

Bond and Lead Breakage

Breakage of a wire bond, separation of the wire from the IC pad or the lead frame, or breakage of a TAB lead attachment are, fortunately, problems that occur very infrequently. Defects due to these internal package problems are usually at the parts per billion level.

The internal assembly of fine pitch packages does push our knowledge of reliable assembly to the limit. For example, to enable the very thin (<2.0 mm thick) TSOP and QFP packages, the wire bonds are swept at a lower angle than used for thicker packages. A comparison of low bond sweep versus the higher conventional bond is shown in Figure 2.32.

Bond

Low bond sweep angles may increase the number of postassembly failures due to electrical or environmental malfunction. Electrical malfunction would be caused,

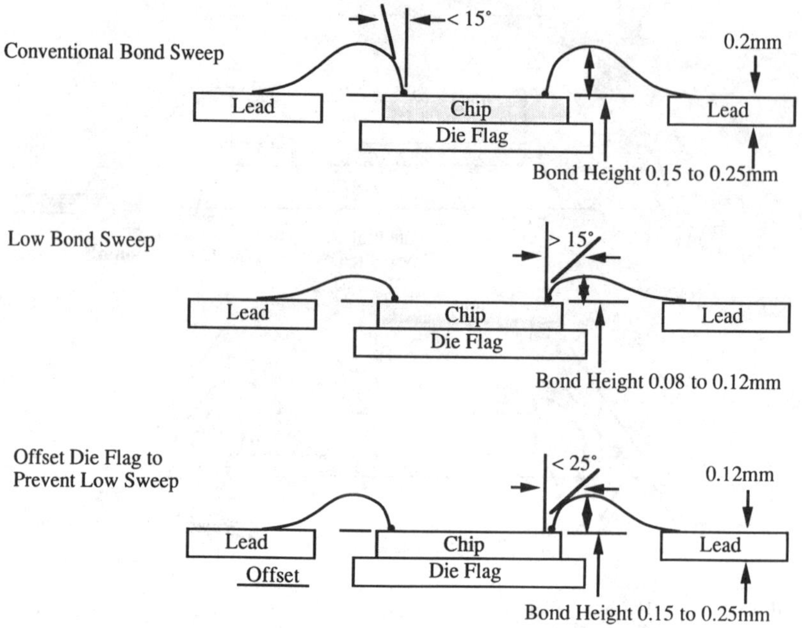

FIGURE 2.32. Thinner packages are the result of advances in wire bonding technology that allow lower sweep angles than those previously thought reliable.

in this case, to the wire bond touching the edge of the chip creating an electrical short. This is shown in Figure 2.33.

The lower sweep angle may induce more stress in the wire to IC bond point. This extra stress may cause the bond to separate when the IC package is exposed to thermal cycling or severe mechanical shock. This is shown in Figure 2.34. To minimize the increase of these potential bonding related problems in the thin packages, the die flag is offset below the plane of the lead (shown in the bottom of Figure 2.32). This allows the bond sweep angle to be less than 25% from the vertical. However, as was discussed in the last section, this construction also

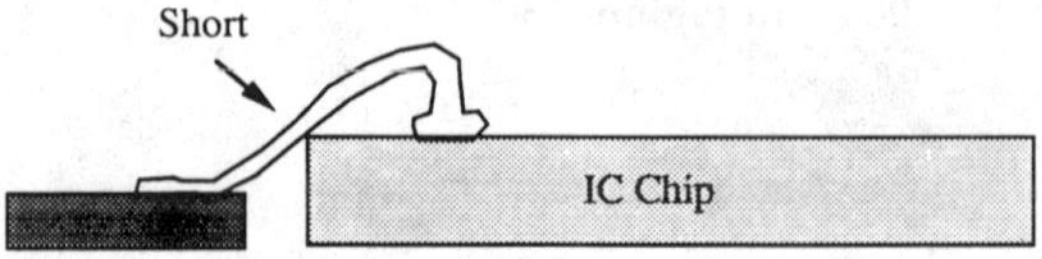

FIGURE 2.33. An example of a failure, a short, due to unreliable excessively low sweep angle.

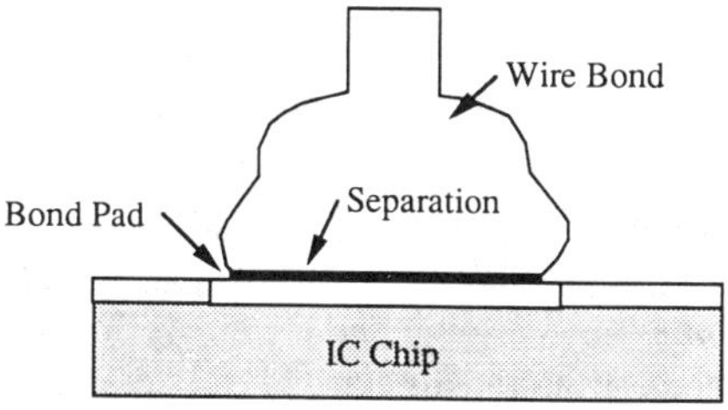

FIGURE 2.34. Wire bond separation from the IC bonding pad due to the stress introduced by thermal or mechanical cycling.

reduces the plastic thickness between the bottom of the die flag and the edge of the package, which may increase the package's susceptibility to cracking.

TAB inner lead bond reliability is receiving significant study as interest in this package assembly technique increases. The strength of the bond of the TAB inner lead to the metalized or bumped IC chip pad is dependent on the bump metals and their quality, the inner lead bond (ILB) temperature, time, and pressure. A generalized graph showing the relative ILB bond strength as a function of bump metal and bonding temperature is shown in Figure 2.35. In general, the test results for a properly inner lead bonded TAB leads show that TAB leads are as mechanically reliable as wire bonds.

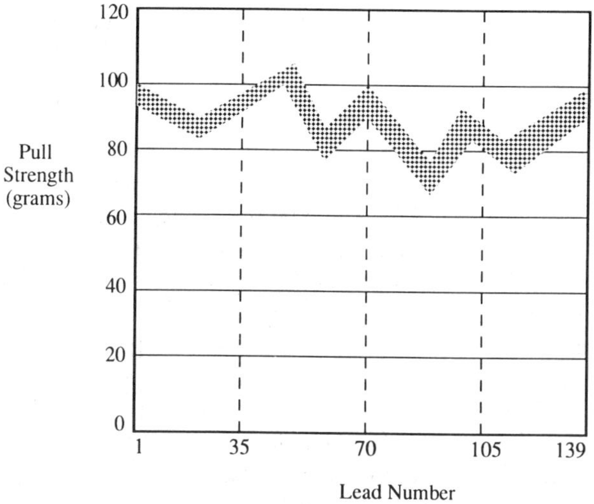

FIGURE 2.35. A generalized graph of the relative Inner Lead Bond strength of a TAB bond, as a function of the bump metal and bonding temperature. (Source: MCC)

Chip Cracking

Chip cracking is a very insignificant problem in today's integrated circuits. However, the thinner fine packages press the reliable process capabilities to the limit. The long thin TSOP package can be subjected to significant package warpage when mounted to a printed circuit board. This warpage puts stress on the IC chip and die flag. If the chip attachment is marginal and the stress due to warpage is large, the chip may crack or separate from the die flag. The cracked chip will fail to function electrically early in its usage. The separated chip attach will generally result in a slower failure due to chip overheating, since the thermal path from the IC junctions to the die flag will be of a higher thermal resistance. To help prevent the occurrence of chip cracking, the die flag is designed with chamfered ends. Also, some suppliers coat the IC chip with a compliant gel that provides a flexible region for the chip, much as bubble packing used to protect sensitive items in shipping boxes.

Thermal Failures

Integrated circuit failures due to overheating or inability to operate correctly at the rated temperature extremes are probably the most common of the various types of failures examined in this section. Both overheating and operational problems may be due, for some failures, to the quality of the package assembly. On the other hand, some failures may be due to chip design or package design and materials. Failures due to poor chip design are beyond the scope of this text. Failures due to package assembly and materials are generally due to high thermal resistance which allows the chip to overheat. When a chip overheats, the transistor switching thresholds, breakdown voltages, and tracking with other components are altered. When the alterations exceed the limits needed for the circuit to function correctly, a failure occurs.

Thermal resistance is a measure of how fast heat passes from one point to another. The generalized thermal resistance of an IC package consists of two resistive analogies: thermal resistance from the semiconductor junctions to the outer edge of the IC package, and thermal resistance of the IC package to the ambient atmosphere. These analogies can be expressed (and shown in Figure 2.36 as θjc (junction to case resistance) and θca (case to ambient resistance). These two terms are usually combined and listed on each IC supplier's data sheet as θja (junction to ambient resistance). Package materials and package design are directly responsible for the θjc resistance of the package. Fine pitch packaging tends to reduce the thermal resistance and thus improve the ability to get the heat away from the chip faster than the older, larger packages. This reduced θjc is shown in Table 2.9; the table also lists the θja of the package in an ambient of still air at a temperature of 25°C. As shown, a dual-in-line package (DIP) operating at 1 W with a copper lead frame has a θjc of 40 to 45°C *above* the ambient.

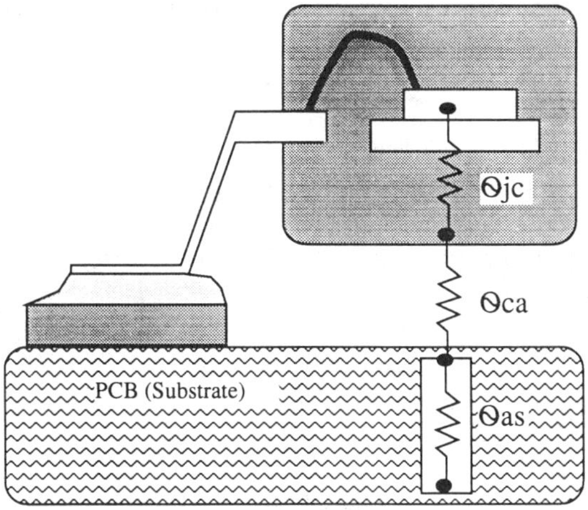

FIGURE 2.36. A diagram of the thermal resistance analogies for a semiconductor package soldered to a printed circuit board.

Therefore, this chip will have junction temperatures of 65 to 75°C. It may be best to limit the safe maximum junction temperatures on a chip to 105 to 125°C, and there is a trend to lower the safe maximum to 65 to 85°C.

If the safe maximum junction temperature is 105°C, for example, then the maximum ambient temperature or more specifically the temperature on the outside of the package is limited to 55 to 65°C. As a result the DIP package in this table will need special cooling to keep the package temperature below the 55 to 65°C limit if the chip dissipates 1 Watt and the maximum junction temperature is 105°C. If the maximum safe junction temperature is 125°C, then the maximum

TABLE 2.9 The Junction to Ambient, Θja, of Various Semiconductor Packages Measured in Still Air. (Source: Signetics-Philips)

Package	Θ_{JA} Range °C/W	Θ_{JC}
Dip	155 - 170 (copper leadframe)	40-45 (copper)
SO-8	200 - 260 (alloy 42 lead frame)	
Dip - 16		
SO - 16	105 - 120 (copper)	32 - 36
SO - 16	85 - 100 (copper)	22 33
PCC - 20	70 - 80	26 - 38
DIP - 44	40 - 47	
PCC - 44	40 - 53	14 - 22
DIP - 64		
PCC - 68	40 - 46	14 - 16
QFP - 84	42 - 50 (TAB bonded)	22 - 24 (TAB)
	31 - 37 (wire bonded)	10 - 13 (wire)

package temperature increases to 75 to 85°C. Now the DIP package is capable of operating without special cooling.

The increase in the number of finely spaced leads in a package brings the benefit of a lower θjc. This is due to the increased metal mass of the leads. If the leads are made of copper, then this metal mass provides a very effective thermal path. Additionally, the thinner packages allow the heat to get to the package edge faster simply because there is less distance for the heat to travel. For example, the wire bonded (2 mil wire) 100 pin QFP package has a θjc of 10-13°C/W. This represents a four to fivefold improvement over the DIP thermal resistance.

Note that Table 2.9 shows the TAB and 1 mil wire bonded QFPs having a θjc of 22 to 24°C/W compared to the 2 mil QFP with θjc of 10-13°C/W. This is due to the reduced metal mass of the thinner leads and wires bonded to the chip. This comparison gives a relative indication of how much heat flows out of the leads as opposed to out of the package body. The traditional approach to package cooling has been to force cool air across the top of the package or to put a large heat sink of a highly thermally conductive metal such as aluminum on the top of the package. Finer pitch packages may also benefit from a cooling method that emphasizes removing the heat from the leads. This will be discussed in greater detail in Chapter 10. The solder joint temperature is one of the prime determinants of solder joint reliability over the service life of the product. In general, the cooler the solder joint and the cooler the IC, the more reliable and longevity of the product.

2.8 ISSUES WHEN SELECTING AN FPT PACKAGE

Too Many Options

FPT presents a myriad of package choices to the circuit designer. To see how many packages can be chosen for a given IC chip, a list was constructed of the various options available commercially to package an ASIC requiring 68 leads. The choices currently available are shown in Table 2.10.

Table 2.11 lists some common specifications for each package. These specifications include lead pitch (center to center) and lead width, and also the overall package size, package thickness, and standoff or clearance height of the substrate. This latter specification is important for cleanability.

One can readily see that little, if any, compatibility exists between the various packaging styles. This means that using a custom IC or ASIC from supplier A in a PQFP package (JEDEC) would not be compatible with the ASIC from supplier B in a QFP package (EIAJ), even if the silicon function is the same. The result is a sole source supplier based primarily upon package type, not silicon. This provides

TABLE 2.10 There are Many Package Options Available for High Lead ICs, Such as the 68 Lead Options Shown in this Table.

Designator	Commercial Name	Technology	Standard Reg.
68 lead PCC	Plastic Chip Carrier	SMT	(JEDEC)
68 lead PQFP	Plastic Quad Flat Pak	FPT	(JEDEC)
68 lead QFP	Quad Flat Pak	SMT/FPT	(EIAJ)
68 lead VQFP	Very Small QFP	FPT	(EIAJ)
68 lead TapePak	TapePak	FPT	(JEDEC)
68 lead PPGA	Plastic Pin Grid Array	TH	(JEDEC)
68 lead COB	Direct Chip on Board	COB	(IPC)
68 lead TOB	Direct TAB on Board	TAB	(EIA/IPC/ASTM)
68 lead Micropak	Micropak	FPT	(Europe/DIN)
68 lead PQFP	Metric PQFP	FPT	(JEDEC)
68 lead Imp	Plastic Quad Flat Pak	FPT	-------
68 lead TQFP	Thin QFP	FPT	(EIAJ)
68 lead CQFP	Ceramic QFP	FPT	(JEDEC)
68 lead CQUAD	Ceramic QFP	FPT	(EIA)
68 lead LCC	Leadless Chip Carrier	SMT	(JEDEC)
68 lead SQFP	Shrink QFP	FPT	(EIAJ)
68 lead GQFP	Guard-Ring QFP	FPT	(EIAJ)

TABLE 2.11 The Nominal Mechanical Specifications for Some of the 68 Lead IC Packages Commercially Available.

	Lead Pitch	Lead Width	Pkg. Area	Pkg. Thickness	Standoff Height
PCC	1.3	.71	625 sq.	4.6	1
PQFP	.64	.3	480 sq.	2.5	.8
QFP	.65	.35	420 sq.	2.8	.13
TQFP	.5 to .25	.35	144 sq.	1.7	.05
TSOP	.5	.25	96 sq.	1.2	.05
PPGA	2.5	.5	900 sq.	4.6	1.0
TAB	.5 to .1	.25	50 sq.	1.0	0
COB	.1	.15	40 sq.	1.0	0

the IC supplier a significant market opportunity to provide uniqueness of silicon function *and* package.

IC suppliers do not purposely limit their package offerings to create a sole source package-hook on a user. But they do limit their package styles because of cost constraints.

The cost to tool an integrated circuit package has always been high. The precision tooling needed for FPT packages has driven this cost to a new plateau. The typical body mold is an estimated $150,000 and the lead separator is $350,000 per lead configuration. Older package tooling costs are about half these amounts.

Package Standards

Package standards define the mechanical tolerances of the package body and its leads. There are no industrywide standards on the material properties or the construction of the package. Although several suppliers use *pseudo-standards* which are the construction standards of one or two of their largest or most quality conscious customers. There are two standards writing bodies for the mechanical tolerances. These are the JEDEC (JC-11) committee of the EIA (Washington, D.C.) and the packaging subcommittee (EE-13) of the EIAJ (Tokyo, Japan). In recent years, the JC-11 JEDEC committee has allowed EE-13 EIAJ standards to flow directly into Jedec for outline registration. While the JEDEC officials will say these are not standards, they can be construed to be for all practical purposes. Interested readers can receive an official copy and updates of the mechanical outlines for fine pitch (including TAB) and every other semiconductor package type by ordering publication 95, "JEDEC Registered and Standardized Outlines for Semiconductor Devices" from the EIA. The EIA offers an update service and a companion publication for all surface mountable passive components, entitled EIA-PDP-100, "Registered and Standardized Mechanical Outlines for Electronic Parts".

Samples of the standards designators for some fine pitch packages are described in Table 2.12.

Users should only use packages that are supplied to the JEDEC or EIAJ standards. This prevents unknown tolerances and many potential problems.

Reliability and Quality

Quality and reliability issues were outlined in Section 2.7; however, common experience shows that fine pitch packages provide a very reliable means for protecting the sensitive integrated circuit. There have been very few field failures reported due solely to package construction; this includes failures due to moisture and package thickness related package cracking. This problem is now well under control, resulting from supplier precautions such as prebaking and thicker packages under the die flag.

TABLE 2.12 The Standards Documents that Define the Allowable Tolerances for sEmiconductor Packages use the Designators Shown in this Table for the Various Fine Pitch Packages.

Package	Jedec Pub. 95
PQFP	MO-069-086,-089
QFP	MO-071
TQFP	- - -
TapePak	MO-094
GQFP	MO-109
CQFP	MO-090,-100, -114
Tape QFP	MO-102
VSOP	MO-117
TSOP	- - -
CQUAD	MO-018
Land Grid Array	MO-062

The largest package related problems tend to be parametric electrical performance and solderability. Both of these problems are discussed in later sections.

References

1. Freeman, M., 1991. Minutes of Meeting No. 2 EIAJ EE-13/EIA JC-11 Package Outlines.
2. Hoffman, P. 1988. TAB Implementation and Trends. In *Solid State Technology*, June, 1988.
3. Holzinger, S. 1988. TAB Types and Material Choices. Paper read at Expo SMT '88, Sept. 1988, Las Vegas, NV.
4. Johnson, B. et. al. 1989. Electronic Materials Handbook, Vol. 1. Materials. Park, OH: ASM International.
5. Kiehna, L. 1991. The Financial Impact of the SMT Process. In *Surface Mount Technology*, July, 1991: 46–47.
6. Mangin, C.-H. 1989. Surface Mount Changes Shop Floor Economics. In *Electronic Packaging and Production*, Sept. 1989: 112–114.
7. Marcoux, P. et. al. 1989. An Introduction to Tape Automated Bonding and Fine Pitch Technology. Lincolnwood, IL: Surface Mount Council, SMC-TR-001.
8. McKenna, R. 1989. Surface Mount Device Package Cracking: An Overview. Journal of Surface Mount Technology 2(4): 20–29.
9. McShane, M., et. al. 1990. Lead Configuration and Performance for Fine Pitch SMT Reliability. Proceedings of Nepcon West 1990 1: 238–257.
10. —1990. Report on the Reliability and Durability of Signetics Surface Mount Devices. Sunnyvale, CA: Signetics Co.

3

Fine Pitch Product Applications

3.0 APPLICATIONS

Fine pitch packages will, in time, be used in all applications incorporating electronics. Today, FPT is associated closely with application specific ICs, high lead count microprocessors and math coprocessors, and high density memory applications. Who begat whom—in this case, does FPT allow more justification for ASICs and memory, or vice versa—is fodder for some lengthy debates, but not in this text.

In this chapter, we examine the present applications where FPT packages are used, including some unique applications as a result of FPTs density capabilities. The forecasted usage of FPT by package type is also estimated.

3.1 PRODUCT APPLICATIONS

The present product applications for fine pitch packages can be listed as:

1. Computers and peripherals
2. Consumer and automotive
3. Communications
4. Memory

Computers and Peripherals

Portability and small size have been two of the prime selling features of personal computers and disk storage devices for the past ten years.

The QFP packaged microprocessor and logic devices, and SO packaged memory devices ushered in the era laptop computer (Figure 3.1) in the mid-1980s. The finer pitch QFP packaged microprocessors and ASIC devices and the TSOP

FIGURE 3.1. Photo of a laptop computer.

packaged memory devices introduced the next computer size down, the palmtop computer (Figure 3.2) and allowed significant increase in the functional capacity of the larger footprint desktop and laptop computers.

These new added features on the larger footprint computers that have been introduced because of (a) extra un-populated board area and (b) greater ASIC capability in a small package. Some of these features are built-in modem, sound generators, math coprocessors, and cache memory. Additional features include on-board color video capability, local area connectivity, and coprocessors to enable running software files composed on different computer operating systems and software applications.

Disk storage devices have made tremendous leaps in memory capacity largely due to improvements in magnetic plating and detection technology, but also in improved electronic densities. This is illustrated in the graph in Figure 3.3 of hard disk drive capacity (in MB) and physical capacity/size, and Figure 3.4 which shows a disk drive.

Consumer and Automotive

This category represents the largest usage of FPT devices. Products in the consumer category include video camcorders, portable TVs, CDs, radios, smart credit cards, and photographic cameras.

FIGURE 3.2. Photo of a palmtop computer.

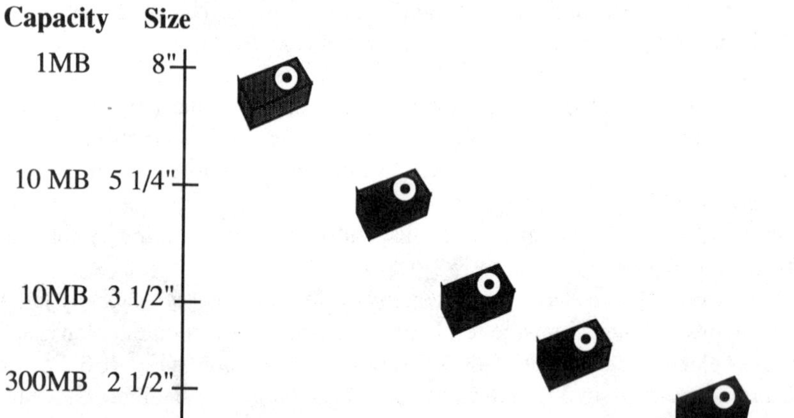

FIGURE 3.3. As circuit densities have increased, so have the memory capacities for disc drives in small sizes.

FIGURE 3.4. Photo of a dense high capacity disc drive memory.

Within the automotive sector are engine and emission controllers, accessory monitors, and entertainment systems.

Communications

The obvious products incorporating FPT in this product area include facsimile machines, cellular telephones, and desktop PBX exchange systems.

The remaining product categories of industrial and automation, and military and aerospace are the smallest user categories due primarily to low product volumes. The significant products in these two areas are shown in the summary of applications by category shown in Table 3.1.

Some of the applications of the finer pitch packages can be expressed by lead count, IC chip size and the lead pitch. These are shown in Figure 3.5. Most current products using FPT packages use high pincount ASIC components and microprocessors.

Memory

Another leading application will be memory cards. These will be used in several of the product categories discussed above. An example of a memory card is shown

TABLE 3.1 FPT is Used in a Wide Variety of Applications and Products.

Automotive Applications
Radio controls
Engine Controls
Emission controls
Accessory monitors
Computer Applications
Portable, transportable, laptop and
Palmtop personal computers
Super computers
Parallel processor computers
Fault tolerant computers
Communications
Cellular telephones
Portable transmitters
Desk-top PBX systems
Consumer Products
Miniature radios
Compact disc players
Video camcorders
Electronic typewriters
Military
All advanced VHSIC programs
Avionics
Smart weapons
Industrial
Portable industrialized computers/controllers
Remote intelligent controllers
Instrumentation
Portable oscilloscopes
Portable multi-channel analyzers
Portable medical monitors/analyzers
Portable medical dispensing/controlling systems

in Figure 3.6. A memory card is the size of a credit card and about twice as thick. Encased in the card is a printed circuit board containing significant memory and decoding circuitry. The information stored in the memory can be of a personal nature such as financial and medical history. It can also serve as alternative storage of computer user files. Memory cards are more versatile than magnetic and optical disks since the card can contain enabling circuits and other functions.

Consumer applications such as VCRs, camcorders, and facsimile machines are

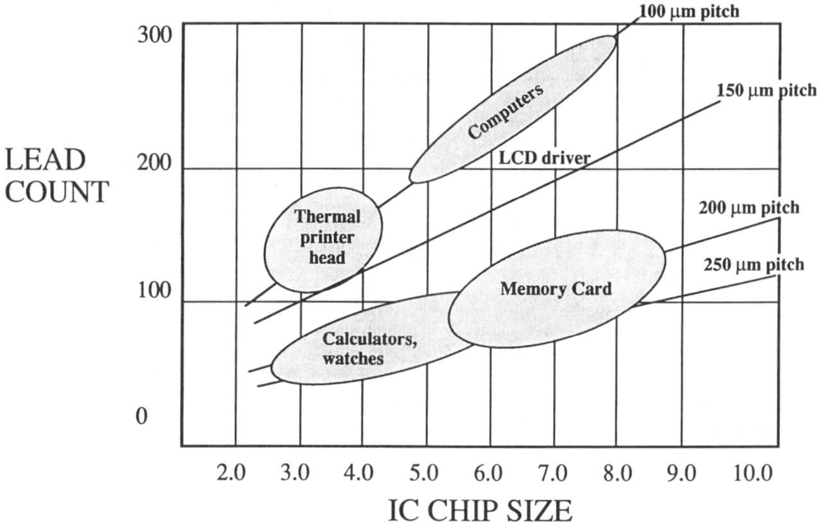

FIGURE 3.5. Common applications of fine pitch IC packages, shown as a function of lead density. (Source: TechSearch International)

currently the highest volume users. Table 3.2 shows the projected usage of SMT and FPT devices by consumer product application. This forecast is from the Japanese Printed Circuit Association. We have added the FPT column based on the expected percentage of FPT packages.

From Table 3.2, it can be seen that electronic cameras are the largest user category. The density advantages of FPT packages has created new camera products such as the autofocus "point and shoot" cameras and the small S-VHS camcorders.

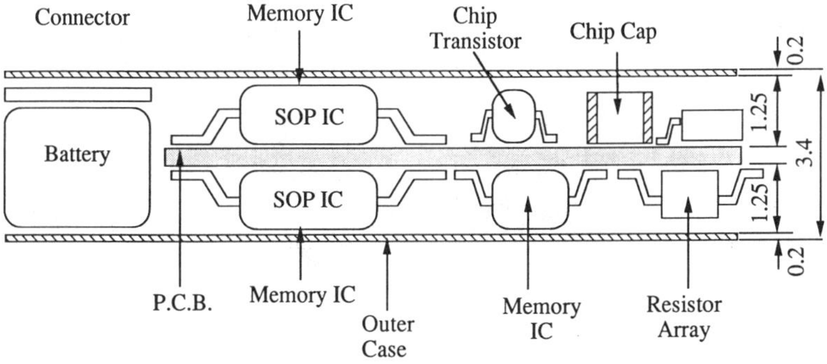

FIGURE 3.6. Cross sectional diagram of a dense memory card. (Source: Mitsubishi Electric)

TABLE 3.2 Forecasted Applications for SMT and FPT by the Japanese Printed Circuit Association. (Source: JPCA and PPM Associates)

	1985 (% SMT)	1990 (% SMT)	1990 (% FPT)
Cameras	86	94	14
Electronic watches/clocks	45	75	11
Video cameras	40	70	10
Radio cassettes	29	48	7
Sewing machines	--	42	6
Stereos	4	41	6
VCRs	9	36	6
Calculators	23	35	5
TVs	12	32	4
Broadcasting	29	58	9
Facsimile	2	45	7
Telephone exchanges	16	43	6
Car electronics	5	39	6
NC equipment	12	37	6
Cable communications	5	37	6
Mini computers	--	32	4
Mini and large mainframe computers	1	20	3

$$\% = \frac{\text{SM devices}}{\text{Total devices}}$$

Over the next five years, small computer applications, particularly the palmtop computers, will grow to the second leading application category.

3.2 PACKAGE USAGE

FPT package usage has grown dramatically over the past few years. FPT and other assembly technology growth rates are illustrated in Figure 3.7. The annual growth rate of fine pitch package usage is increasing at a rate of 55 to 65% per year. This compares with a 30 to 35% rate for SMT, and a declining rate for through-hole packages.

It was estimated that in 1990 approximately 1.2 billion FPT packages were used worldwide (3% of the 38 billion total IC packages). Almost all of these were of the QFP style. In 1995, it is estimated that 7 billion FPT packages will be used. This represents about 10% of the 70 billion total estimated IC packages to be used. By the year 2000, the forecast is for 21% of all of the IC packages to be of the fine pitch category.

The fine pitch packages, representing only 3 to 10% of the total components on a typical printed circuit board, will however represent the largest portion of the unit cost in both component and assembly cost. In several cases, the half dozen or fewer

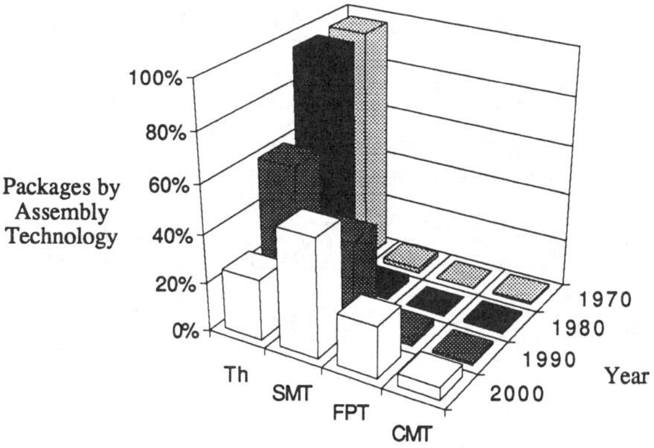

FIGURE 3.7. Estimated package growth rates segmented by assembly technology.

FPT packaged integrated circuits on a printed circuit board can represent 60 to 90% of the entire product's cost.

In 1995, the majority will still be the quad flat pak (QFP) packages, but with a trend toward the finer pitch (< 0.5mm) styles as lead count requirements continue to increase. By the year 2000 the 0.3 mm QFPs are expected to be the dominant packages.

References

1. Aburaya, K. 1989. Examples of the Latest SMT Installation Techniques in Audio and Video Equipment. Proceedings of ISHM "89, Oct. 1989, Baltimore, MD.
2. Marcoux, P. 1989. *Fine Pitch Technology—A Market Study*. Saratoga, CA: Electronic Trend Publications.
3. Vardaman, E. J. 1990. New TAB Developments in the United States and Japan: A Market/Technology Comparison. In *Semiconductor World*. June, 1989.

4

Printed Circuit Boards for Fine Pitch Technology

4.0 BASIC PCB PROCESSING

The term printed circuit board (pcb) is used to describe the substrate that provides the interconnections from one IC package to all the other components and IC packages in the circuit. A substrate is also called, by many, a printed wiring board (pwb), but, this text will use the more common term *printed circuit board*. The substrate must meet several requirements to fulfill this task.

The ideal substrate or printed circuit board will:

- Provide adequate mechanical support
- Allow undisturbed electrical interconnection
- Meet the allowable size, cost, and weight budgets
- Exceed the reliability expectations of the product
- Not impede the package to substrate assembly and test processes

The vast majority of printed circuit boards are laminates of copper, a bonding resin such as epoxy, and a reinforcing material, such as woven fiberglass. The materials of a printed circuit board as they relate to manufacturing, electrical properties, and reliability will be the focus of this chapter.

Basically, to form a printed circuit board using the subtractive process to remove copper on fiberglass reinforcing material, the fabricator usually starts with a laminate that is surfaced with thin sheets of copper foil on one or both sides of the fiberglass reinforcement. The basic subtractive printed circuit board fabrication process is shown in Figure 4.1. The copper is bonded to the reinforcement, by a laminator, using various epoxy resins. This laminate is purchased by the fabricator.

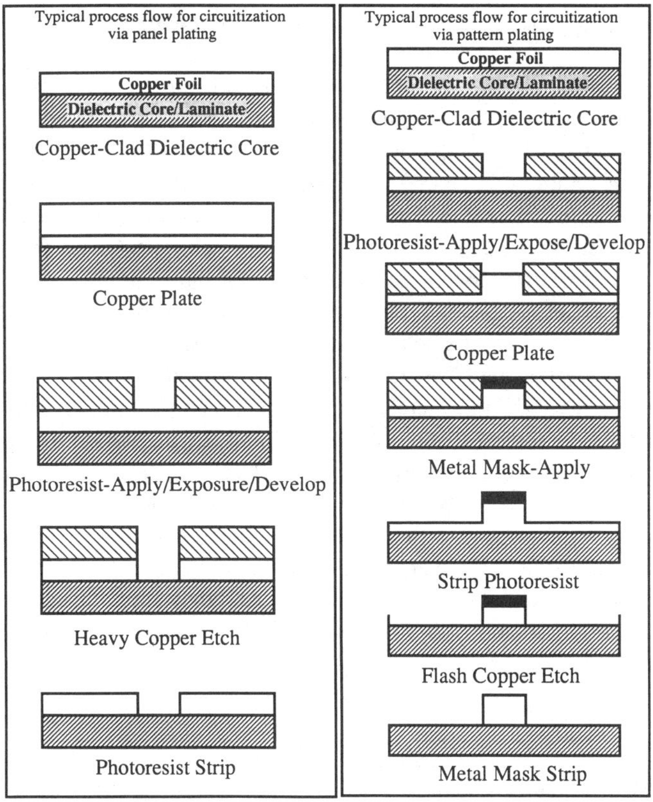

FIGURE 4.1. The steps used in subtractive printed circuit board fabrication. (*Adapted from PC Fab, May, 1990*)

The fabricator first drills the pin and via holes in the laminate. Next the copper is coated with a photosensitive material and a photoimpression of the trace, land and via features is imprinted onto the material by exposing it to light (normally ultraviolet) through a phototool mask, or these impressions are drawn on using a light or laser beam. The photoresist material over the nonactive copper areas is washed away, exposing the copper. The exposed copper is acid-etched, leaving only copper interconnect patterns. This is repeated for the multitude of layers that form the interconnection on the board, with the exception of the outer two sides. The layers are then laminated together under pressure and heat to form a laminate sandwich. Fabricators may also use a vacuum or autoclave to assist the process. The outer layers are masked with photoresist, but before the copper is etched it is electrolessly and electroplated to deposit copper in the holes to form the through the board interconnects. The exposed photomaterial is removed and the copper is coated with tin to allow etching of these thicker traces.

In contrast to the subtractive fabrication process is the additive process. This is illustrated in Figure 4.2. The additive process starts with a treated bare laminate material that is free of copper. The necessary holes are drilled or punched into the laminate. The laminate is coated with a photoresist that becomes a permanent feature on the board after it is exposed and the trace areas are cleaned. Copper is then electrolessly deposited in the trace and through-hole areas. The copper is coated with a semi-permanent resist that masks the copper in the trace and through-hole areas. After the unwanted copper and resist are removed, a soldermask is applied and the board is solder dipped and leveled.

The correct design and specification of the printed circuit board is key to achieving the required form factor, performance, manufacturability, quality, and reliability of the product. Meeting these needs and staying within the cost budget of the product can be challenging and often elusive. A large portion of printed circuit boards produced worldwide are fabricated in captive factories, i.e., they serve one customer. Because their needs are derived from one customer, the captive fabricator should be in an excellent position to adopt new technologies in a timely fashion to meet the customer's needs. And they should be able to readily acquire the funding to add equipment and staff. However, this is not the case, the

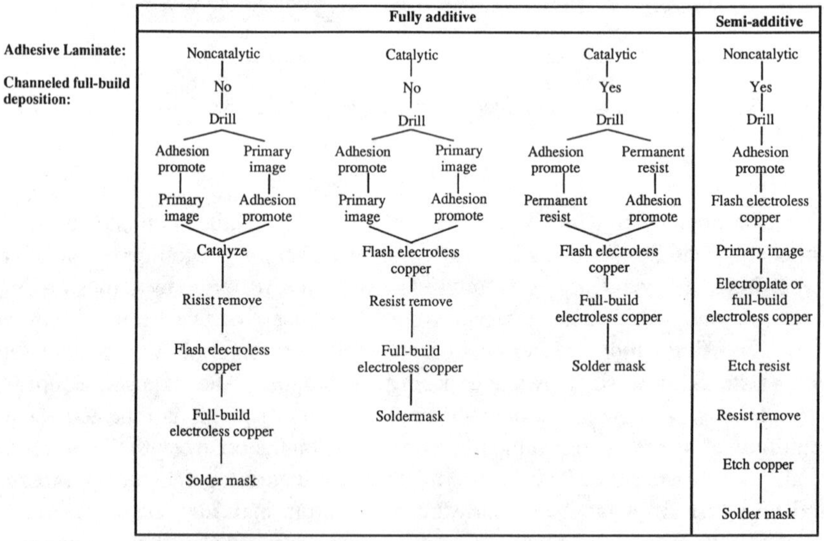

FIGURE 4.2. The steps in processing fully or semi-additive printed circuit boards. (*Adapted from Electronic Materials Handbook, Vol. 1. ASM International*)

large captives while serving one company really serve several diversified divisions within that company, each with different needs. And as a result, these factories are often slow to add new board technologies, or to meet the needs of new assembly technologies such as fine pitch. As a result, a growing number of users of printed circuit boards use external contract fabricators.

Currently, approximately 40% of the printed circuit boards fabricated are done by noncaptive, contract fabricators. Contract fabricators serve many customers. The trend over the past several years has been for end equipment manufacturing companies (OEMs) to deemphasize and in some cases to abandon their captive fabrication facilities in favor of using contractor services.

There are several possible reasons for this trend toward contract fabricators. For example, the effect of changing corporate direction can render a large captive fabrication facility obsolete in a short period of time. The large captive facility is sometimes viewed as too inflexible or too expensive to change to newer and nonmainstream technologies. Large captive factories often lack the people-push needed to make internal changes as quickly as a contractor will tend to do. Figure 4.3 illustrates this trend. From 1989 to 1990, the top 10 contract fabricators enjoyed a 9% increase in sales in an overall flat market. During the same period the top 10 captive fabricators witnessed a 10% decrease in annual sales.

Individual contract fabrication companies are usually not as well financed as the captives, which means they are usually more financially constrained in adding new equipment, staff and technology. However, as a body, contractors form a fluid aggregate of new technology developments. Improvements are driven by the

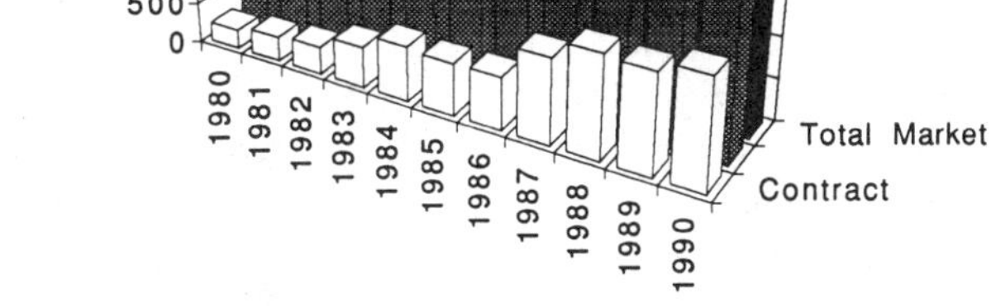

FIGURE 4.3. Sales of printed circuit boards of independent vs captive suppliers. (Source: Kirk-Miller Associates and the IPC)

survival instincts of the individual companies within the body. Those contractors with a strong desire to survive, manage to find the customers and capital necessary to keep pace. Those that don't, fade away.

Locating the capable contract fabricators can be a time consuming and expensive chore. Monitoring the contractor for changes or deficiencies in the printed circuit board process capability and quality can be another expensive task. Often, these chores are three to ten times more expensive than the initial amount invested in designing the printed circuit board.

4.1 MATERIAL OPTIONS AND PROPERTIES

There are several material options available simply to provide mechanical support and electrical interconnection for components. However, only a few of these materials provide all the desired properties, with the overwhelming favorite being copper laminated on epoxy-fiberglass. This material offers low cost, flexible size, multiple interconnect layers, and satisfactory electrical performance. If designed and fabricated correctly, it also provides for excellent manufacturability and reliability.

Other viable materials for special applications include different resin systems, such as, polyimide, cynate ester, high temperature epoxy, and injection molded plastic. Other viable reinforcements include quartz, aramid fiber, ceramic, insulated steel, anodized aluminum, and copper bonded to flexible substrates (such as mylar or or polyimide).

Reinforcing Materials

Over the years several alternative materials have been explored to replace fiberglass. But none have been able to displace fiberglass. This is due in part to the significant variation in their relevant properties and cost. Some of these materials and their properties are listed in Table 4.1 and discussed below.

Fiberglass is the most commonly used reinforcing material for printed circuit boards. Fiberglass is used in over 95% of the boards produced worldwide. This is due to its low cost, high availability, and resultant manufacturability. There are two types of fiberglass that may be used in electronic boards, E-glass and S-glass. Of the two E-glass is used almost universally. Both glasses are grades of soda lime fibers that are bundled into a yarn and then woven into a fabric blanket. The E denotes it as electrical class, and S denotes a high strength class. There is a third class that may be of future interest, D-glass, which offers a lower dielectric constant but is not as strong as E- or S-glass.

The number of glass strands in the yarn and how densely the yarn is woven determines the surface smoothness, how many layers are laminatable, and the coefficient of thermal expansion in the z-axis (change in thickness). The glass

TABLE 4.1a Properties of Some Various Reinforcing Materials. (Adapted from Electronic Materials Handbook, Vol. 1. ASM International)

Material	Form	Modulus		Dielectric Constsnt @ 1 MHz	Specific Gravity	CTE, ppm/°C		Relative Cost
		GPa	MSI			Long	Radial	
E-Glass	Fiber	1.5	10.5	5.5	2.54	5.5	5.5	1
S-Glass	Fiber	1.8	12.5	4.5	2.49	2.6	2.6	
D-Glass	Fiber	1.1	10.0	3.6	2.16	3.1	3.1	
Quartz	Fiber	1.45	10.0	3.8	2.2	0.54	0.54	
Kevlar 29*	Fiber	2.0	12.0	4.1	1.44	-4.0	60	14
Kevlar 49*	Fiber	2.68	18.5	4.1	1.44	-5.0	60	
e-PTFE	Fiber	0.19	28.0	2.0	2.54	70	120	

*(aramid)

TABLE 4.1b

Material	Thermal Conductivity W/mK	Moisture Absorption (%)	Bulk Resistivity $10^6 \Omega$-cm	Voltage Breakdown KV	Softening Point °C
E-Glass	.32	0	10^3 to 10^4	>40	841
S-Glass	.32	0	10^3	>40	860
D-Glass	.32	0	10^6	>40	771
Quartz	.30	0	10^6	>40	1670
Kevlar 29*	.25	0.5-2.0	10^6	>40	>260
Kevlar 49*	.25	0.5-2.0	10^3 to 10^6	>40	>260
e-PTFE	.26	0.4	$>10^3$	>40	>260

*(aramid)

fabrics are produced in various styles as shown in Table 4.2. Close-up diagrams of the cloth weave of some of the common fabrics are shown in Figure 4.4.

Though there are many different glass cloth styles available, the industry uses only a few. The preferred and recommended single or combined fabrics for different multilayer applications are:

Application (Nominal Thickness)	*Fabric(s)*
Military (0.05mm)	1080/106
Commercial (0.05mm)	2113
All (0.15mm)	2116/106 or 2113/1080
All (0.25mm)	2116/2113
All (0.2mm)	7628

Quartz reinforcement has been used in some applications where low dielectric and low CTE are required. Quartz forms a very good bond with epoxy and polyimide resin resulting in a strong board with higher frequency capability then fiberglass. Unfortunately, quartz is more costly than fiberglass. The quartz cost to

TABLE 4.2 Some Properties of Common Fiberglass Fabrics. (Source: ADI/Isola)

Style	Count (Per/Inch) Warp x Fill	Yarn (U.S. Nomenclature) Warp x Fill	Nominal Thickness (Inches) For reference only	Weight (OSY)
104	60 x 52	D900 1/OxD1800 1/0	.0011	0.55
106	56 x 56	D900 1/OxD900 1/0	.0013	0.72
108	60 x 47	D900 1/2xD900 1/2	.0024	1.40
1080	60 x 47	D450 1/OxD450 1/0	.0021	1.38
112	40 x 39	D450 1/2xD450 1/2	.0036	2.08
2112	40 x 39	E225 1/0xE225 1/0	.0032	2.04
113	60 x 64	D450 1/2xD900 1/2	.0034	2.39
2113	60 x 56	E225 1/OxD450 1/0	.0031	2.28
2313	60 x 64	E225 1/OxD450 1/0	.0033	2.36
2116	60 x 58	E225 1/OxE225 1/0	.0037	3.01
7628	44 x 32	G75 1/OxG75 1/0	.0068	6.03

Notes: Above values are industry targets. Yarn count tolerance is +2. All styles listed above are plain weave. Constructions with identical yarn counts, but having yarns with different filament diameters shall be designated by changing the first digit of the style number to next lowest available number ie. 2116 vs 1116

FIGURE 4.4. Close-up views of three common fiberglass fabric weaves used in circuit board reinforcing laminates. (Courtesy of ADI/Isola)

fiberglass is in excess of 14 to 1. Quartz fibers also increase drill wear during fabrication which adds to the board cost.

Aramid is the generic name for DuPont Kevlar. It is a fiber of aromatic polyimide. Kevlar is an attractive reinforcement for printed circuits because of its negative CTE (− 5.0 ppm/°C) in the x-y axis. Unfortunately the CTE is high in the z-axis (60 ppm/°C), which may introduce microcracks into plated through-holes.

Expanded Polytetrafluoroethylene (e-PTFE) was developed by W. L. Gore. The fabric e-PTFE is also called expanded teflon. This material features a very low dielectric constant ($\epsilon_r \sim 2.0$) and provides an excellent board for high frequency circuits, especially microwave applications. Early problems with the material hampered its use. The biggest problem has been its low dimensional stability, it shrinks after copper etch by 1 to 2%.

Bonding Resins

The copper foil is bonded to the reinforcing material by the uncured epoxy resin in the laminate. The epoxy does double duty, as it functions as the adhesive and the dielectric. The available resins are epoxies, triazine (BT), cynate ester, and polyimide.

The bonding resin imparts other important attributes to the final board beyond simply the attribute of bonding the copper foil to the reinforcing material. The resin determines the board's glass transition temperature (T_g) which can directly affect the way the board is handled during reflow soldering. The resin-richness also affects the relative dielectric (ϵ_r) constant, the loss tangent (tan δ), the thermal conductivity, and the coefficient of thermal expansion (CTE).

Epoxy resin is the most widely used and best understood of the bonding systems. The most common epoxy is FR-4 type, which is a brominated thermoset epoxy with a dicyandiamide hardener. The bromine is added as a fire retardant. FR-4 type epoxy, also called difunctional epoxy, has a low relative dielectric constant (ϵ_r = 3.6) and a relatively low glass transition temperature (T_g = 125°C).

The difunctional epoxy has served the electronics industry well for the past three decades by offering low cost, stable performance for through-hole, wave solder assembly, and operating frequency requirements up to 40 MHz. Fine pitch technology presses the viability of difunctional epoxy, since it requires a higher T_g and higher operating frequency range. A desired T_g of 190°C enables easier handling during reflow. And lower dielectric constant and loss tangent values allow circuit operations at frequencies up to 75 MHz. A dielectric constant of ϵ_r = 3 or less is a desired figure by many leading edge users.

The transmission speed of a digital voltage pulse is inversely proportional to the square root of the dielectric constant, ϵ_r:

$$t \propto \frac{1}{\sqrt{\varepsilon_r}}$$

As the pulse propagates along the trace from one component to another, some of its electrical energy is absorbed and some is radiated. The lost electrical energy is a function of the resin material bonded to the trace. This is expressed as the loss tangent, tan δ, and is usually specified at 1 MHz. To be relevant, both the dielectric constant and the loss tangent need to be measured at the product's maximum transition frequency, which is different than the circuit's clock rate. This will be discussed further in the Section 4.3.

To increase the glass transition temperature, fabricators add to the difunctional epoxy, or substitute completely, higher functional epoxies. These more complex bonding epoxies are known as multifunctional and tetrafunctional epoxy. This results in an increase from $T_g \approx 130 - 140$°C (difunctional) to $T_g \approx 150 - 170$°C for multifunctional and $T_g \approx 180°C$ for tetrafunctional epoxy resin. These T_g values, along with ϵ_r and tan δ, are shown in Table 4.3.

Bismaleimide Triazine (BT) resin is a blend of bismaleimide resin and triazine, which is also known as cyanate ester. BT was introduced by Mitsubishi Gas Chemical Company in the late 1970s. The BT blend is made up primarily of triazine and exhibits a Tg of 220 to 250°C. Other benefits of BT compared to epoxy are a lower dielectric constant ($\epsilon_r = 2.9$ vs 3.6) and a lower loss tangent (tan $\delta = 0.001$ vs 0.032).

When bonded with E-glass material the BT/E-glass composite offers significant benefits as an FPT compatible printed circuit board without a prohibitive cost penalty (see Table 4.3), especially if the BT is blended with FR-4 type epoxy.

Cyanate ester resins were introduced in the early 1970s for various applications. They were considered for printed circuit fabrication but developed low interest because of their tendency to absorb and retain relatively high amounts of moisture. Further refinements in resin purity beyond that of BT have reduced the moisture absorption property to where cyanate esters have become of interest again as a possible replacement for epoxy in high performance printed circuit boards.

Refined cyanate esters, beyond the triazine ester, offer slightly higher Tgs (260°C) but no significant improvement in dielectric constant or loss tangent. Also, their coefficient of expansion is the same as that of epoxy below the glass transition temperature. Consequently, the higher cost of a cyanate ester/E-glass composites limit its interest to only higher performance applications.

TABLE 4.3 Some Properties of Common Bonding Resins for Printed Circuit Board Fabrication. (Adapted from Electronic Materials Handbook, Vol. 1. ASM International)

Resin	Generic Type	Supplier	Dielectric Constant or Relative Permittivity	Loss Tangent	Specific Gravity	Glass Transition Temperature Tg °C
G-10FR(a)	FR-4 epoxy	Norplex	3.6	0.032	1.35	125
RSM-1212	FR-4 epoxy	Shell	3.6	0.032	-	125
Quatrex 5010	Tetra Func. epoxy	Dow	3.6	0.032	1.22	180
RSM-1151	Tetra Func. epoxy	Shell	3.6	0.032	-	180
G-200(a)	BT/epoxy	Norplex	3.2	0.012	1.35	185
BT2110F	BT	Mitsubishi	3.1	0.003	1.35	250
BT2060BG	modified-BT	Mitsubishi	3.9	0.001	-	220
Kerimid	Polyimide	Rhone Poulenc	3.2	0.020	1.4	275
RSM-1206(b)	Polyimide	Shell	-	-	-	-
RDX 64826	Cyanate ester I	Interez	3.1	0.005	1.21	260
XU-71787	Cyanate ester IV	Dow	2.8	0.004	1.18	260
UDEL	Polysulfone(c)	Union Carbide	3.0	0.004	1.24	192
ULTEM	Polyetherimide(c)	General Electric	3.1	0.0065	1.27	260

Notes: Func. = functional, (a) These designations refer to laminates available from Norplex Oak. The tabulated data are for the resin blends employed in these laminates. (b) This non-MDA polyimide is usually provided as a blend with an epoxy. (c) Thermoplastic polymer

Polyimide resin is a plastic that forms an adhesive and dielectric fluid when mixed with a ketone. Its high glass transition temperature (T_g>275°C) has made it popular as a tape for several high temperature applications. Polyimide is the commonly used resin for tape automated bonding (TAB) tape and for flexible circuits. It has also been used for several years as a soldermasking tape sold under the trade name of Kapton (Dupont).

Soldermasks

In the final phases of fabrication, the circuit pattern, exclusive of pads, lands, and other to-be-soldered interconnect areas of the board, is coated with an insulating material. The insulating material, called a *soldermask*, serves primarily to prevent solder from joining to places where it does not belong. The various forms of solder mask are shown in Figure 4.5.

In wave soldering, a soldermask prevents solder shorts between pads and traces. During reflow soldering, a mask prevents shorts due to runout of the solder paste, and solder starvation. Solder starvation occurs when the solder wicks down the trace or a via hole, pulling it away from the joint between the package lead and the board land (see Figure 4.6). The adhesion of the soldermask to the copper trace provides enough of a deterrent to the solder so that it will not flow down the trace and away from the land.

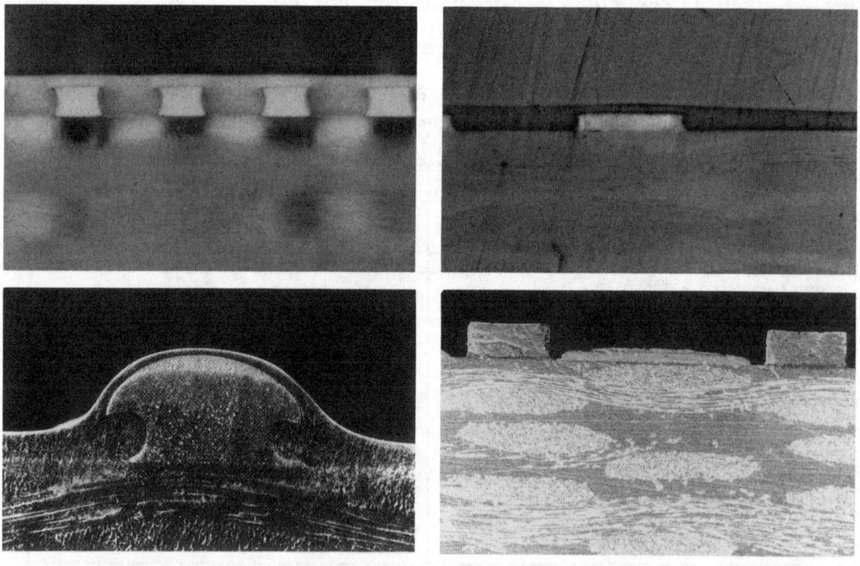

FIGURE 4.5. (a,b,c,d) Photos of cross sectional views of traces covered with soldermasks. (Photos a and b courtesy of Dupont) (Photos c and d courtesy of Dynachem, Morton Intl.)

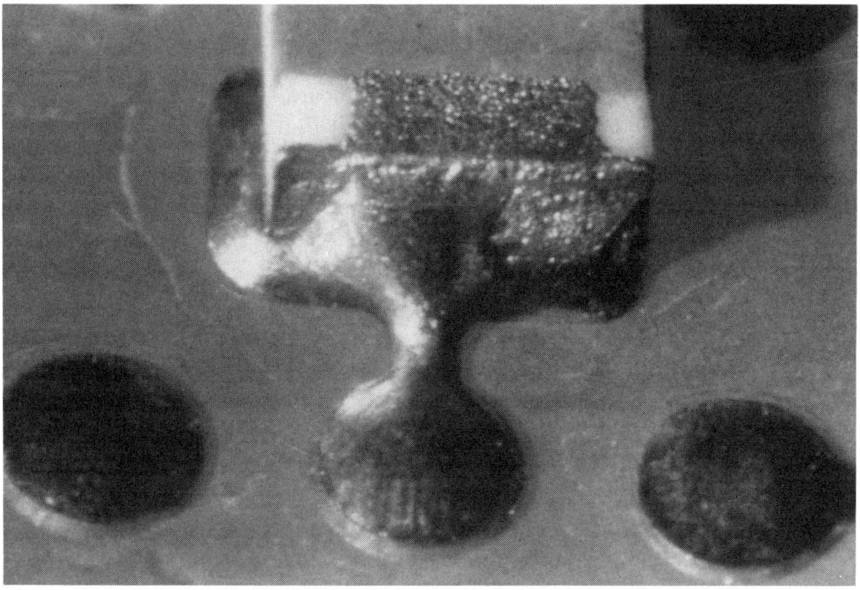

FIGURE 4.6. The effect of solder wicking on a trace that is insufficiently covered with soldermask. This wicking may create starved solder joints. (Photo courtesy of the IPC)

The soldermask serves other purposes as well. It protects the board surface from cuts and scratches during handling. It prevents entrapment of flux and chemicals between fine spaced traces. This entrapment could otherwise cause electromigration or corrosion. Soldermask provides a barrier to moisture and chemicals which could cause current leakage, corrosion, and crosstalk. And the mask also increases the board's resistance to flammability and thermal shock.

During wave soldering, the soldermask reduces the possibility of solder pot contamination. When reflow soldering this is not a concern. Soldermask does add slightly to the board cost and increases the dielectric constant ($\epsilon_r \approx 2.8$ to 3.6) on the top of the traces. If no soldermask is used, the dielectric constant would be that of air ($\epsilon_r = 1$).

The ideal soldermask for fine pitch technology is one which:

- Can coat between land spacings of 0.1 mm (0.004″)
- Does not require oversizing of openings by more than 0.03 mm (0.001″)
- Adheres uniformly to the board and all contours such as 0.1 mm (0.004″) trace spacings
- Does not trap solder, forming solder balls
- Forms a conforming versus encapsulating surface even over copper traces having significant height variations of 2 to 1
- Securely covers (or tents) via holes

These attributes are illustrated in Figure 4.7.

The choice and application of a soldermask can have an effect on the cost, solderability, and reliability of the printed circuit board. Soldermasks are generally grouped in three different varieties: screenable, dry film, and liquid photoimagable masks. However, there are several newer soldermasks that appear as combinations of the basic three. As astutely stated by one soldermask material representative, "No other area in printed circuit board manufacturing has more types of materials and application methods than soldermask. You can screen it, roll it, curtain coat it, spray it with or without electrostatics and slap it on wet or dry. Then you can cure it with ultraviolet light or all sorts of thermal cycles."

Screenable Masks. Thermally or UV cured liquid polymer's have long been the material of choice for soldermasks. These liquids are squeegeed through a mesh

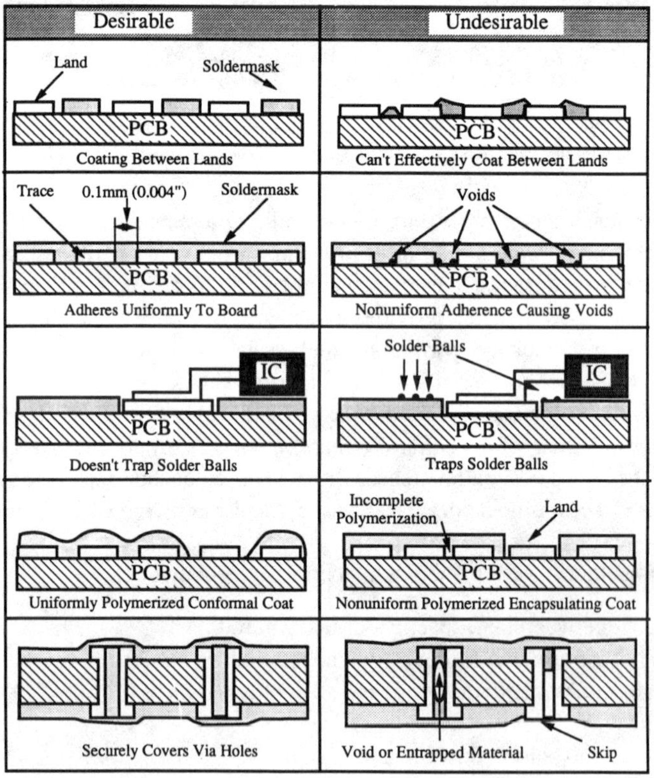

FIGURE 4.7. The desirable and undesirable attributes for soldermasks that are used on printed circuit boards.

stencil directly onto the board and trace surfaces. Most of the masks in this category are epoxy based and are available in either one- or two-part systems. A cross section of screened-on or wet soldermask is shown in Figure 4.5a.

Surface mount and especially fine pitch densities have exceeded the capabilities of all of the commonly used screenable masks. The screenables have difficulty uniformly coating traces spaced less than 0.2 mm (0.008″) apart. The stenciling process is difficult to control without smearing and misregistration. Therefore, it is customary to design the mask openings 0.25 mm (0.01″) larger than the actual opening. This amount of oversize distance is intolerable for fine pitch spacings.

Another drawback to the screenables is the setup time and cost. Each circuit requires a custom stencil. This stencil needs to be attached to the printer and registered to the printed circuit board. This consumes time and, even more costly, pilot circuit boards (boards to test the registration and setup). Many fabricators manually squeegee on the mask. This introduces great variability depending on the operator, the time of day, and the operator's mood.

Dry Film Masks. A dry film soldermask (Figure 4.5b) is a polymer sheet with photosensitive characteristics. It is sandwiched between a thin plastic support sheet and a thin release sheet. The release sheet is peeled away as the film is laid on the

FIGURE 4.8. Photo of a lifted, or tombstoned component. (Photo courtesy of Omni Training Corporation)

board. Prior to lamination using heat and vacuum, the plastic support sheet is removed. A mask containing the circuit lands and openings is placed over the film and exposed to UV light. The unpolymerized film over the lands and openings is chemically removed, exposing the copper. Dry film is able to register very closely to the openings, so oversizing is not necessary. A typical oversize specification is 0.05 mm (0.002″). Dry films are usually thicker than the copper traces, which improves the films' ability to tent via holes, but may excessively increase the height of overplated copper traces relative to underplated land areas. This latter situation causes small chip components and flush mounted FPT packages to "teeter-totter", rendering some of the terminations as unsoldered fence posts pointing toward the sky. This problem is referred to as tombstones, the Manhattan effect, or drawbridging.

Dry films, because of the lamination process and greater thickness, offer relatively poor surface adhesion and uniformity of coverage. This leads to voids, breaks, and film delamination. The voids and breaks usually occur along the side edges of traces. The void can act as a capillary to draw in cleaning fluids and other chemicals which, if left unremoved, cause corrosion and electromigration. Delamination occurs when a trace is too close to a large opening, such as a land. This situation also invites entrapment but is usually considered a cosmetic defect. Dry film voids, breaks, and delamination are shown in Figure 4.9.

Photoimageable Masks. Photoimageable masks are liquids containing acrylic monomers, polymers, fillers, photoinitiators, and dyes. When exposed to UV light, the monomers form crosslinked hydrocarbon chains that are strong and resistant. Some of the commercially available photoimagables require drying before imaging, while others allow wet imaging. A cross sectional view of a printed circuit board with a photoimageable mask is shown in Figure 4.5c.

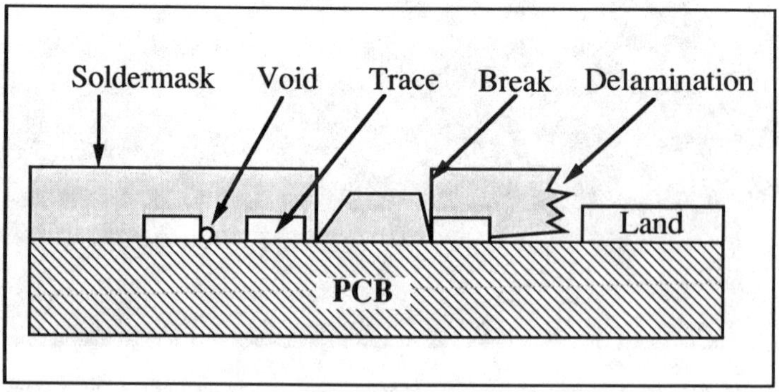

FIGURE 4.9. Possible void defects due to poorly laminated dry film soldermask.

The dryable photoimageable masks can be processed on some of the same equipment as the dry film masks. They can also be double-coated to allow stronger via hole coverage. More advantageously, both sides of the board can be coated and etched at the same time saving redundant steps. The dryable image masks are preferred over the wet image masks for fine pitch and surface mount boards, since the coating conforms to the board features and cures more uniformly than the encapsulating coatings. Encapsulating wet masks have the problem of thickness dependent cure times. Frequently this results in uncured mask material that can either delaminate or outgas during reflow soldering. Delaminated mask invites the entrapment of contamination. Outgassing of the mask material results in poor or completely unsolderable lands.

The liquid can be applied in thin layers to the board using roller coating, curtain coating, screen coating, or spray.

Roller coating applies a very thin and uniform layer of mask on the board. This is a time-tested and well understood process. The liquid is first deposited onto a grooved roller and then transferred onto the board surface. The roller height must be set high enough to deposit a sufficient coating of mask on the highest and lowest traces. This adjustment is usually manual and accomplished by first measuring the highest trace, and setting the roller higher than that trace by approximately 0.025 mm (0.001″).

The curtain coating method dispenses a fountain of mask material out of a slit shaped nozzle. The board is rapidly passed under the nozzle so the material deposits on the board much as a table cloth being laid on a table. The wet film coats the traces and vias. This is very dependent on the viscosity of the mask material and the speed of application; if either is outside tolerance, a significant problem with embedded bubbles and holes is the result.

Bubbles and holes are also a problem with another mask application method, screen coating. Screen coating uses a mesh screen with and without masked areas. The liquid mask is squeeged through the screen and transferred onto the board. If the screen has masked areas, then registration becomes as much of a problem as with the previously described wet film soldermask.

Two methods relatively new for soldermask application but old in practice are electrostatic spray and electophoretic bath. These methods are used to apply paint to cars, appliances, and other complex surfaces. Both methods are capable of depositing very thin and uniform coatings that conform to a wide variety of surfaces. Which approach is best is the topic of several debates. Both approaches require expensive deposition and filtration systems. Because of this expense, potential users will continue to debate until clear advantages or customer demands arise.

To coat a board using electrostatic spray, the board is attached to a high electrical potential of an opposite polarity from the spray. This causes the spray to be attracted to and bond securely to the board due to the electromotive force. This method requires ultraclean conditions, as any and every dust particle is also attracted to the board. Electrostatic spray is sensitive to the variation of the electrical field across the board; as a result there will be thicker deposits at the corners and cutout areas.

Electrophoresis is a bath coating process. The board, when immersed, acts as one electrode in the circuit. The bath contains resin micelles that are attracted and adhere to the board surface when a voltage is applied to the circuit. The coating thickness is very thin (approximately 0.01 to 0.02 mm); once the micelles attach to the surface, they lose charge and no additional coating occurs over that area.

There appears to be a difference in the coating properties if the board acts as the cathode or anode of the circuit. In paint spraying, coatings applied to cathode functioning metal results in better corrosion protection. Printed circuit board coating systems slightly favor an anode functioning system. The reasons are not clear but involve coating tackiness as well as electrode and bath life.

The liquid photoimageable masks offer superior registration and reduced thickness compared to wet and dry film masks. Unfortunately, this reduced thickness causes potential problems when a photoimeagable mask is used to tent via holes. Untented via holes may cause excessive air leaks when vacuum hold-down fixtures are used during paste stenciling or electrical test. This problem can be overcome by using a combination mask. A combination mask is a photoimageable liquid mask covered with a thin dry film. Both masks are exposed and developed at the same time.

A comparison of these four mask methods is shown in Table 4.4.

TABLE 4.4 Properties of the Four Major Soldermask Selections Used on Printed Circuit Boards.

	Screen Printed Wet Film	Dry Film	Photoimageable	Combination Photoimageable Dry
Over-protection & Strength	High	High	High	High
Registration Quality	Low	High	Highest	High
Thickness	Medium	High	Low	Medium
Adhesion Quality	High	Medium	High	High
Application Variability	High	Low	Low	Low
Coverage (or tenting) Ability	High	High	Medium	High
Affect on Solderability	High	High	Low	Low
Relative Overall Cost	Lowest	Highest	Medium	High

Exposed Copper Coatings

Prior to the implementation of surface mount assembly in the early 1980s, fabricators applied the soldermask over tin or tin-lead coated copper traces and pads. Soldermask adheres reasonably well to the coated traces, and tin-lead provides and excellent solderability preservative for long periods of time. However, tin and tin-lead coated traces cause many problems for small surface mount components. Typical problems are solder wicking away from the joint, wrinkled and raised soldermask, and shorts under the soldermask. Eliminating the tin-lead coating under the soldermask solves these problems but raises a new problem, that of preserving the solderability of the exposed copper.

The copper needs to be treated prior to soldermask application to enhance the mask's adhesion.

Temporary Coatings. Many users want to use only the exposed bare copper for attachment of fine pitch and small surface mount components. For these users, the fabricator sprays on a temporary copper protective coating. This can be a water or chemical soluble lacquerlike coating that is removed by the user just prior to paste printing. This coating may also be a material that breaks down with heat, such as flux, during the reflow operation and is washed away during cleaning (or it can be left on the board if it's a "no-clean" flux).

This temporary copper protective coating works well for users who do not expect long storage times for the boards, say, less than 6 months prior to assembly. Where long term storage (>6 months) is required it is best to coat the copper with tin-lead.

The conventional approach for tin-lead coating is one of the most severe processes the printed circuit board experiences. The board is dipped into a molten 60/40 tin-lead solder pot at 250°C (500°F). A high tin solder (70 to 80%) concentration is used in some applications due to the environmental concerns with the lead metal. As the board is withdrawn from the pot it is sprayed with a focused jet of hot air from multiple directions. This hot air "knife" is to remove the excess solder and to leave a flat coated surface. This operation is referred to as *hot air solder leveling* (HASL). Earlier HASL boards received unidirectional air knife treatment which left slanted solder deposits such as those shown in Figure 4.10. The multidirectional air knife minimizes the unevenness and thus eliminates some of the package solderability problems due to the uneven coating. The multidirectional leveling also increases the thickness of the coating which minimizes the copper concentration and its effect of reducing solderability (See Figure 4.11)

The thin tin-lead coating has been found to provide a solderable coating after exposure to harsh storage and mild solder fluxes. After storage for up to two years in high humidity (> 50% RH) most HASL boards are still readily solderable using R or RMA (now known as type L) fluxes.

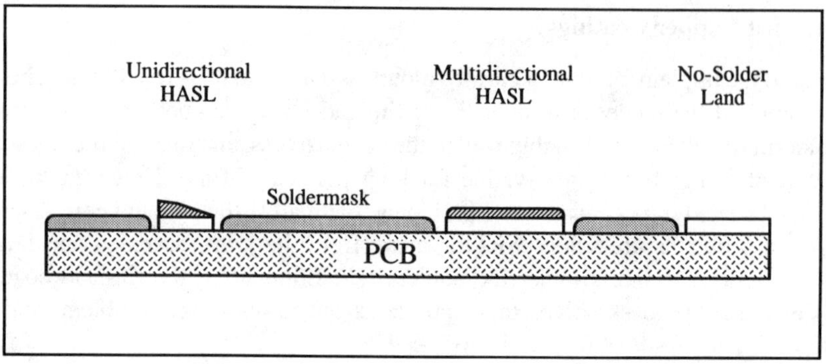

FIGURE 4.10. The cross sectional appearance of hot air leveled solder, or HASL, on a copper land area.

A technique that avoids the shortcomings of HASL boards and placing components into wet solder paste is called *pre-reflowed* solder. Pre-reflowing is simply solder paste that is stenciled onto the exposed copper areas of the board by the fabricator. The paste is immediately reflowed without the components. This forms a solder crown (see Figure 4.12) which can be coated with flux by the assembler at the time of assembly. The package is placed on the fluxed solder and reflowed to form a joint.

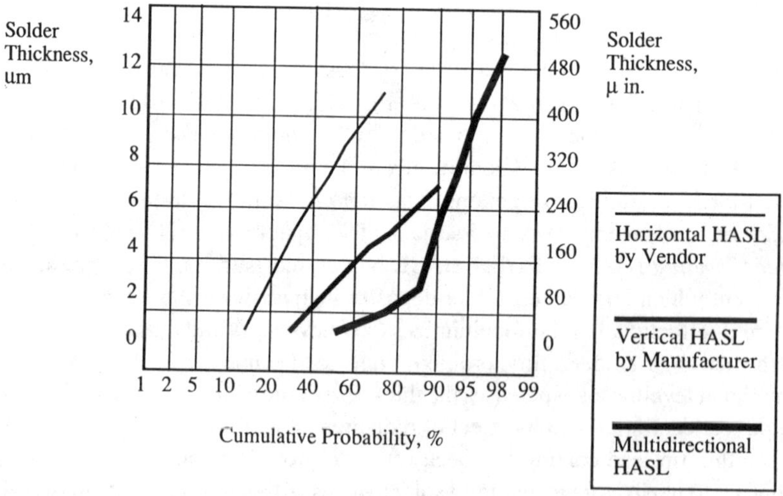

FIGURE 4.11. Comparison of Hot Air leveled solder capabilities. (Source: ATT, Bell Laboratories)

FIGURE 4.12. Solder crown height as a function of land width.

4.2 FABRICATION AND ASSEMBLY ISSUES

As the reader will recall, the resin, reinforcement, and copper are pressed together to form a laminate. To form the laminate the reinforcing fabric is impregnated with the bonding resin. The available combinations and main attributes are shown in Table 4.5 (Properties of Composites). As previously mentioned, the most common combination is difunctional epoxy, E-glass, and copper. This combination is known as FR-4. The users of multilayer printed circuit boards will continue to favor FR-4 because of cost and availability; however, there may be some growing usage of multi-functional epoxy with E-glass and triazine or BT with E-glass. This forecasted growth (Figure 4.13) is a result of the need for higher glass transition

TABLE 4.5 Selected Properties of the Composite Laminates that may be used for High Speed Applications. (Adapted from Electronic Materials Handbook, Vol. 1. ASM International)

Typical properties of some laminates with potential for high-speed applications

Laminate reinforcement/resin	Supplier	Dielectric Constant	Loss Tangent	Water uptake mg	Composite CTE ppm/°C x, y axis	z-axis @ T = 25°C
E-Glass/Epoxy	Many	4.7	0.021	10	12 - 16	80 - 90
E-glass/BT-blend	Many	4.5	0.010	15	12 - 16	80 - 90
E=glass/polyimide	Many	4.5	0.018	25	11 - 14	60
E-glass/cyanate ester	Nelco	3.9	0.003	-	12 - 16	80 - 90
E-glass/XU71787	Norplex	3.8	0.003	6	~12	~80
S-glass/XU71787	Norplex	3.6	0.003	6	~8	~40
Quartz/polyimide	Mica	3.6	0.010	25	6 - 12	34
Kevlar/epoxy	DuPont	3.7	0.030	10	4 - 7	80
Kevlar/polyimide	DuPont	3.6	0.008	25	3 - 8	83
e-PTFE/epoxy	Gore	2.8	0.012	10	24	261
e-PTFE/polyimide	Gore	2.8	0.010	25	11 - 14	60
e-PTFE/XU71787	Gore	2.4	0.003	6	–	–
Ceramic/PTFE	Rogers	2.9	0.0012	1	–	–
Nonwoven glass/PTFE	Rogers	2.2	0.0008	1	~8	~105

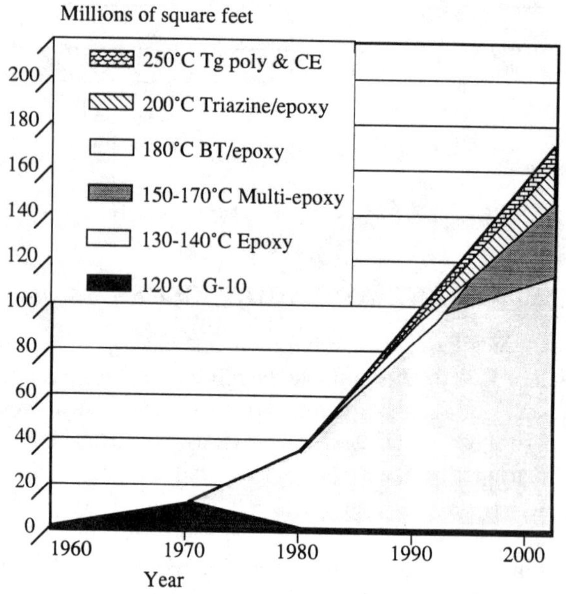

FIGURE 4.13. Forecasted use of the various composite laminate materials. (Adapted from "Trends in Electronic Substrate Technology. In Electronic Packaging and Production, Sept., 1990)

temperatures to minimize board handling problems during reflow. The higher Tg is also to minimize potential solder joint problems due to plastic deformations of the FPT joints as a result of being formed during the "above Tg" CTE phase.

To etch the thin copper traces precisely, fabricators must use a copper foil of consistent properties. There are several different copper foils fabricators can select from. With the proper etch-resist *thickness and width* most of these form satisfactory fine lines, down to 0.08 mm (0.003″). The thicker outer layer copper is usually masked with tin instead of photoresist and the exposed copper is etched in an alkaline and sulfuric peroxide solution. Since the etch works both vertically and laterally (Figure 4.14), the width of the tin etch resist must be wider than the final trace width. As will be discussed further in Chapter 13, it is important in the successful implementation of FPT to allow the fabricator some freedom with the circuit artwork. This way adjustments to the trace width can be made *in specific* areas to achieve the fine traces within the specification desired.

Board warpage is a significant problem for FPT, especially on large boards. Board warpage during solder paste stencil and component placement result in open and shorted joints, or weak solder joints. When the board is laminated during fabrication, the fabricator must assure a uniformly thick multilayer laminate that is free of voids, and with minimum warpage. To help assure this, the preferred

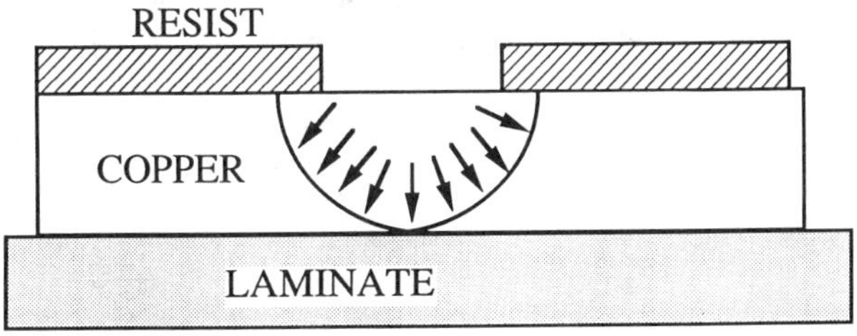

FIGURE 4.14. The isotropic etch patterns during copper etching.

lamination technique is vacuum autoclave. The vacuum autoclave units help remove any moisture, bubbles, and resin buildup while the board is pressed together. The typical pressure ranges from 175 psi to 350 psi. While the layers are pressed, heat (depending on the resin system used) is applied to harden the epoxy. The layers may swim and loose registration during the lamination and subsequent processing. The swim during lamination is due to the reduced viscosity of epoxy prior to hardening (see Figure 4.15) Most fabricators characterize the swim rates for their process and materials, and adjust the photo artwork to match their specific needs. An illustration of the amount of dimensional change experienced by polyimide and a military grade E-glass such as style 106 is shown in Figure 4.16.

In order for all the layers to line up properly or register for interconnection, the via hole pads must be large enough to allow the drill to be within the

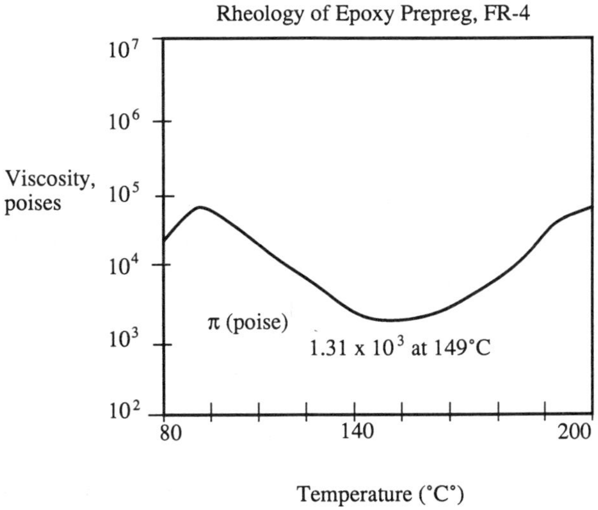

FIGURE 4.15. The variation of epoxy viscosity (rheology) during thermo-curing.

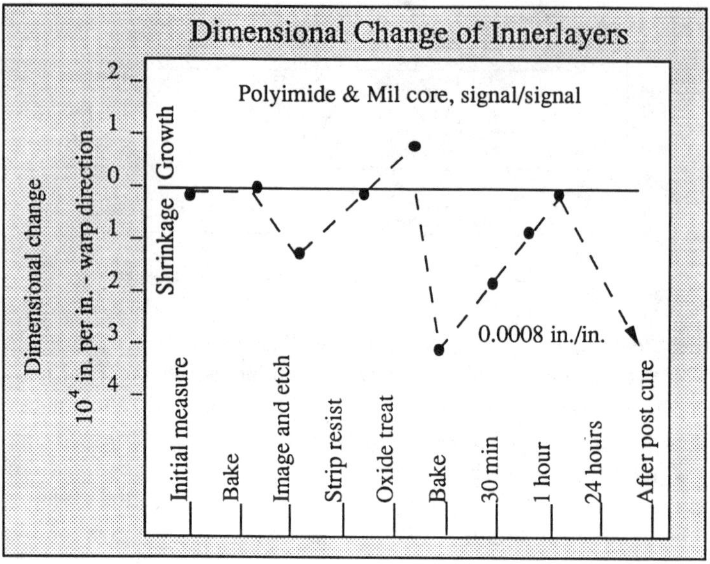

FIGURE 4.16. The dimensional change experienced by the innerlayers of a multilayer board during the various process steps of fabrication. (Source: Hinton PWB Engineering)

perimeter of the pad, with no part of the drill breaking out. To prevent making these pads so big as to waste board area, the fabricator must know the dimensional and swim characteristics of the materials and processes. Also , the designer must know the material aspects to meet product cost and electrical performance budgets.

Thinner glass using 113, 2113, 2313, 116, 2116, 2316 styles, usually exhibit less dimensional change and are thus favored by fabricators for multilayer boards. Thicker glass styles such as 7628 are prone to greater swim and are favored for single layer or low multilayer applications.

The lamination pressure affects the laminate results. The pressure strongly influences the resin richness and the glass/resin press-out. The resin richness plays a very key role in the bonding strength of the copper to the glass. The less resin, the weaker the bond. One problem with fine pitch devices is the difficulty of desoldering all the solder joints uniformly to remove a package. If too much heat is applied or if the resin bond is weak when heated, then some of the metal lands can be removed along with the package. This presents a slight problem if the lands are the only features removed, and not the plated barrels of adjacent via holes. Wire jumpers from the remaining traces or the via holes form easy substitutes to the missing lands. If, however, the barrels are pulled out or if too many lands are lifted, then the replacement task is complicated since several layers will need to be reconnected.

Another reason to limit the lamination pressure is to avoid over pressing the glass fibers. The glass is not a ductile material as is a metal such as copper. When pressed, the glass simply stretches out. During lamination the epoxy bonds to the glass in this pressed-out condition. When the pressure is removed, the glass returns to its prelamination size. In some areas, the resin will lose its bond on the glass and form voids, which are called *measles* because of the laminate's physical appearance.

Another method that some have used to minimize the dimensional change during lamination is to put those layers needing heavier copper in the inner layers and the thinner copper on the outer layers. The heavy layers are usually the power and ground interconnections because of the need for higher current densities. These layers are usually 1 oz. (nominal thickness =. 03 mm) or 2 oz. (nominal thickness = .06 mm). Any signal traces requiring heavy copper should be routed in the power or ground layer or a separate layer. This allows the outer signal layers to use a uniform 0.5 oz (nominal thickness = 0.015mm) copper thickness.

To plate the drilled through holes after the photoresist is etched, a process called Pattern Plating is used. Pattern plating means the copper is electroplated only on the exposed areas and not on the areas masked by the photoresist. This additional copper plating also increases the thickness and, hence, the electrical current handling density. However, because of variations in current density resulting from circuit pattern variations, when the extra copper is pattern plated on the outer layers, it results in different thicknesses of copper on the surface. This causes uneven surfaces. These uneven surfaces create problems when the solder bearing paste and the protective soldermask are applied.

4.3 COMPOSITE ELECTRICAL PROPERTIES

The increasing operating speed and frequencies of integrated circuits is not only limited by package performance but board performance as well. The board materials, trace sizes and spacings, and board coatings have electrical properties that degrade and hamper the transmission of an electrical signal at the higher frequencies. Some of these properties act as "thieves" and rob energy from a signal. Others act as discontinues that divert the signal, forcing it to radiate noise, reflect signals, or to crosstalk with other traces.

The clock frequency of a digital circuit is not the maximum speed parameter of a circuit. The clock is the device that creates a regular heartbeat of pulses. The other circuits perform various functions set by the pace of the clock. Some circuits are edge triggered, which means they start their tasks during the rising or falling edge of the clock pulse. So the speed of the rise or fall of the clock pulse is the critical speed parameter. It is not uncommon for a digital system operating at a 10 MHz clock rate to have some signals needing fast, say, 2 or 4 nanosecond rise times. In these cases, a copper trace on a board must act as an electrical transmission line. If the travel time of the signal down the trace approaches one-half the signal's transition (rise or fall) time then the circuit's performance is degraded. Degradation appears as erratic race conditions, missing logic states, or nonfunctionality.

Signal travel time on a typical copper-clad FR-4 board is about 7 psec/mm. The maximum trace length to assure proper functionality for a 2 nsec transition time pulse is estimated to be:

$$\text{Trace Length max} = \tfrac{1}{2}\,(\text{transition time})\,(\text{travel time})^{-1}$$
$$\text{Trace Length max} = \tfrac{1}{2}\,(2\ \text{nsec})\,(7\ \text{ps/mm})^{-1}$$
$$\text{Trace Length max} = 143\ \text{mm}$$

To avoid problems, a designer needs to keep the trace lengths short for critical traces and to understand the board's material properties.

To understand these material properties, it is useful to describe the fabricated board in terms of an electrical circuit. For this purpose, a four-layer board will be used, with two signal layers, a ground, and a power layer. A cross sectional view of this board is shown in Figure 4.17. The electrical model (Figure 4.18) represents a three-dimensional view of the electrical parasitics along and at the intersection of the traces.

Along the length of the trace the electrical signal encounters a small series resistance, inductance, and parasitic capacitance. The trace resistance reduces the voltage and produces a small amount of heat as the voltage travels along the trace. Of greater concern is the effect the resistance has when lumped with the parasitic capacitance that occurs along the periphery of the trace. If the series resistance, R_S, is large compared to the output impedance, R_O, of the circuit driving the signal down the trace, then the combined series resistance, the parasitic trace capacitance,

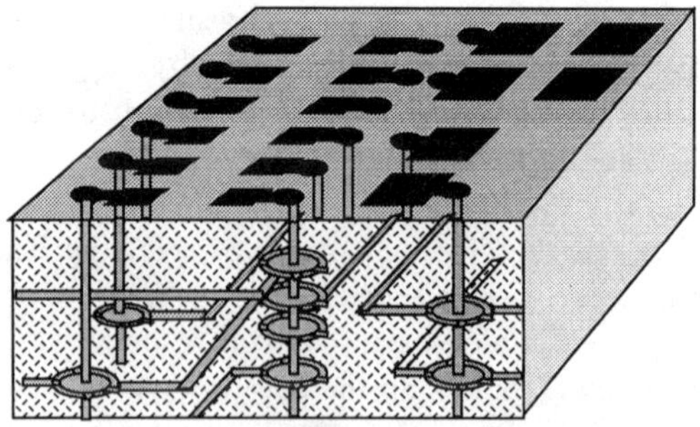

FIGURE 4.17. A cross sectional view of a 4-layer printed circuit board.

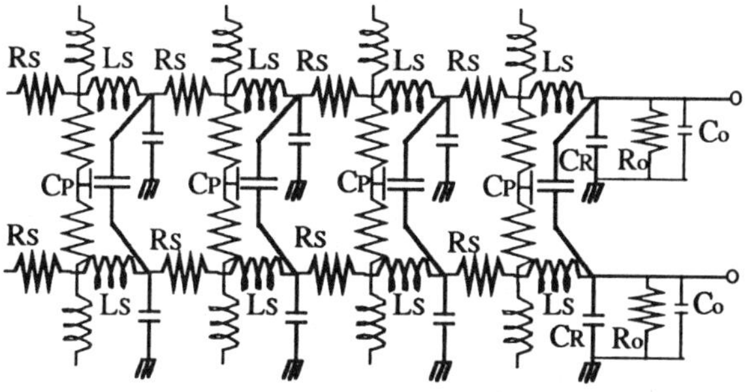

FIGURE 4.18. Circuit diagram analogy of a 4-layer printed circuit board.

and the input capacitance of the receiver combine to impede the rise and fall times of the signal. This results in time delays whenever the signal changes value. The time delay, t_D, can be approximated from the equation:

$$t_D \ 5 \ \ k \, fC_p \, (R_O \, 1 \ \ ½ \, R_S) \ 1 \ \ C_O \, R_O \ 1 \ \ C_R \, (R_O \ 1 \ \ R_S)g$$

where

- R_S 5 the trace series resistance
- R_O 5 the driver circuits output resistance
- C_P 5 the parasitic trace capacitance
- C_R 5 the lumped load capacitance of the receivers and any other capacitances
- C_O 5 the package capacitance of the output driver lead
- k 5 is a value from 1.0 to 1.6 that is dependent on the value of R_OC_R.

The inductance element creates a tiny magnetic field that radiates from the trace. If other traces adjacent to, above, or below the radiating trace are close enough to intersect some of the magnetic flux, then an electromotive force (EMF) is induced in these traces. The electromotive force causes current changes in the affected traces as the signal in the originating trace changes. The current changes appear as voltage spikes and noise. Since the power traces would be most affected by these spikes, designers attach capacitors to these traces. These capacitors dampen the amplitude of the spikes and filter out some of the noise.

The major parasitic parameter limiting a circuit's performance is capacitance. Capacitors act as loads to weigh down a signal, and as filters to shunt signal energy. The lower the capacitance, the lower the time delay, t_D.

Capacitance is created by bringing two conductors of differing potentials into close proximity with a dielectric material in between. The negative potential conductor will store an electrical charge in an amount that is a function of the facing area of the conductors, the distance between the conductors, and the relative dielectric constant. The capacitance value can be expressed as:

$$C = \frac{\varepsilon_R A \times 10^{-6}}{4.45d}$$

where

C = the capacitance in microfarads
ϵ_r = the relative dielectric "constant"
A = the area of the conductor facing an adjacent conductor
d = the distance between the conductors

Air has a relative dielectric constant of 1; FR-4 material ranges from 4 to 5 depending on several factors to be explored shortly. When interconnecting fine pitch devices the traces are very close, which reduces the value d in the capacitance equation. This reduction of spacing increases the value of the capacitance loading the signal. The increased capacitance does more than load the signal, it also allows signals to cross over to other traces; this is called crosstalk. Since the trace's face area and space is constrained by the package and fabricator constraints, lowering the parasitic capacitance is best done by lowering the overall trace area and the dielectric constant. To lower the overall area, designers need to take care during board layout and physical design and make critical traces small, thin, and short. Also traces should not run parallel for long distances.

To lower the dielectric constant means changing the board material. Table 4.5 shows the composite properties, which include the dielectric constant of different board materials. The dielectric constant value listed is for the material at a frequency of 1.0 MHz. The dielectric constant is not a constant; it varies with frequency, absorbed moisture, and temperature. It also varies with the composite reinforcement to resin ratio. The reinforcement to resin ratio effect on the dielectric constant of epoxy-fiberglass (FR-4) is shown in Figure 4.19.

The greater the amount of resin volume surrounding the laminate the lower the dielectric constant. Figure 4.19 also shows the ϵ_r at 1 MHz and 100 MHz. The ϵ_r of epoxy fiberglass is very frequency dependent, with the highest ϵ_r being at lower frequencies. This frequency dependency must be taken into consideration by a designer. As shown in Figure 4.20 the range in variation of ϵ_r is from 4.9 at 1 KHz to 4.0 at 1 GHz. This means that a designer using the nominal ϵ_r listed in Table 4.5 at 1.0 MHz would experience a 6.6% error for calculations made on signal traces operating at 100 MHz ($\epsilon_r = 4.2$)

Some alternative board materials such as polyimide/aramid and PTFE/glass

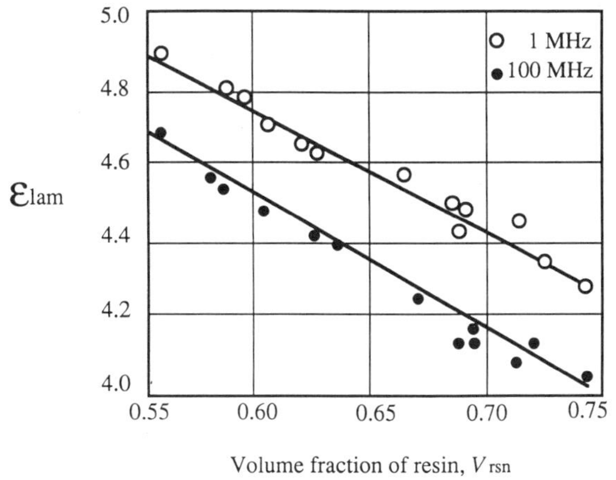

FIGURE 4.19. The greater the volume of resin surrounding the fiberglass, the lower the dielectric constant, ϵ_r. (Adapted from *Electronic Materials Handbook*, Vol. 1. ASM International)

display far less change in ϵ_r with frequency. However, their cost and availability usually over-shadow their improved ϵ_r.

Temperature and absorbed moisture also change the value of ϵ_r. An increase in either temperature or absorbed moisture increases ϵ_r.

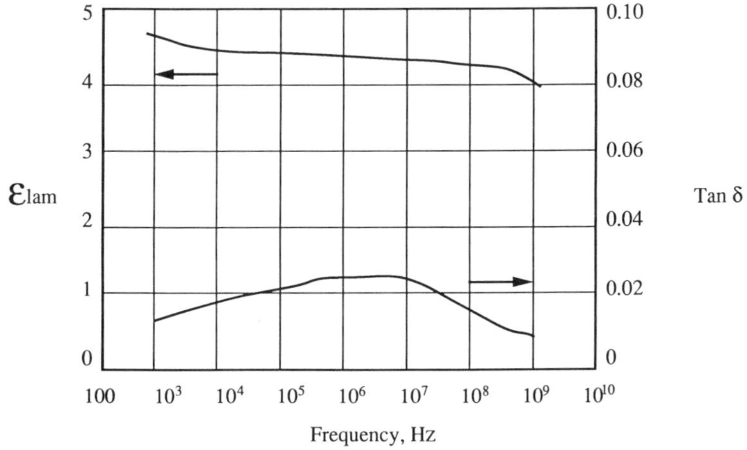

Typical frequency dependence of relative permittivity, Ɛ lam and loss tangent, tad δ, of an FR-4 laminate. Sample on which measurements were performed had a volume fraction of resin of 0.724.

FIGURE 4.20 . The dielectric constant, ϵ_r, is not a constant with frequency variations. At higher frequencies ϵ_r decreases. (Adapted from Electronic Materials Handbook, Vol. 1. ASM International)

Signal energy is lost from the signal as it travels down a trace. The signal attenuation by the dielectric surrounding the trace is approximated by:

$$A_D = \frac{27.3\,(\tan\delta)\varepsilon_r^{1/2}}{\gamma_O}$$

where

A_D = the attenuation in dB per unit length.
$\tan\delta$ = the loss tangent
ϵ_r = the relative dielectric constant
γ_O = the free space signal wavelength

Tan δ, the loss tangent (tan δ) is usually specified at 1 MHz by most material vendors, but as is shown in Figure 4.20 it varies with frequency in FR-4.

When the sensitive signal traces are embedded in the sublayers of the board the assumed value of ϵ_r is that of the laminate. If the traces are on the board surface and covered with soldermask, then the assumed ϵ_r is the weighted sum of the ϵ_r for the soldermask and the laminate.

4.4 COMPOSITE THERMAL PROPERTIES

Fine pitch packages present a challenge for thermal management. The devices in the FPT packages can generate large amounts of heat that must be removed to avoid component failure. The FPT packages also have small, easily oversoldered leads which form rigid joints on the printed circuit board. If the thermal coefficient of expansion of the package material is not closely matched with the board then solder joint reliability problems due to solder joint fatigue and creep occur.

Removing the Heat

Fortunately, fine pitch packages are usually large in surface area. Even though more of the device heat exits the package leads than was typical with a through-hole package, a significant amount of the heat rises to the top of the package. Removing this heat uses the same heat sinking techniques developed for through-hole packages. This means gluing heat sinks on the package and blowing cool air over the assembly.

The thermal expansion properties of a printed circuit board become very important with the use of the leadless chip carrier (LCC) package. The LCC has no leads, but is attached using deposited terminations called *castellations*. The deposited terminations are made of various metals to form a contact with the ceramic package, the outer metal being gold to preserve solderability. Since the LCC lacks

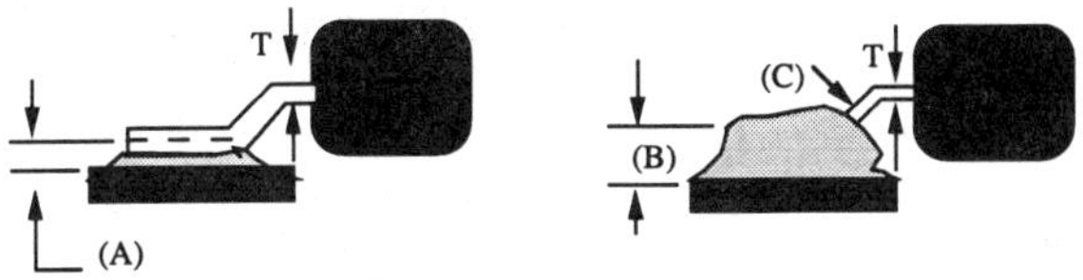

FIGURE 4.21a. A diagram of an oversoldered Gull Wing lead. (Source: PPM Associates Workmanship Manual)

a lead, it mounts flush to the board, forming a constrained union of the ceramic package with the printed circuit board. When the assembly experiences temperature changes, the materials change size according to their coefficients of thermal expansion (CTE).

The generalized problem with the LCC is that ceramic has a CTE of approximately 5 ppm/°C and FR-4 in the x and y axis has a CTE of 12 to 16 ppm/°C. If the ceramic package is large and the thermal change is also large, then the solder joints may be stressed to a degree that some of them fracture. This problem is potentially acute on 44 pin and higher lead count LCC packages mounted on FR-4 and operating in a temperature environment that varies from − 55 to 125°C. This problem has been widely reported and researched and is discussed further in Chapter 10. Many users feel it is limited to large ceramic packages on FR-4. Unfortunately, the small leads on fine pitch packages make them almost as susceptible to the CTE mismatch problem as the LCC. Not only are the FPT leads small but they are easily oversoldered. Oversoldering the lead by allowing the solder to go over the mid-point of the bend in the gull wing lead, as shown in Figure 4.21b, results in a stiff constrained assembly.

Engelmaier has developed, from extensive studies, two models to predict solder joint failure as a result of CTE induced fatigue. These models are discussed in Section 10.1. If the lead has flexibility, then these models predict that it will exhibit a much lower probability of failure. The same is true if the CTE mismatch between the package material and the board are reduced. Table 4.6 lists the nominal CTEs and cost ratios for selected composite materials. The CTE of ceramic is generally

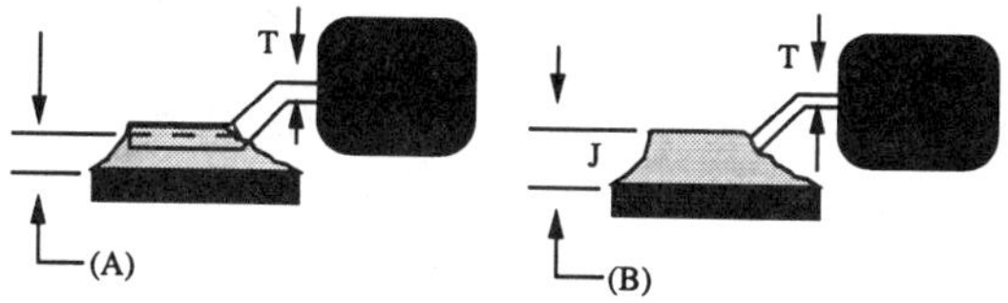

FIGURE 4.21b. The preferred solder appearance for a gull wing lead. (Source: PPM Associates Workmanship Manual)

TABLE 4.6 The Coefficient of Thermal Expansion, CTE, for Commonly Used Printed Circuit Board Laminates. (Adapted from *Electronic Materials Handbook*, Vol. 1. ASM International)

Laminate reinforcement/resin	CTE ppm/°C Composite z-axis @ T = 25°C	x, y axis	Cost Factor
E-Glass/Epoxy	80 - 90	12 - 16	1
E-glass/BT-blend	80 - 90	12 - 16	1.1 - 1.2
E-glass/polyimide	60	11 - 14	2 - 3
E-glass/cyanate ester	80 - 90	12 - 16	1.2 - 1.3
Quartz/polyimide	34	6 - 12	14
Kevlar/epoxy	80	4 - 7	3 - 7
Kevlar/polyimide	83	3 - 8	4 - 8
Nonwoven glass/PTFE	~105	~ 8	14

described as between 5 to 7 pm/°C. The CTE of copper is approximately 17 ppm/°C.

The spread in CTE of FR-4 (E-glass/epoxy) is due to the reinforcing fiberglass. Changing to a different reinforcement/resin system may be the best approach for critical applications. In general (see Section 10.1), to optimize reliability, package leads should be flexible, the lead should have limited solder mass to assure flexibility, the ΔCTE between the board and package should be small, and the solder joint temperature should be held as close to room temperature as possible.

References

1. Gray, F. 1990. An Overview of Commercial and Military Packaging. In *PC Fab*, Jan. and Feb., 1990.
2. Hinton, P. The High-Yield Challenge in Laminating MLBS. In *Electronic Packaging and Production.* Jan., 1990.
3. Lynch, L. et. al. Electronic Materials Handbook, Vol. 1. Materials. Park, OH: ASM International.
4. Marcoux, P. 1992 SMT/FPT Workmanship Guidelines. Rev. 3.2. Sunnyvale, CA: PPM Associates.
5. Murray, J. So Many Solder Masks. In *Circuits Manufacturing*. Feb., 1990.
6. Senese, T. Trends in Electronic Substrate Technology. In *Electronic Packaging and Production*. Sept., 1990.

5

Solder and Application Methods

5.0 SOLDER

Solder is a soft metal alloy that bonds to metal surfaces to form a joint. Solder has been used in electronic products used for decades to provide a reliable mechanical joint and an excellent electrical conductor. Tin-lead alloy solder has been the standard. However, double sided SMT assembly, environmental concerns with lead, and the demands of fine pitch technology are causing some users to explore alternatives.

For fine pitch applications the ideal solder is one that:

- Has a low melting temperature ($T_m < 200°C$)
- Is highly conductive ($\sigma < 0.001\ W/mm^3$)
- Is extremely reliable in all applications
- Satisfies required visual appearance criteria

For fine pitch and surface mount assembly, the solder must be preapplied to the board in small, finite quantities before the component packages are placed. This is different from through-hole assembly where the package leads are inserted first and then passed over a molten solder pot that contains an "infinite" amount of solder. The amount is infinite compared to the amount needed to solder the assembly.

The application methods to preapply sufficient solder for fine pitch surface mount applications include:

- Hand-tinning each pad with a soldering iron and solder wire
- Electroplating solder onto the land areas
- Syringing solder dots using a syringe and solder paste
- Stenciling solder paste using a stencil and printer

The two latter application methods use solder blended into a paste as the means of depositing the solder onto the board. Today's solder paste for fine pitch applications is a much more refined material than the surface mount grade solder paste of the 1980s. Solder paste is basically a blend of solder alloy particles and flux. Mixed in with the flux are solids, activators, solvents, and thixotropic agents. Solder in paste form continues to be the most economical vehicle for depositing the small quantities of solder at the precise locations of each lead, termination, via, and joint on a fine pitch and surface mount assembly. This is shown in Figure 5.1

However, several issues face the user in using the paste successfully. These issues can be summarized below:

- Precise deposition, i.e., minimal misregistration
- Adequate volume to form a reliable joint
- Easy printability in small areas
- Solderability during storage and use
- Flux removal

5.1 SOLDER ALLOY CHOICES

Solder paste contains the metal alloy that is electrically conductive and mechanically bondable to the metals to be joined. For the past three decades the preferred

FIGURE 5.1. A close-up photo of solder paste on land areas. (Photo courtesy of ESP Solder Plus)

alloy has been tin-lead in concentrations ranging from 60% tin—40% lead to 63% tin—37% lead. Tin-lead alloy has met all of the criteria previously mentioned that are desired in a solder for electronic applications. Tin-lead has a low melting temperature (183 to 189° C); it is highly conductive (less than 0.1 microhm/cm); has been extremely reliable in long term field usage; and it produces an appealing smooth solder joint with a satiny luster. Tin-lead solder does have some potential shortcomings for fine pitch technology. One shortcoming is its relatively high melting point during the reflow of the second side of a board. This can affect the first side components and their solder joints if remelted. In very demanding applications tin-lead solder alloy may offer less than optimum yield and tensile strength, too high a melting temperature, excessive elongation and creep, and reduced thermal fatigue life. The importance of these shortcomings is influenced strongly by the user's volume needs, cost sensitivity, application, and reliability demands. For the vast majority of applications (>95%) tin-lead alloy will continue to be the alloy of choice.

The most common are 63%/37% tin-lead (Sn63) and 62%/36%/2% tin-lead-silver (Sn62). 60%/40% tin-lead is commonly used in wave soldering, and infrequently in solderpaste since there is no price savings and Sn63 offers slight advantages.

In the early years of surface mount technology the addition of silver was popular. This is the 62/36/2 tin-lead-silver alloy in Table 5.1. A small amount of silver (2%), is enough to prevent the leaching or scavenging by the tin-lead of the

TABLE 5.1 The Various Metal Solder Alloys Commonly Used for Electronics.

	Solidus Point °C	Liquidus Point °C	Tensile Strength lb-f/in² @ 25° C	Shear Strength lb-f/in² @25°C
63% tin - 37% lead	183	183	6120	4130
60% tin - 40% lead	183	189	6080	4070
62% tin - 36% lead - 2% silver	179	179	6120	3980
43% tin - 43% lead 14% Bismuth	144	163	5260	N/A
52% Indium - 48% tin	118	118	~1720	N/A
42% tin - 58% Bismuth	138	138	N/A	N/A

silver from silver or palladium-silver termination components such as chip capacitors. Leaching is the removal of the metalization on the termination through solution in other metals. In this case, the silver is readily dissolved or leached from the termination by the tin in the tin-lead alloy. Silver termination components are now coated with a nickel barrier and a solderable outer coating metal to prevent leaching. This barrier prevents tin-lead leaching and eliminates the need for the silver addition to the solder.

Many believe the silver adds to the tensile strength of the alloy, but as can be seen from Table 5.1, this is not the case. The silver does, however, yield nice satiny joints, but at extra cost.

The various other metal alloys that are used for electronic soldering are also listed in Table 5.1.

Ternary and quaternary solders are being recommended by some suppliers as better than tin-lead alloy. Ternary and quaternary solders are those with additional metals added to the alloy. For example, bismuth may be added to lower the melting temperature and indium may be substituted for tin to form ductile joints when gold is present. Gold in tin-lead forms brittle joints.

The bismuth (Bi) and indium (In) bearing solder alloys are used primarily for soldering gold plated junctions. In recent years these alloys are also being increasingly used because of their lower melting temperatures (118 to 163° C vs. 183 to 189° C). They are being used by some for the second side solder on double side assemblies. This allows the assembly with the soldered primary side to go through a second solder step without the primary side reflowing. This avoids weakening the solder joints, the possibility of the components falling off the board, and the exposure of sensitive packages to excessive thermal energy.

The alloy 46Sn/46Pb/8Bi is favored by some users because of its ability to draw off of the soldermask or laminate when this paste is stenciled on as a continuous stripe instead of in individual land deposits. Another alloy, 54Sn/26Pb/20In, is used for touchup since it can solder open leads at a lower melting temperature without reflowing the previously reflowed adjacent Sn63 or Sn62 joints.

Major drawbacks with the bismuth and indium alloys are the higher cost and the gray, cold appearance of the resulting solder joints. 46Sn/46Pb/8Bi is slightly duller than Sn63, while 42Sn/58Bi is very dull and extremely brittle at room temperature. The 54Sn/26Pb/20In alloy is very ductile but significantly higher in price than Sn63 solder.

Of the alloys listed in Table 5.1, the preferred alloy for dense electronic assembly is 63/37 tin-lead, which is also called *eutectic solder*. The eutectic of an alloy is the unique blend of the two or more metals that produces the lowest liquidus-solidus point. This is shown in the metallurgical equilibrium diagram of tin-lead (Figure 5.2). The lowest liquidus-solidus point is about 183°C and occurs when the alloy is 61.9% tin and 38.1% lead (which is the real eutectic of tin-lead). This melting temperature is much lower than that of the individual melting points

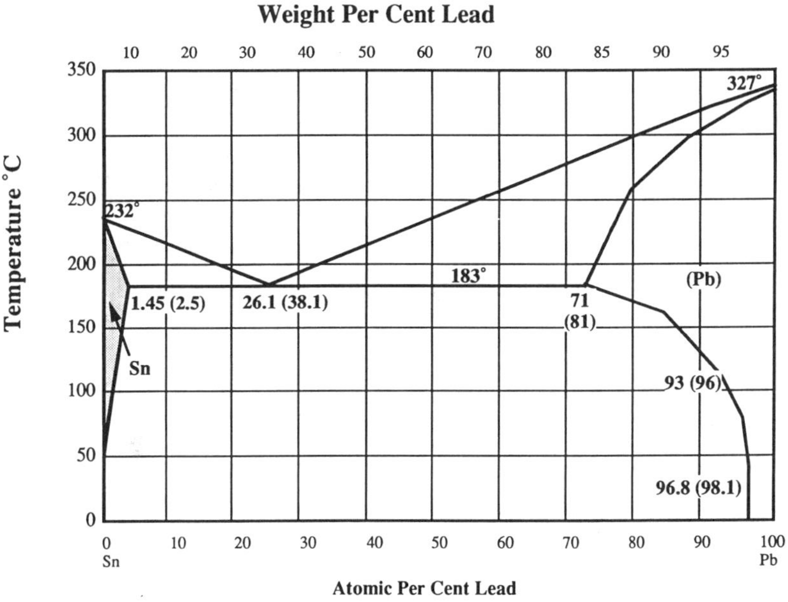

FIGURE 5.2. The metallurgical equilibrium for tin and lead alloy.

of tin (232°C) and lead (327°C). Because of inaccuracies in early analysis, the eutectic was set as 63% tin and 37% lead instead of the real eutectic of 61.9%/38.1%. Even though the errors were later discovered, the commonly accepted name for the eutectic of tin-lead is and will remain 63/37. The eutectic mixture also yields, in most cases, the strongest joint of these metals since there is minimal formation of dendrites of either of the primary phase metals. These primary phase regions are usually lead rich regions. These are of concern in joints subject to high mechanical and high thermal stress since they have been shown to be points of joint fracture.

In high stress environments joints have been found to crack and fail at the boundaries in the area of grain growth resulting from the high stresses. These regions show as dark areas in Figure 5.3. Optimum solder shows small dendrite regions and voids, as is shown in Figure 5.4. Optimum solder joints like that illustrated in Figure 5.4 are the result of using quality solder, careful storage, proper preheat and rapid post reflow cooling.

To enhance the fluidity and to possibly prevent tin phase transformation, suppliers add traces of other elements to solder, a common element being antimony. The amount of primary alloy and trace elements is carefully controlled and specified by an industry approved standard. The industry standard for solder alloys, which sets controls on the purity and amount of the metals to be used, is the

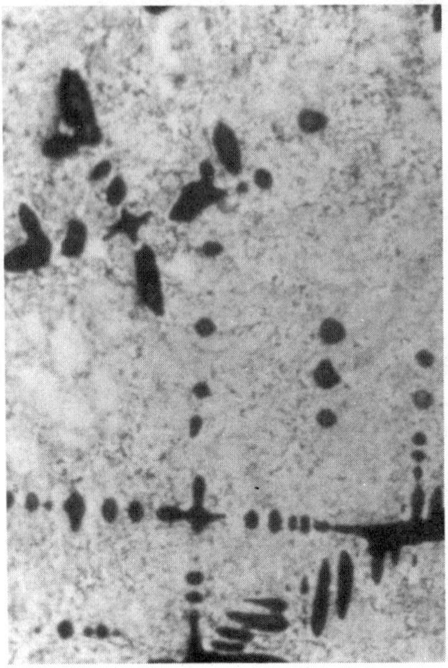

FIGURE 5.3. An extreme close-up view of tin-lead alloy solder with excessive grain growth. (Photo courtesy of Alpha Metals)

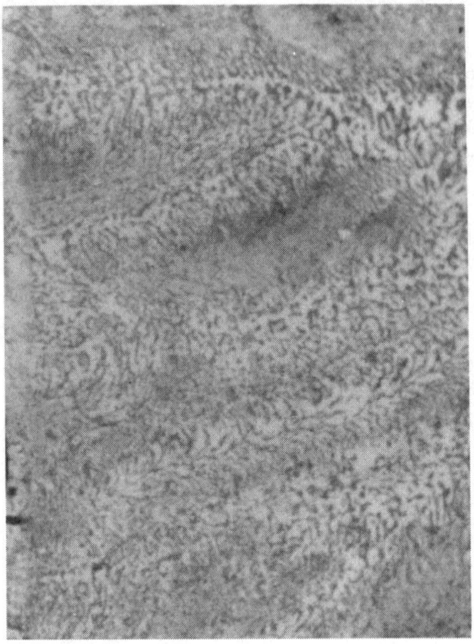

FIGURE 5.4. A close-up view of optimum solder joint appearance. (Photo courtesy of Alpha Metals)

QQ-S-571. It specifies the allowed ingredients in the solder alloy and the flux. It also specifies the test methods for evaluating the compliance and quality of a solder.

5.2 FLUX

Flux is a highly essential chemical needed to achieve excellent solder joints. It serves many functions. The prime function, however, is to be the cleaning agent that removes the metal oxides that hinder solderability. Additionally, the flux aids stencil printing and extends the storage life of solder in paste form. Electronic grade flux is a mild acid at elevated temperatures and relatively inert at room temperature.

For fine pitch assembly, the ideal flux is one that:

1. Removes oxides from lands and leads in a short period of time and at relatively low temperatures (approximately 125°C)
2. Disintegrates after reflow or leaves only a transparent, inert substance after reflow, so no cleaning is required

3. Leaves no byproducts that encourage dendrite growth, corrosion, leakage, or electromigration
4. Remains tacky for the required time until reflowed
5. Has minimum slump from ambient until reflow temperature

Ten years ago, the dream flux envisioned above was nonexistent. Severe environmental damage, caused by the chlorofluorocarbon (CFC) containing cleaning agents used to clean the rosin based fluxes provided the necessary incentive to develop fluxes closer to this ideal. Today, no-clean and water soluble fluxes provide excellent solderability, reliability for electronic applications, and environmental safety.

Rosin Fluxes

The original flux, and most of today's flux, is rosin based. Rosin is a naturally occurring product of pine trees. Rosin's ingredients of interest are abietic acid, isopimaric acid, neoabietic acid, pimaric acid, and dihydroabietic acid. These acids can all be lumped into the organic carboxylic acid group consisting mainly of pimaric and abietic acids. Their chemical structures are shown in Figure 5.5. These acids can be readily derived into refined esters. The esters allow suppliers to tailor the flux to meet the specific time and temperature requirements of the user while providing sufficient chemical activity to clean the metals to be joined. The way the

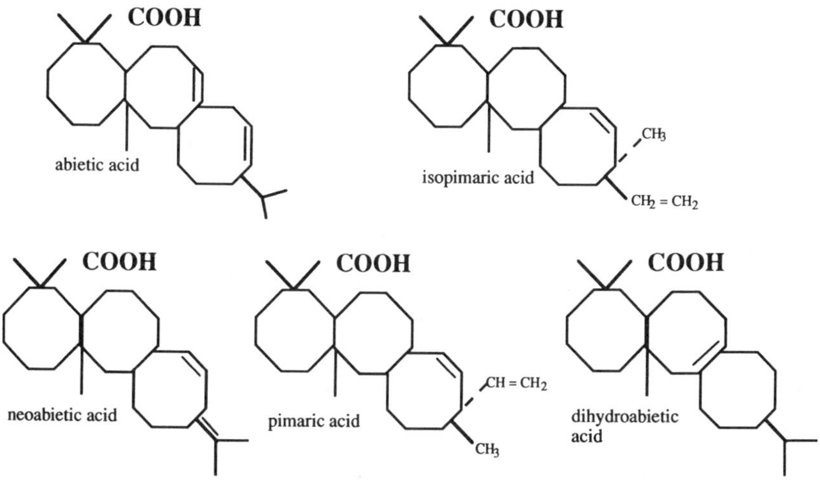

FIGURE 5.5. The chemical structure diagrams for the primary acids in rosin flux. (Source: Petrofilm, Inc.)

acid esters clean the metals is through a thermal energy aided chemical reduction reaction. Without thermal energy, the rosin ingredients are basically inert to 200°C. With the addition of thermal energy (temperature × time) the acid esters are able to break the molecular bonds of the oxidized metals. This releases the basic metal and oxygen. The oxygen may join as gaseous oxygen, or combine with hydrogen to form water. The generalized equations equations for this reaction are:

$$x\,M + y\,O_2 \rightleftharpoons Mx\;O_{2y}$$

$$MxOy + y\,H_2 \rightleftharpoons xM + y\,H_2O$$

$$x\,M + y\,CO_2 \rightleftharpoons MxOy + y\,CO$$

To regulate the strength and storage life of the acid esters, suppliers add activators, crystallization and oxidation inhibitors, and flow modifiers. The activators increase or decrease the acid strength when exposed to thermal energy. Flux is prone to crystallization and oxidation with time and temperature, which reduces its strength. Inhibitors slow or prevent this. The viscosity of flux over the reflow temperature spectrum determines its ease of runout. Excessive runout (shown in Figure 5.6) causes solderballs and shorts. By adding modifiers to slow the runout of the flux these problems are minimized.

Rosin from pine trees can be extracted from three sources on the tree. The first is from the gum, which is the thick liquid that lies between the bark and the cord wood. Next is from the stump, and the third is a byproduct of the wood pulping process used in making paper. The rosin from each source has a different characteristic. Gum rosin mixes well with alcohol solvents and is slower to crystallize while in storage. It also has the mildest odor during reflow. Tall oil rosin, the byproduct of wood pulping, is more stable over temperature than gum rosin, which means it is less likely to char during reflow. Uncharred flux is easier to clean after reflow than charred flux. However, tall oil rosin is more prone to crystallization during storage unless an inhibitor is added to the paste. Rosin from pine tree stumps is called wood rosin; it has a heavier odor and less stability than gum or tall oil rosin.

Rosin is graded by color. For electronic applications, the lightest color, i.e. water white, is the preferred color grade.

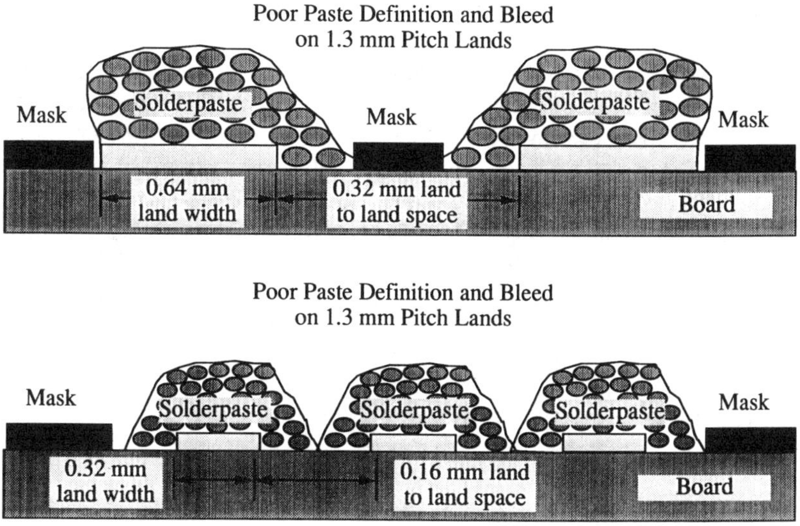

FIGURE 5.6. Excessive paste runout from the lands can cause shorts, solder balls, and weak joints. For wider surface mount packages, it is possible to use soldermask between the lands to block some of the solder bleed and reduce shorting. In fine pitch, this isn't possible.

Synthetic Resin Fluxes

Modern chemistry has developed synthetic fluxes that act like pine rosin. The synthetics can be tailored to remove undesired rosin properties such as odor and residue color. Additives to the synthetics can make them more or less active than natural rosin. An important feature of the synthetics is that they are used in the manufacture of water soluble and no-clean fluxes.

Formulating the ideal flux just to perform the oxide reduction reaction on a variety of base metals (copper, tin-lead, gold, nickel, etc.) requires a great understanding of physical chemistry. Add to this the additional requirements that the flux must aid stencil printing, help heat transfer during reflow, and be easily cleaned (or be safe when left as a no-clean product). It is easy to arrive at the realization that solder flux is a very complex material requiring a user's through understanding and careful process control. Users are advised to consult the references at the end of this chapter for a better understanding of flux chemistry.

The 1992 revision of the QQ-S-571 solder specification combined the European and U.S. flux classification methods. Prior to this change, fluxes for use in Europe were graded by chemical type and in the U.S. by flux residue activity level. The newer classification is based on flux composition loosely grouped by activity level. The flux classes include resin or nonresin, rosin or nonrosin, organic or inorganic. To be classified as a specific composition, a minimum of 51% of the nonsolvent portion of the flux must be either resin or nonresin, rosin or non-rosin, organic or inorganic material.

The nonsolvent (or nonvolatile) portion of a flux may be dissolved or dispersed in a suitable nonhalogented solvent such as isopropanol. The classification table from QQ-S-571 is shown in Table 5.2.

To classify the activity level of these fluxes, the specification uses a subjective "L for low, M for medium, H for high" approach which is listed in the notes section rather than in the main body of QQ-S-571, as before. This is because the previous R (for rosin), RMA (for rosin mildly activated), RA, and RSA (for rosin super active) system did not provide an accurate method of classifying flux activity. The new activity level classes equate to the old levels roughly as follows:

L0 type fluxes—All R, some RMA, and some low solids no-clean fluxes
L1 type fluxes—Most RMA, and some RA fluxes
M0 type fluxes—Some RA, and some low solid no-cleans

TABLE 5.2. The Flux Classifications Defined by the IPC-SF-848 Standard (Previously Numbered as the QQ-S-571). (Source: IPC)

Symbol	Base Material	Activity Designator
0 or 0	No Flux	
A	Rosin	L0
B	Rosin	L1
C	Rosin	M0
D	Rosin	M1
E	Rosin	H0
F	Rosin	H1
G	Resin (other than rosin)	L0
H	Resin (other than rosin)	L1
I	Resin (other than rosin)	M0
J	Resin (other than rosin)	M1
K	Resin (other than rosin)	H0
L	Resin (other than rosin)	H1
M	Organics (other than resin)	L0
N	Organics (other than resin)	L1
P	Organics (other than resin)	M0
Q	Organics (other than resin)	M1
R	Organics (other than resin)	H0
S	Organics (other than resin)	H1
T	Inorganics	L0
U	Inorganics	L1
V	Inorganics	M0
W	Inorganics	M1
X	Inorganics	H0
Y	Inorganics	H1

M1 type fluxes—Most RA fluxes
H0 type fluxes—Some water soluble fluxes
H1 type fluxes—All RSA, and most water soluble synthetic activated fluxes.

A low solids flux is defined as a liquid flux made with 5% or less nonvolatile contents with minimal postsoldering residues. These are considered by many to be very inert over the operating temperature of a product and therefore are not required to be removed, except possibly for test probe access, high frequency or high temperature applications, or esthetic reasons. No-clean flux is a flux class that may also be applied to the low solids fluxes. However, there are some higher solids content fluxes supplied whose residues are inert and do not degrade product reliability or require cleaning or removal.

No-Clean Fluxes

The advantage to using a no-clean flux is enormous. Eliminating the cleaning of the flux after reflow saves cleaning cost, may improve the product's reliability, and most important, helps the environment. Cleaning with solvent or water-based cleaning agents uses expensive equipment and costly, polluting agents. To clean under fine pitch devices, a cleaner needs several high pressure, specially designed nozzles plus multiple cleaning chambers. Solvents have become increasingly expensive, and in some cases, banned. This is due to environmental and disposal problems. Even water processing has become a major expense item in certain geographical areas (California, for example).

The high pressure spray cleaners can easily damage the fragile leads on some of the fine pitch packages, especially the tape automated bonded, or TAB packages. Studies have found it beneficial to leave the inert flux residue on a board. These studies conclude that the product's reliability is not degraded and in fact may be improved by the conformal coating effect of inactive solids around the solder joints. This conformal coating encapsulates the activator and prevents the absorption of moisture which would otherwise encourage corrosion of the board's traces, current leakage, or dendrite growth.

The environmental benefits of eliminating cleaning are the preservation of the air quality of our earth and also the ground water quality. All world-class and environmentally conscious electronic manufacturers treat their cleaning agent emissions before releasing them. However, the treatment systems are not perfect, in that they cannot catch every possible pollutant. And, treating cleaning emissions is certainly expensive.

A no-clean flux is not much different from the fluxes that are cleaned off the board. To qualify as a no-clean flux, the material must:

1. Leave no corrosive residues
2. Leave a nontacky residue that does not collect dust

3. Be safe for personnel and must not degrade equipment
4. Allow penetration of probe pins for electrical testing
5. Have a low light reflectance (for easy AOI inspection)
6. Provide excellent solderability

To prevent any remaining flux residues from reacting or corroding, the reactive elements need to be insulated from the board. The reactive elements in flux are the organic acids. The solids portion of the flux comprises the nonvolatile substances which may include the organic acids. In "cleaning-required" L1 flux, the solids content is as high as 40%. In no-clean flux, the solids content is reduced to 1 to 5% of the total concentration. The activators in no-clean fluxes have changed from chloride-containing halides to carboxylic and dicarboxylic acids. Some use only fatty acids, and amino acids. These sublime during the reflow soldering process.

The steps to successfully utilizing a no-clean flux in a solder paste for fine pitch surface mount applications are:

1. Use only highly solderable components and boards
2. Maintain good storage and inventory use procedures
3. Establish a stable reflow solder thermal profile
4. Purchase a solder paste that has a consistent and tightly controlled amount of low-solids flux
5. Choose a supplier who notifies customers of any changes in the recipe of the paste or flux
6. Use the least amount of flux possible so as to minimize any residue

5.3 SOLVENTS AND THIXOTROPES

The other constituents of solder paste fluxes are solvents and thixotropes. These additions regulate the rheology, or flow characteristics of the paste.

There are three areas where the flow needs to be regulated:

1. In storage
2. During printing
3. Just before and during reflow

During storage it is undesirable for the solder alloy particles to separate from the flux. If separation occurs than the particles may require remixing. A thixotropic additive minimizes this problem. When printing the paste the thixotropic agent helps the paste flow into the stencil openings easily and completely. After the stencil is removed, the thixotrope keeps the solder particles in contact with the flux so they do not slump.

Solvents are added to keep the other solder paste flux ingredients in solution and in reaction with the thixotropes. Degraded flux activity due to storage is not a significant problem. Crystallization and charring of the flux during reflow are problems. These problems cause flux residues that may or may not result in corrosion, but they are considered by most users and customers to be visually rejectable. Generally, the solvents are high temperature boiling solvents, such as butyl carbitol and beta terpinol.

During the preheat phases of reflow the solvents evaporate from the paste. Other added solid ingredients prevent flux breakdown as the solvent evaporates. These high boiling point solvents must be carefully selected to assure their complete evaporation in the reflow soldering process using the no-clean type of fluxes. If they remain on the board they can be absorbed into the interface between the soldermask and the copper. Once absorbed they may transport activators which may cause corrosion or electromigration.

5.4 SPECIFYING A PASTE

Regardless of the alloy, the solder is prepared in a paste form by mixing particles of the metal alloy with solder paste flux (see Figure 5.7). The alloy, as previously mentioned, forms the mechanical and electrical joint, so its purity and quality cannot be emphasized enough. The flux system contains the active and transport ingredients that act as a print vehicle, a binder, and a cleaner.

Specifying the Alloy

The choice of alloy is usually Sn63, 63/37 tin-lead. The use of 60/40 tin lead is also acceptable but its higher surface tension limits its use in some applications. Although Sn60 solder may be less expensive than Sn63 paste, the amount of solder used per board, including very dense two sided boards, is a very small fraction of the product's cost. Silver bearing solder offers no clear advantage when the silver terminations of components have a nickel barrier. The lower melting point alloys containing indium and bismuth allow lower temperature second side soldering, and reflow soldering at temperatures closer to the glass transition temperature of FR-4 boards. However, these alloys introduce several extra costs. First is the cost for an extra reflow oven (to avoid changing profiles in the primary side oven), and the extra cost of the alloy. Also the extra cost of field repairs, since technicians will be more apt to over-rework a low temperature solder joint using a solder iron set to a temperature appropriate for eutectic solder. Inspectors and customers will need to be retrained in accepting the gray "cold-joint" appearance of the bismuth and indium alloys.

Alloy Particle Shape. The shape and size of the alloy powder is critical to successful fine pitch stenciling. If the particles are too large and uniform in size

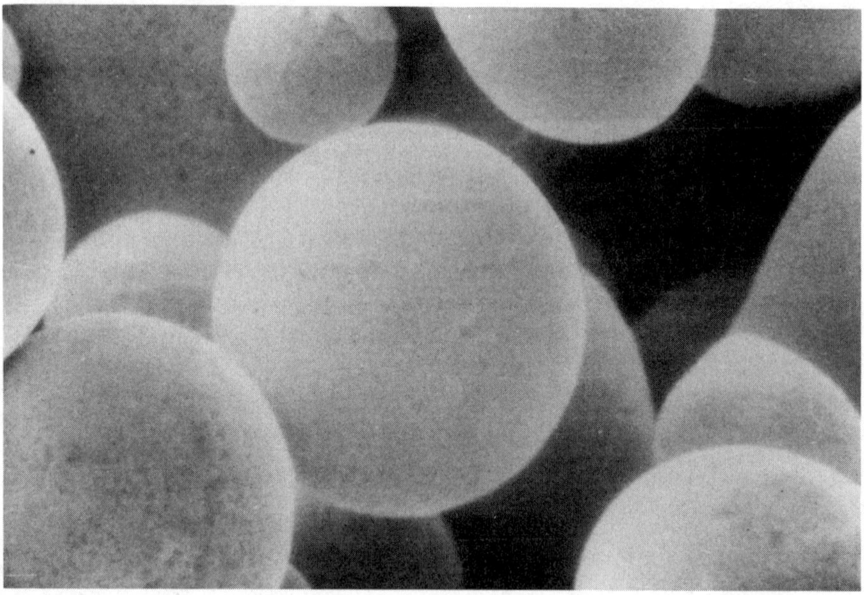

FIGURE 5.7. (a,b,c) A scanning electron microphoto of a solder alloy spheres at various magnifications. (Photos courtesy of Alpha Metals)

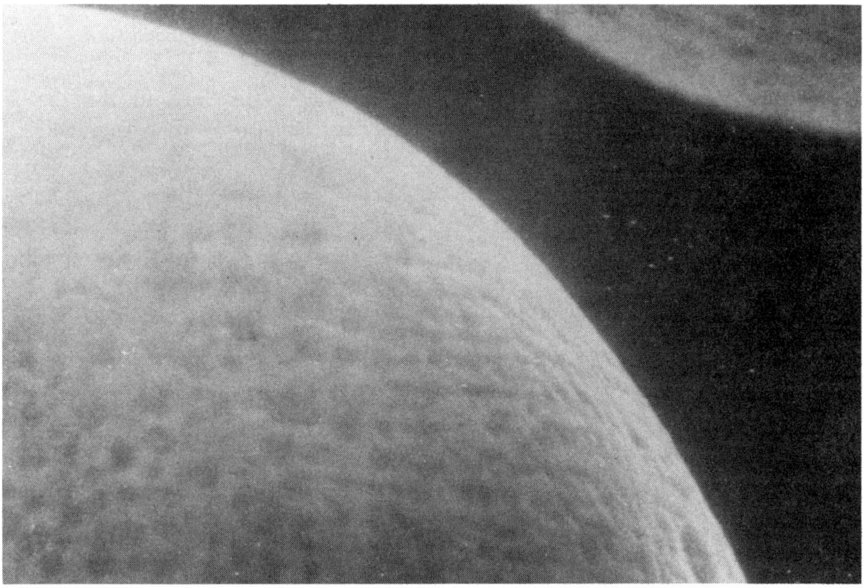

FIGURE 5.7. *(Continued)*

high metal percentages cannot be used. Also, they have low binding energy and will run out during preheat. This runout causes the paste to flow off the land and onto the soldermask and possibly to join with other paste deposits, as illustrated in Figure 5.6. After reflow, the runout areas can leave embedded solderballs in the soldermask, or worse, solder shorts between joints. Smaller particles of variable size have higher binding energy during preheat, resulting in less runout, solderballs and shorts. The particles should also be approximately spherically shaped to provide minimum surface area and hence minimal oxide. The amount of oxide on the particle contributes to the formation of solder balls.

Particle Formation. Spherically shaped solder particles are usually formed by either injecting molten solder alloy under pressure into a cooling inert gas atomization system, or by dropping the molten solder onto a solder wetted spinning wheel. Both methods give excellent results. A photo showing the resultant spherical shape in shown in Figure 5.7.

After particle formation, the solder particles are graded into various size categories. Originally, the grading was performed using mesh screens. The particles were vibrated through a stack of mesh screens. Each lower screen had more wires per inch thereby restricting the larger particles from falling through to the next screens. For surface mount technology, the common particle size is − 200 to + 325 mesh. This means the particle is smaller than 200 mesh wires per inch, but larger than 325 wires per inch. In physical size −200 to +325 particles range in size from 75 to 45 μm.

With newer grading techniques, a revised grading system was adopted that is slowly replacing the mesh specification. This newer classification system is used in specification IIW/ISO and is shown in Table 5.3.

Paste suppliers are permitted to use smaller size particles as fillers, which helps to decrease the runout properties of the paste. The filler particles displace extra flux and increase the binding energy of the large particles. The amount of smaller particle filler is controlled by the IIW/ISO specification, since the smaller particles (<20 μm) may cause solder balls.

The size of the particle limits the stencil and screen opening size. If the particles are too large, they will not pass through the opening, or they clog the opening without sticking to the land site. For good printing, the size of the particle should be 4 to 4.5 times less than the stencil opening and 2 to 2.2 times less than the screen opening. Table 5.4 lists the particle size by type and the minimum recommended stencil and screen opening for that paste type. Table 5.4 also shows the relationship between the package lead pitch, the stencil opening, and the powder type.

From Table 5.4 one can see that a type 2 powder provides an adequate stencil opening to powder diameter ratio for 1.3 mm (0.05″) and 1.0 mm (0.04″) pitch packages. Type 3 is adequate for lead pitches down to 0.4 mm, provided the power particles are spherical; it is marginal if the particles are nonspherical. A type 5 powder is marginally adequate for a 0.25 mm pitch.

Users can specify to the supplier the amount of metal by % weight. For stencil print applications, 88 to 91% metal weight is appropriate. For syringing the range of 85 to 88% works well.

TABLE 5.3 The Size of the Solder Alloy Particles Tabulated by Power Size Type. (Source: IPC-SF-848 Minutes)

Type	1% W/W maximum, larger than:		90% W/W minimum, between:		10% W/W maximum, less than:	
	μm	mil	μm	mil	μm	mil
1	150	6	75 - 150	3 - 6	75	3
2	75	3	45 - 75	1.8 - 3	45	1.8
3	45	1.8	20 - 45	0.8 - 1.8	20	0.8
4	35	1.4	20 - 36	0.8 - 1.4	20	0.8
5	25	1.0	15 - 25	0.6 - 1.0	15	0.6
6	15	0.6	5 - 15	0.2 - 0.6	5	0.2

Based on IIW/ISO specification.

TABLE 5.4 The Minimum Size Opening in a Stencil or Screen for a Given Powder Size Type. (Source: IPC-SF-848 Minutes)

Paste Type	Largest Particle Size, μm	Minimum Stencil Opening, mm (in)	Minimum Screen Opening mm (in)/mesh
Type 1	150	0.68 (0.027)	0.33 (0.013) / 80
Type 2	75	0.38 (0.013)	0.17 (0.006) / 105
Type 3	45	0.20 (0.008)	0.10 (0.004) / 150
Type 4	36	0.16 (0.006)	0.08 (0.003) / 180
Type 5	30	0.13 (0.005)	0.06 (0.002) / 200
Type 6	20	0.08 (0.003)	0.05 (0.001) / 250

Specifying the Flux

In general, specify the least amount of the lowest activity flux. And use one that is compatible with the chosen cleaning method. The low active fluxes are the type L fluxes which are equivalent to the older R and RMA classification. Use rosin and synthetic fluxes with solvent cleaning systems, use water soluble fluxes for aqueous systems, and use no-clean fluxes when no cleaning is required. Using the least amount of flux reduces runout, shorts, and solderballs. And if the flux is a no-clean type, minimizing its volume improves the safety of not cleaning the assembly after reflow soldering.

Table 5.5 lists the % volume of flux for the corresponding % of metal by weight for 63/37% tin-lead solder.

One element of flux that causes corrosion is the ionic halide additive. The effect of halide additives in flux can be seen in Figure 5.8, which shows the copper corrosion rate in microinches per year versus the halide additive content. For reference, the industry cleaning standard allows a maximum corrosion rate of 200 microinches per year. Copper in air-only corrodes at a rate of 30 to 100 microinches per year (depending on whose air it is exposed to.)

5.5 STORAGE

Solder paste is shipped in jars ranging from 150 grams to 2 kilograms in size, in cartridges of 400 ml to 1.5 kilograms, and in syringes of 10 to 100 grams.

TABLE 5.5 The Percentage by Volume and Weight of Flux And Metal in a Solder Paste Mixture.

% Metal by Weight	% of Metal by Volume	% of Paste Flux by Volume
80	33	67%
85	41	59%
88	48	52%
90	53	47%
91	56	44%
92	59	41%

Storing the paste in a cold, dry area is beneficial. The cold temperature slows the rate of chemical reaction and reduces separation by significantly increasing the viscosity. The storage area should have a low controlled humidity (RH $< 50\%$) and the jars of solder paste should be tightly sealed. A refrigerator with a moisture

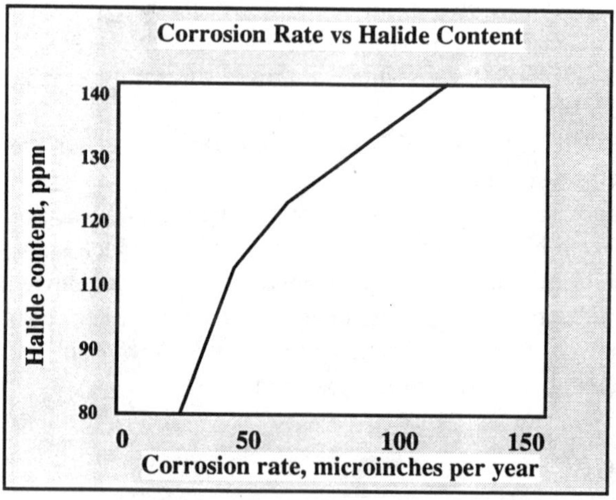

FIGURE 5.8. This graph shows the amount of copper that is corroded by the specified flux after a year of exposure.

absorbing desiccant works well for this purpose, unless specified otherwise by the paste supplier.

Reusing solder paste, from a previously opened jar or the excess from the printer, is acceptable if the paste is not separated and oxidized after exposure to moisture on the factory floor and subsequent storage. Storing the paste during floor use in a dry box containing a nitrogen atmosphere and a desiccant may be beneficial. The reused paste should be tested for solder balls, slump and tack time before reusing.

5.6 PASTE QUALITY CONTROL AND TESTING

Proper specification, supplier selection and storage are three of the quality control procedures required to assure optimum solder paste. The consistent quality of the solder is of utmost priority since the paste is the only material that provides the mechanical and electrical connection of the components to the interconnecting board. There are several other common tests that should be conducted at incoming (or receiving) inspection of the paste from the supplier. Some of these tests should also be performed before and during the board printing operation. The battery of tests used for testing and monitoring solder paste includes:

1. Solder ball
2. Viscosity
3. SIR (surface insulation resistance)
4. Tack time
5. Slump
6. Flux activity
7. Dryness
8. Acid value (for rosin pastes)
9. Solids level
10. Particle size and shape
11. Wetting balance

These tests are listed in no particular order or priority. Some users may want to perform all of these at incoming inspection. However, only some of the first five are performed during the printing process, depending on the user's use requirements.

The solder ball test is the most universally utilized test because it is an easy test to perform. The solder ball test is used to detect oxidized metal, crystallized flux, and excessive moisture in the paste. A small quantity of paste is placed on a nonwetting substrate (alumina, glass or FR-4). A volume 0.2 mm thick by 6 to 7 mm in diameter is adequate. The substrate containing the paste is heated to 210° C using a hot plate. As the paste heats, the time it takes for the solder to melt is monitored. As the paste heats the flux will rise to the surface and spread around the periphery of the sample. Any violent bubbling of the flux is an indication that the paste has absorbed moisture, which may cause the paste to be pushed away from the joint during reflow. After approximately 20 to 30 seconds, the solder alloy melts and, ideally, will melt into one large solder ball. (This sequence is shown in Figure 5.9).

If the melt leaves several uncoalesced solder balls as shown in Figure 5.9, Categories 3 and 4, then the paste should be discarded because of excess solder balls.

The viscosity test is important to assure proper printability, especially in the small fine pitch openings. Viscosity testing has been abandoned by many users of SMT since the stencil openings are relatively large and undemanding. FPT, being more difficult, renews the need for this test. It is important to know that the

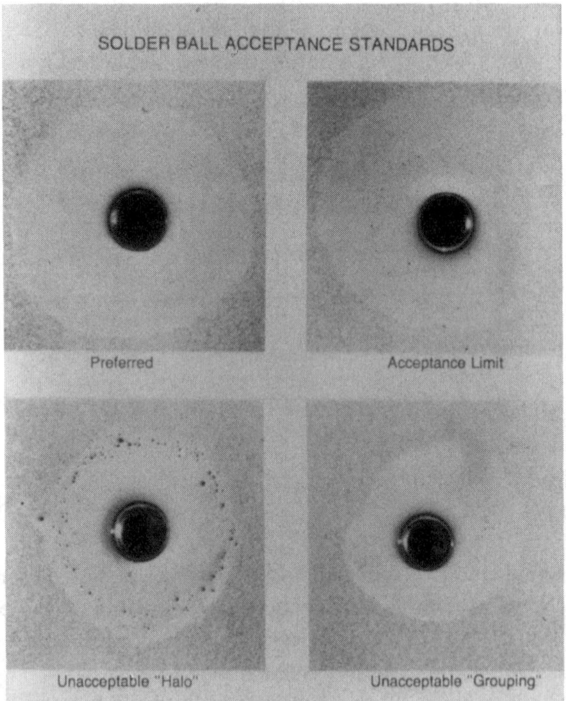

FIGURE 5.9. The sequence of events in the popular solder ball test. As the solder paste is heated it melts into one solderball. If the paste is excessively smeared or oxidized, then several tiny solderballs occur which implies that the paste is are unacceptable for production usage. (Source: ATT, Bell Laboratories)

commonly used viscosity tests are subjective, so the user must be consistent in test method (time, temperature, and equipment) in order to get meaningful test results.

The viscosity test uses a viscosity meter which is basically a motorized paddle with a sensor and meter to indicate the resistance encountered by the paddle as it rotates through a viscous material. The shape of the paddle, its rotation speed, the paste temperature, and the test time are all critical. They should all be the same each time the test is conducted. For example, the paste should be tested at the temperature at which it is printed, usually 25°C. For testing, the temperature should be within ± 0.5°C (≈ ± 1°F) A typical solder paste will change viscosity at a rate of 20,000 to 30,000 poise/°C in the temperature range of 20 to 30°C.

For SMT applications, typical viscosities are 600,000 to 800,000 centipose. For FPT, higher viscosities of 900,000 to 1,400,000 centipose are desired, again to assure printability into small stencil openings.

The surface insulation resistance (SIR) test is used to test the conductivity of the flux, any flux residues and the board surface. This test can be used in several process points other than its customary postcleaning point. The SIR test uses an interdigitated trace pattern as shown in Figure 5.10. The resistance between the two traces is measured using an ohmmeter, with readings typically greater than 50 megohms.

This test should be used at the incoming and paste print process to test the resistance, especially if no-clean fluxes are used. A measured reading of greater than 100 megohms (per IPC-SF-818) indicates an inert region between the traces. A good test for the inertness and corrosiveness of the flux between the traces of the

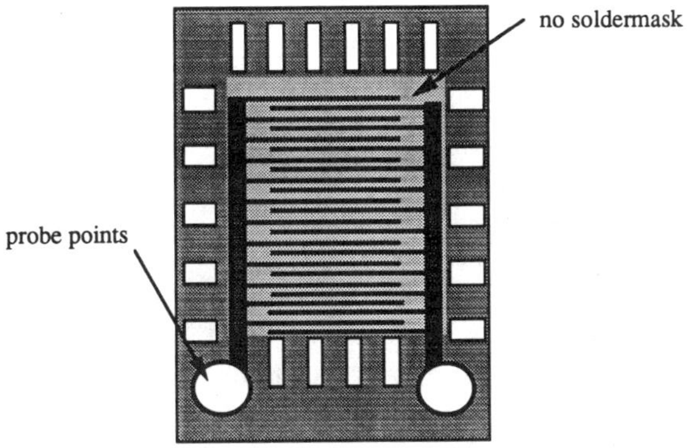

FIGURE 5.10. The surface insulation test, or SIR pattern that is designed onto test coupons and open board areas. Electrical testing of the SIR pattern provides an easy and effective means to test cleanliness.

SIR pattern is to subject the test sample to 85% humidity at 85°C for 7 to 28 days. After 85/85 testing dry the sample and test the resistance. Also examine it for evidence of corrosion (green or blue green precipitates) and evidence of dendrite growth.

The tack time is the length of time the paste will be sticky after printing. Loss of tack will result in parts falling off the board during or immediately after placement. Commercially produced tack testers use the basic method of pressing a probe into a paste sample and measuring the resisting force necessary to remove it. An equivalent method, although less quantitative, is to place the leads of a component into a paste sample. Then turn the sample upside down and observe that if the paste is adequately tacky, it holds the component in place.

The slump test is a measure of the amount of spread or runout a pillar of paste will develop over time and temperature. This test can be conducted by stenciling paste onto a printed circuit board and monitoring the slump or spread of the paste deposit over time at ambient temperature and 150°C. The quick and dirty tack test can be done on the same board. The test part is placed into the paste after a predetermined period of time has passed. The time delay is normally equal to the minimum time necessary to place the parts. As the paste dries less slump occurs.

The other tests listed include the flux activity through wetting balance tests. These tests are best done on a sample basis at incoming, or accepted as guaranteed by the supplier. A description of the test methods of these tests is given in the QQ-S-571 specification.

5.7 APPLYING SOLDER PASTE

Solder paste for fine pitch applications is best applied to the printed circuit board by using a stencil, a stencil printer, and a squeegee. For low volume and repair situations, a syringe applicator may also be acceptable, but far less controllable. Screens, which are common to some SMT users, are seldom used in FPT. The small solder particles needed to pass through the small wire openings of a screen oxidize too easily. Larger particles simply clog the screen.

A stencil printer is basically a board holder with a frame holder and a squeegee mechanism. A basic stencil printing system is shown in Figure 5.11. When selecting an appropriate printer, the user will find a wide ranging array of printer costs depending on options. The practical options to an FPT grade printer include alignment aides, squeegee pressure and speed adjusters, a board holder/with a nesting fixture, and a cleaning mechanism for cleaning the bottom of the stencil. The importance of these options will vary with the user's throughput needs and budget. For very low volume, a manual system is marginally adequate. For high volume or high accuracy, an automated system with vision assisted alignment and automatic stencil cleaning is important.

FIGURE 5.11a. A typical printer, stencil and squeegee system for printing paste onto the land areas. (Photo courtesy of MPM Corporation)

The common problems with printing paste in closely spaced, small openings are bridging, smeared paste (border violations), shorts, misregistration, and opens. The numbers and type of possible defects a user could experience is shown in the Pareto chart in Figure 5.12.

Stencil Alignment

Typical alignment aides include drilled location holes, etched fiducial marks, and video cameras. Drilled location holes with micrometer adjustments are the least desirable means of alignment because of the inaccuracy and variation in hole tolerance.

Fudicial marks etched into the board and stencil allow a highly accurate alignment. Typical fiducial marks are shown in Figure 5.13. For automated systems, a video camera views the location of the fudicial mark on the board and aligns the stencil mark to the board. Alignment accuracies of ± 0.03 mm (± 0.001″) across a 30.5 cm (12″) board are typically required.

FIGURE 5.11b. A close up of the squeegee, stencil and paste dispensing nozzle of an FPT grade stencil printer. (Photo courtesy of DEK, USA)

The Squeegee

A squeegee is used to push the paste across the surface of the stencil at a speed that allows the paste to roll into and fill the stencil openings. A side view of a squeegee and printer setup is diagramed in Figure 5.14. The typical FPT squeegee is a piece of hard plastic, such as polyurethane or teflon, shaped in a diamond or rectangle. Metal squeegees such as those made of steel are used by some to achieve the desired hardness and uniformity. The squeegee is held mechanically with a fixture that assures secure grip at an even steady force. The force, speed, stiffness, angle, and length of the squeegee are critical.

The printer should have a monitoring gauge to assure consistency over time. The squeegee force supplies the horizontal and vertical shear force necessary to push the paste across the stencil and down the hole. If the squeegee force is set high and the squeegee is soft, then a force vector is created that develops a scooping motion. The magnitude of the force vector is a function of the force on the squeegee and the hardness of the squeegee. If the force is large, then the vector will be

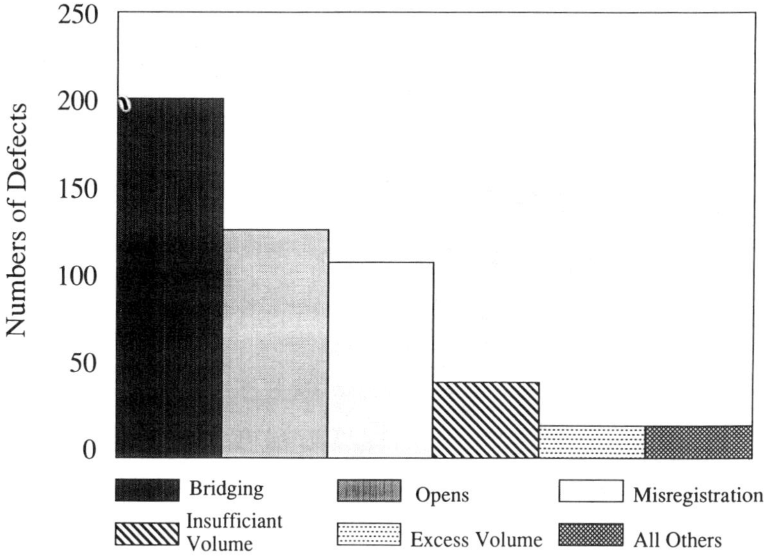

FIGURE 5.12. A Pareto chart showing the common defect causes and rates experienced in paste printing. (Source: MPM)

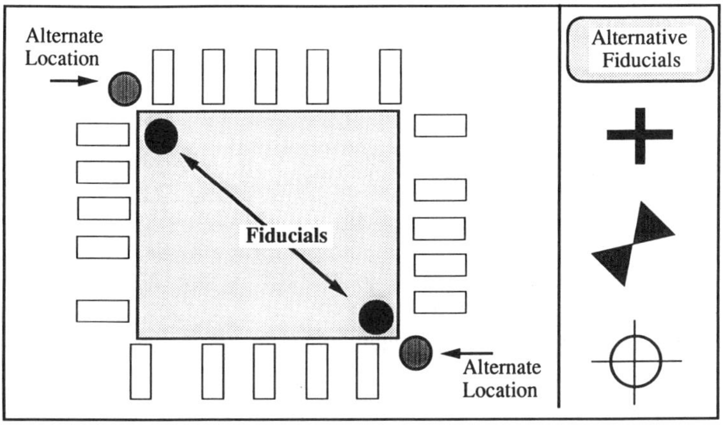

FIGURE 5.13. Fiducial marks are marks etched into the copper foil, they allow a vision system to accurately align the land areas so optimum registration of the stencil to the board is achieved.

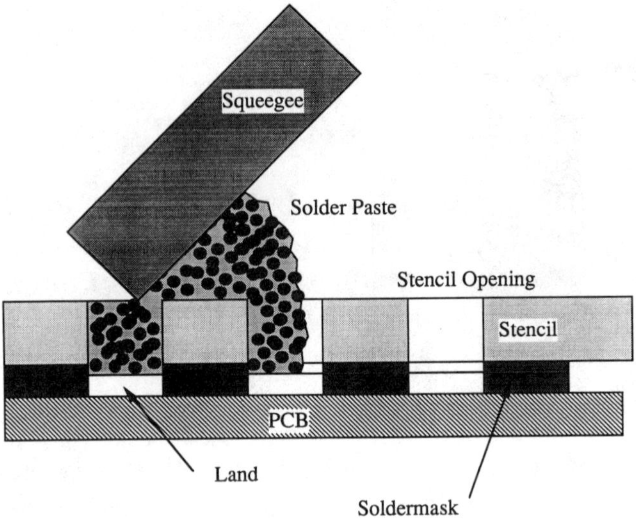

FIGURE 5.14. The side view of a squeegee and stencil displaying the suggested stand-off distance and contact force.

stronger in the horizontal direction. If the squeegee is soft, then the vector will favor the vertical direction. This is illustrated in Figure 5.15.

The force setting on the squeegee should be adjusted so as to produce a clean stencil surface on a single pass. Greater force results in excess paste scooping. Less force leaves uneven paste deposits and smearing on the lands due to paste bleed under the stencil. The typical force is between 3 to 10 kg (7 to 22 lb), depending on the stencil height (also known as the *snapoff height*) above the board.

The advantages of a noncompliant squeegee are that it:

- Resists solder paste stick
- Prevents solder paste scoopout (or scavenging)
- Can be machined in very straight even blade lengths

For SMT applications, a polyurethane squeegee with a hardness range of 75 to 85 durometer (shore A scale) is very adequate. For FPT, a squeegee hardness of 90 to 100 is more desirable. A noncompliant material such as polycarbonate or metal offers sufficient hardness and durability. The squeegee experiences wear during use, so the blade edges should be rotated regularly.

A slow squeegee speed of about 2 to 2.5 cm per second is needed to give a paste of viscosity of 900,000 to 1,400,000 centipoise sufficient time to roll into and to fill the opening of a fine pitch land. As illustrated in Figure 5.16 a slower speed

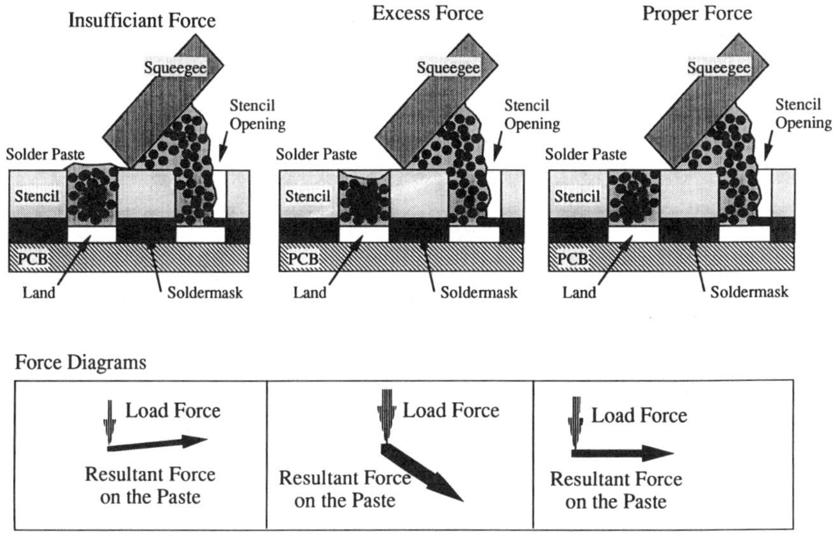

FIGURE 5.15. This diagram shows the effects that insufficient, excessive, and proper squeegee force have on resultant paste print volume.

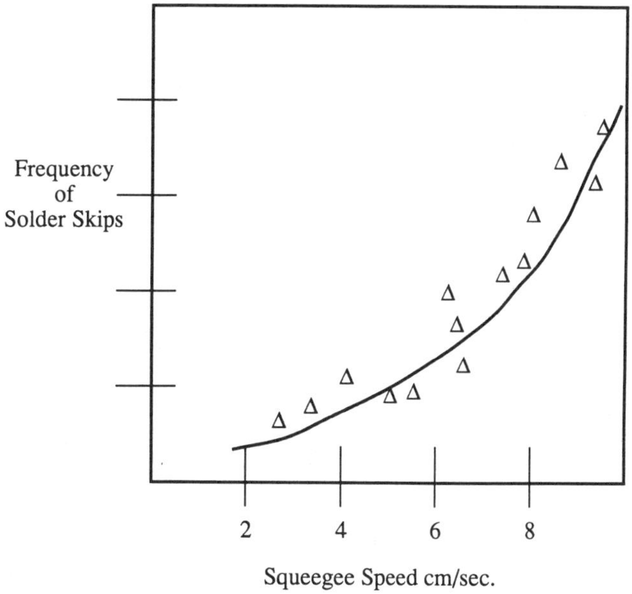

FIGURE 5.16. Squeegee speed is an important control factor, along with squeegee force. This graph shows the possible number of solder skips that occur as a function of squeegee speed.

does not improve solder paste deposit yield, while, a faster speed causes the paste to skip some openings, particularly in those openings that are parallel to the direction of the squeegee travel.

The squeegee needs to be longer than the print width. A length of 120 to 125% of the print width is usually sufficient, depending on the stencil's mounting method. The extra length minimizes squeegee bowing and unevenness.

If the squeegee lacks flatness, then streaks and misses occur. The squeegee is not the only source of unevenness, the board is another source. The board may be held under the stencil with a vacuum plate or in a fixture. Regardless, the board should be "nested" so the squeegee is evenly supported across the print width. The concept of nesting is best illustrated in the Figure 5.17. Nesting on a vacuum plate is done by placing strips of FR-4 material on either side of the board. The strips are the same thickness as the board.

A flat, even squeegee pressure and a flat board enables the stencil to conformally press against the openings forming a gasket. This gasket effect prevents paste bleed (Figure 5.18), and smear under the stencil. Paste bleed if left on the board can cause solder shorts or solder balls. If the paste bleed stays on the stencil it needs to be removed to avoid smearing the next board. Some printers are equipped with an automatic cleaning mechanism that wipes the stencil clean after each pass. The cleaning is usually done with alcohol and lint free cloth or paper.

Offsetting the stencil from the surface of the board rather than allowing it to sit directly on the board helps improve the print definition and minimizes potential smear and bleed. Setting the offset height at approximately five times the stencil thickness provides enough standoff for most stencil sizes. This technique is called *offset printing*. Its counterpart is *direct contact printing*, where the board and

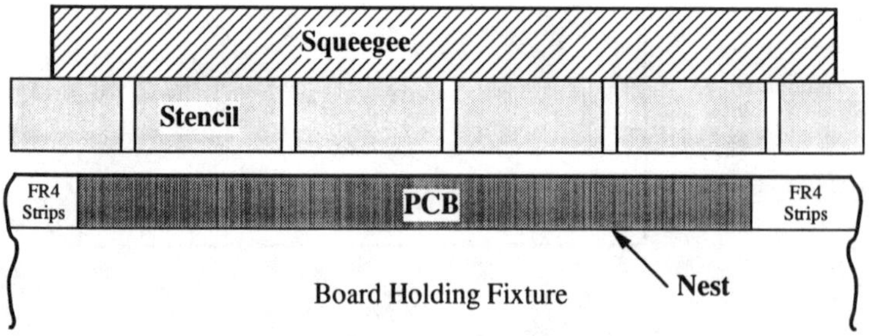

FIGURE 5.17. To prevent solder print problems the squeegee needs to be level along its full distance. Since the squeegee is wider than the board width, the outer squeegee width is supported by nesting the board down into a holding fixture or by taping scrap board material beside the board.

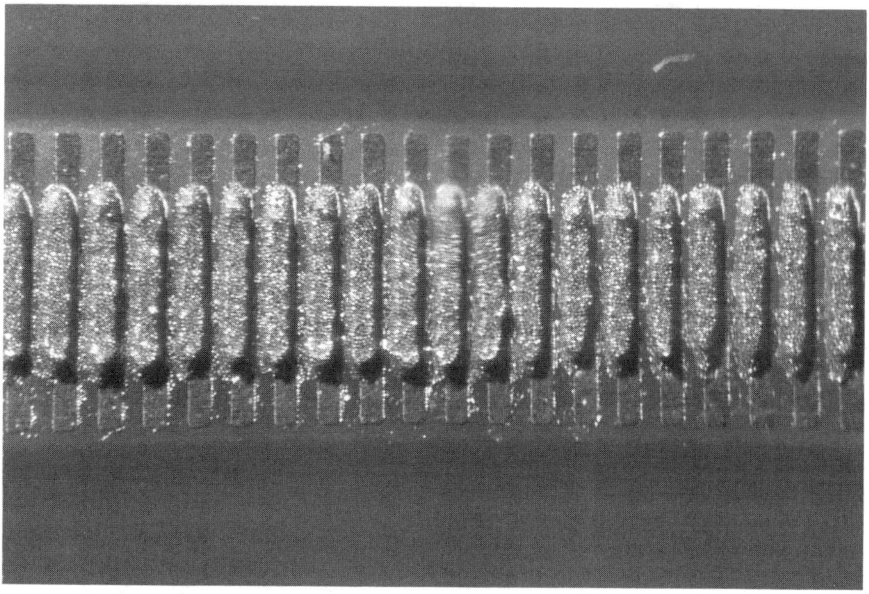

FIGURE 5.18. If the squeegee isn't flat, if the stencil is registered or seated properly, then solder paste smear occurs. The result of smear is solder shorts and solder balls. (Photo courtesy of DEK USA)

stencil are flush together. Direct contact printing works more effectively if the printer raises and lowers the board to the stencil; this is in contrast to the more common method of raising of the stencil holder.

The Stencil

The stencil is an etched piece of thin metal, usually stainless steel, brass, or alloy 42 (steel-nickel). Other metals have been used such as molydenum, but they tend to have undesirable features, such as brittleness, high cost, or etching difficulty. Of the common metals, brass is the least desirable for FPT even though it has been the most common for SMT. Brass is too soft, which means it deforms under the higher printing pressure. When deformed the small interconnects between lands allow paste to bleed and bridge between lands and they are easily broken. These interconnects can also easily break if the brass stencil is cleaned in a high pressure cleaner.

Stainless and alloy 42 steel are better than brass as they offer, respectively, approximately 5 times and 3 times stronger tensile strength than brass. However, neither is as easy to etch as brass. Of the two, alloy 42 is easier to bond resist to and is somewhat easier to etch than stainless steel. These are the main reasons for

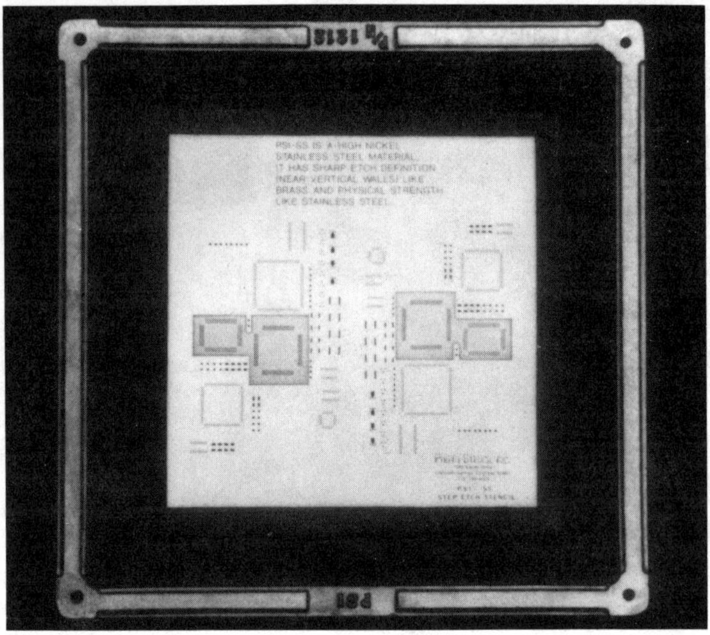

FIGURE 5.19. A photo of a typical stencil mounted on a metal frame. (Photo courtesy of Photo Stencil, Inc.)

its use over the more difficult to bond and etch "300 series" stainless steel. However, the lower tensile strength and the tendency to rust of alloy 42 makes a quality etched stainless steel stencil the preferred alternative.

The etch properties of a stencil are limiting factors in setting the smallest size opening that solder paste will release from. The ability to get paste into an opening is controlled by the size of the paste particles, and to a smaller degree, the squeegee force and evenness. If the stencil side walls are too close together and too rough, the paste sticks to the sides and clogs the opening. The roughness of the edges (shown in Figure 5.20) is inherent in some etching methods but this has not been a major concern with the wider pitch assemblies. Fine pitches result in smaller openings and the opening area to the sidewall area ratio is much less. This causes some or much of the paste to remain stuck to the sidewalls rather than to release and deposit onto the lands. This causes solder opens or starved solder joints that are weak in field use.

Stencil Etching

Stencil openings for fine pitch applications are best formed using chemical etching followed by a micropolishing, or deburring, prior to a final plating in a slippery

FIGURE 5.20. (a,b) The smooth side wall appearance (a) of a stencil assures paste release. If the sidewalls are rough and burred (b), then paste will clog the stencil openings and starve the solder joints on the board. (Photo courtesy of Photo Stencil, Inc.)

coating. Chemical etching is most economical as it is a fast mass production process. To chemically etch a thin steel plate, the plate is coated or laminated on one or both sides with a photosensitive emulsion. For a double side etch, two pieces of identical artwork are aligned to each side. This artwork defines the stencil openings. The photoemulsion is exposed and developed. Unexposed areas of resist are thus opened to allow the etchant to reach the metal plate in the appropriate areas. The plate is then sprayed with the chemical etchant long enough to form the approximate sized openings.

Interest in laser etching is developing as cutting techniques and results improve. With the ability to directly guide the laser using the board's data files eliminates a significant source of accuracy error. For very small openings a CO_2 laser may be used to improve the accuracy of the opening sizes and to smooth the sidewalls. The CO_2 beam is programmed to traverse the perimeter of the critical holes and to burn away the uneven edges and projections. Laser cutting is used because during chemical etching there is both lateral and vertical cutting. This etching proceeds laterally 40% as far as it proceeds vertically. For example, after etching vertically 0.2 mm (0.005″) the lateral etch distance, A, will be 0.4 × 0.2 mm or 0.08 mm (see Figure 5.21). To control this lateral distance, the stencil maker reduces the area of the opening photographically. The amount of area reduction is a function of the metal's grain structure and etch reaction. As a rule of thumb, the artwork is reduced by an amount of 40 to 50% of the plate's thickness for a single side etch. The resultant shape of single side etch opening is hyperboloid, as shown in Figure 5.22. A double sided etch produces an hourglass shape when viewed cross-sectionally. The sidewall variation is slight for a well etched stencil, with a variation of less than 10% of stencil thickness T.

This variation is a function of the fabricator's process control and the geometries to be etched. It can be controlled accurately enough to provide adequate stencils for package pitches down to 0.5 mm (0.02″). Underetched stencils exhibit

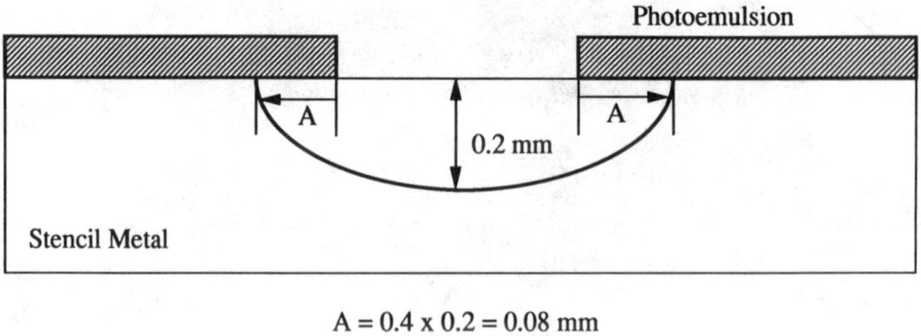

FIGURE 5.21. The lateral etch distance in a metal stencil during single side etching. (Source: Screen Manufacturing Technologies, Inc.)

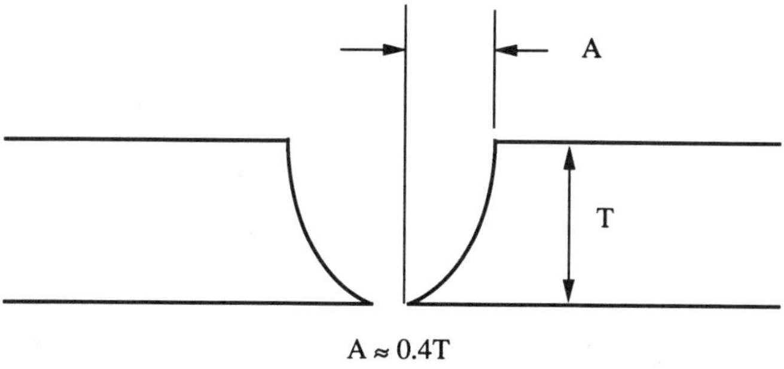

FIGURE 5.22. The resultant shape of the stencil after single-side etching. (Source: Screen Manufacturing Technologies, Inc.)

a more pronounced sidewall variation. Overetched stencils have a reversed hour glass shape (see Figure 5.24). Misregistered resist (Figure 5.25) and lifted resist (Figure 5.26) are quality problems that cause unacceptable printing.

A double side etch is preferred over single side etch. The single side etch clogs easily if the wider opening is squeegee side up. If the wide opening is facing down, then paste smearing and bleed may occur. The wide opening face down approach may work adequately if the wide opening area is tailored to match the required paste area. The paste area is not the same as the land area, but usually 20 to 50%

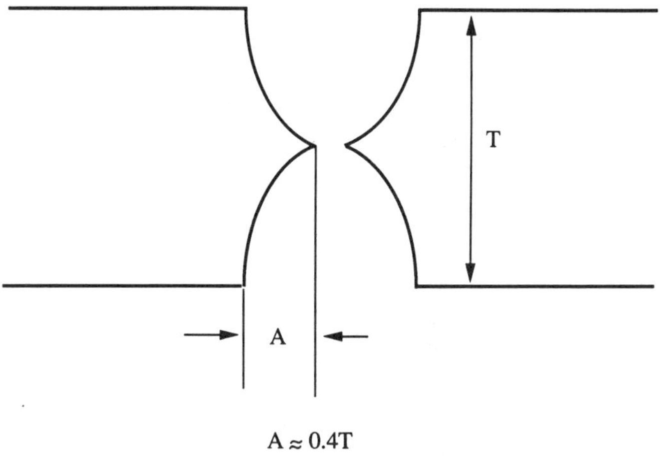

FIGURE 5.23. The resultant shape of the stencil after double-side etching. (Source: Screen Manufacturing Technologies, Inc.)

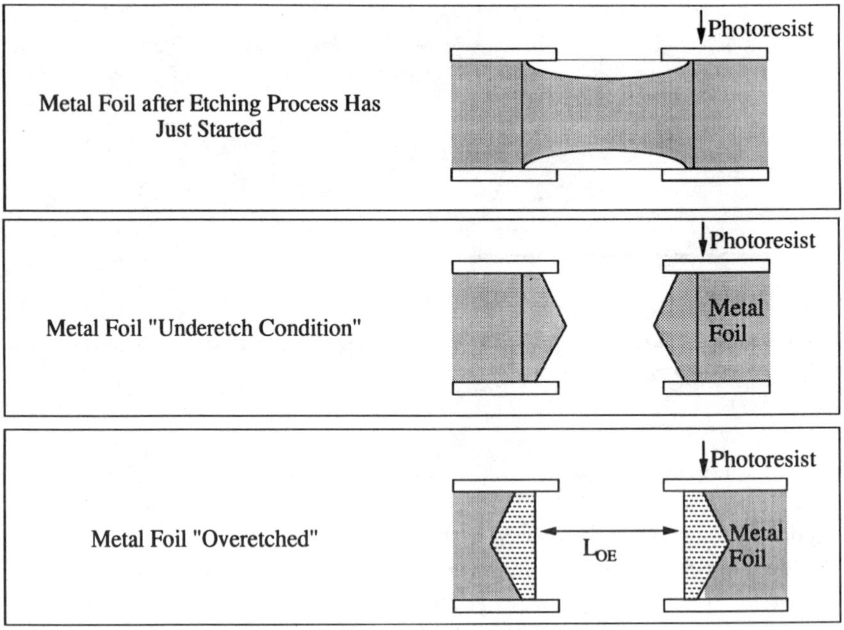

FIGURE 5.24. The appearance of under and over etched stencil walls. (Source: Photo Stencil, Inc.)

smaller to minimize excess solder and bridging. These reduced area stencil openings and other types of successfully used openings are shown in Figure 5.27.

Because of the lateral etch rate, any image with right angles in the artwork will exhibit rounded corners in the stencil after etching (see Figure 5.28). The corner radius is directly proportional to the vertical distance of the etch; e.g., the corner radius of an 0.2 mm thick plate will be 0.2 mm after a single side etch.

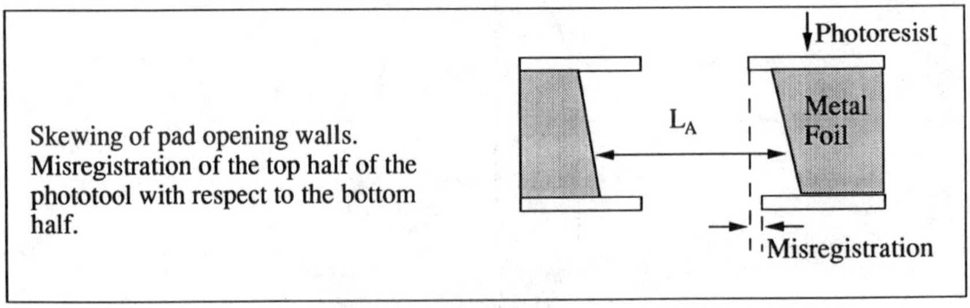

FIGURE 5.25. If the etch resist is misregistered during the stencil fabrication, then proper paste deposition and rebake is difficult to achieve. (Source: Photo Stencil, Inc.)

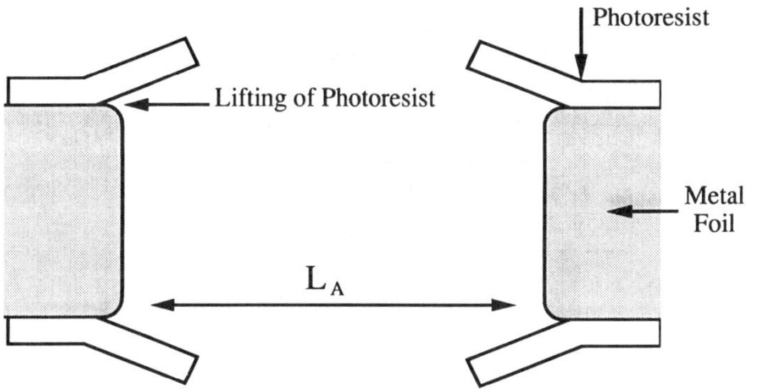

FIGURE 5.26. Lifted resist caused over etching and rounding of the stencil openings. This results in paste bleed and smear. (Source: Photo Stencil, Inc.)

One way to reduce the corner radius and the lateral etch distance is to form the artwork so the etch area of each opening is as small as possible and similar for each opening. In other words, only etch a thin channel around the perimeter of the larger openings rather than etch the entire opening. This technique recognizes that the stencil material's etch rate is a function of the width of the artwork openings. Different width openings will etch at different rates. If the plate is left in the etchant long enough to open a small area completely, then the largest openings will overetch.

To balance the etch rates on a stencil with large area differences in the openings requires other than traditional artwork. The traditional approach is to define an aperture of area equal to the desired opening area and to place that aperture over the appropriate place on the stencil. The balanced etch artwork uses a perimeter line aperture around all openings larger than the smallest land width. The perimeter width is the same as the smallest land width (0.25 mm in Figure 5.29). This is illustrated in Figure 5.29.

Stepped Thickness Stencils

Since the area of a fine pitch opening is much smaller than an SMT opening (>2 : 1 area difference) and since too much solder volume on fine pitch leads causes solder shorts, stepped thickness stencils are sometimes utilized. A stepped thickness stencil is one that is thinner over the FPT parts and thicker over the others. This is illustrated in Figure 5.30.

Stepped stencils effectively reduce the volume of solder paste and reduce the overetch potential of the fine pitch openings. However, stepped stencils are more fragile than uniform thickness stencils, thus requiring special handling and clean-

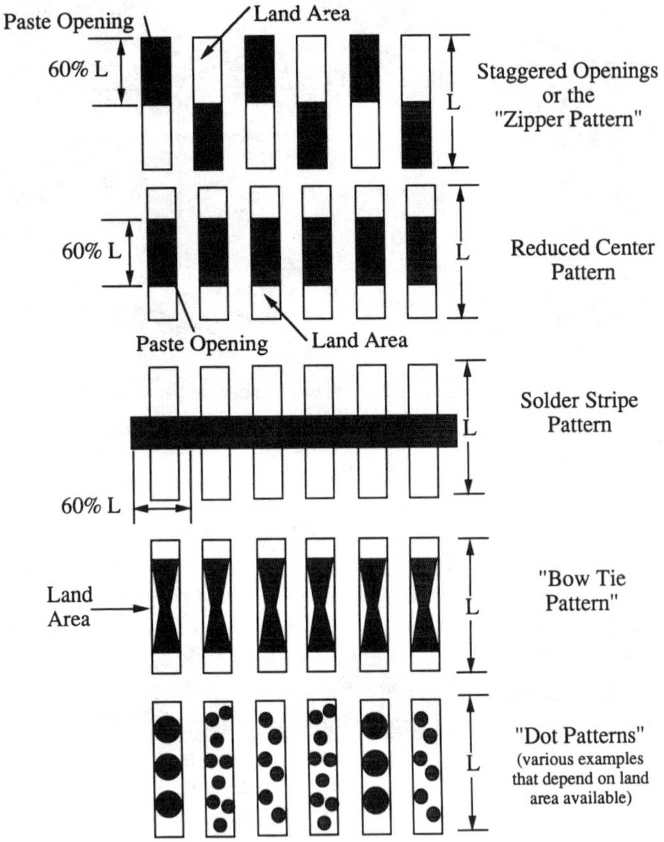

FIGURE 5.27. This figure shows the various stencil opening designs that have been used for successful paste printing for fine pitch assembly.

ing procedures. Additionally, the designer must be careful not to place any components near the step, and the fine pitch openings should be spaced at least 6 to 7 mm from the step. This 6 to 7 mm space allows room for the squeegee to sit level over the fine pitch opening, and it allows proper conformance of the stencil maker's photoresist to the stepped stencil surface.

Stepped stencils, even designed as specified above, are still problematic with the harder (durometer/reading >80) squeegees. The squeegee may not conform to the stepped stencil contour and the steps can create points of wear on the squeegee, which results in a grooved squeegee. If the stepped stencil buckles it is easily ripped by the squeegee, which ruins the squeegee as well as the stencil. Also the reduced thickness accentuates the problem of open joints with nonplanar leads.

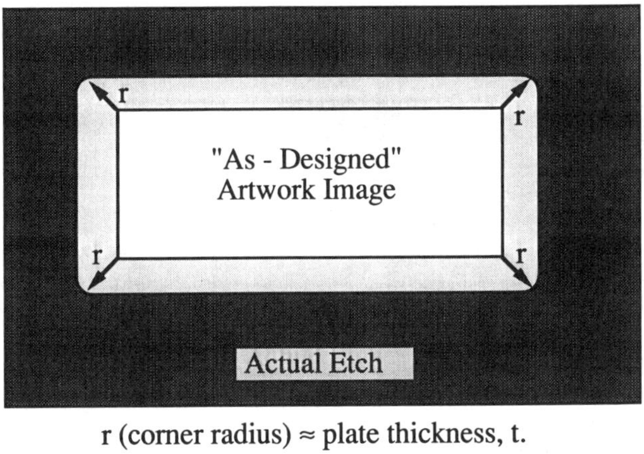

FIGURE 5.28. The actual stencil etch appears different than the artwork design due to the etch factor. This figure shows the rounded corners that occur when square artwork is used.

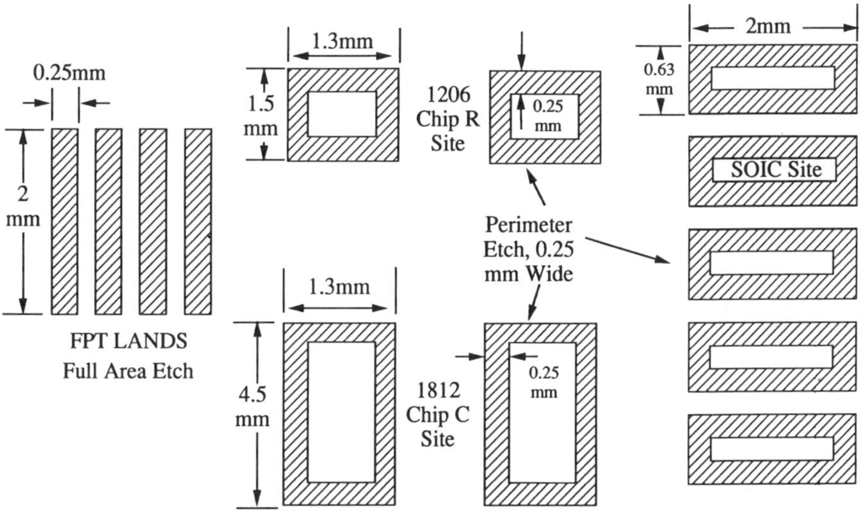

FIGURE 5.29. The etch rates of larger stencil openings can be balanced with smaller openings by designing narrow etch channels around the perimeter of the large opening, this rather than trying to etch the entire large opening as well as the small opening. (Source: Screen Manufacturing Technologies, Inc.)

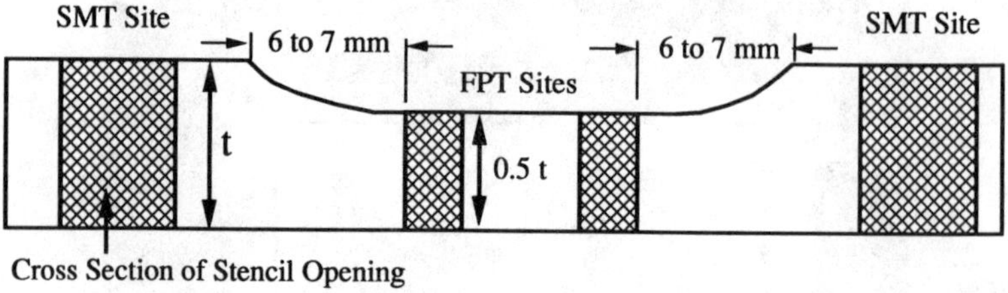

FIGURE 5.30. A stepped stencil, one with certain areas etched thinner than the bulk stencil, is used by some to limit the solder volume to fine pitch land sites. This may help minimize solder problems.

Additional Stencil Design Tips

A border pattern needs to be added that etches holes in the edge of the plate. These holes allow epoxy pass through for attachment to the stencil frame of fabric that attaches to the frame. There are several different border patterns favored by different stencil makers (see Figure 5.31).

The best way to assure an optimum quality stencil from a quality stencil maker is to send the stencil maker the CAD data files (Gerber format or whatever electronic data format is readable). The stencil maker can then assume the responsibility of modifying the opening sizes and setting the border pattern. After all, it is his responsibility to provide the user with the correctly sized and securely attached stencil.

If the user insists on providing the stencil maker the photopositives he should find out from the fabricator the etch factor, border patterns, and locator marks needed to do a quality job.

Attaching the Stencil

The stencil plate is attached to a frame that, in turn, is attached to the printer. The stencil plate may be directly attached to the frame, but this usually results in a plate that has no tension control and is not suitable for direct contact printing because of its lack of flexibility or deflectability.

To enable flexibility while maintaining tension, the stencil is attached to a tensioned fabric which is attached to the frame. The choice of fabric determines the stencil's tension. Typical fabrics are coarse polyester mesh (mesh count 30 to 32 strands per centimeter) or stainless steel. Polyester provides a nominal tension of 22 to 30 newtons/cm^2 (depending on the frame size). The stainless steel mesh provides higher tension, but it lacks memory and can lose its original tension with use much faster than polyester. Polyester is the preferred fabric since it is highly durable and maintains its "memory", i.e., original tension, for a long time during use.

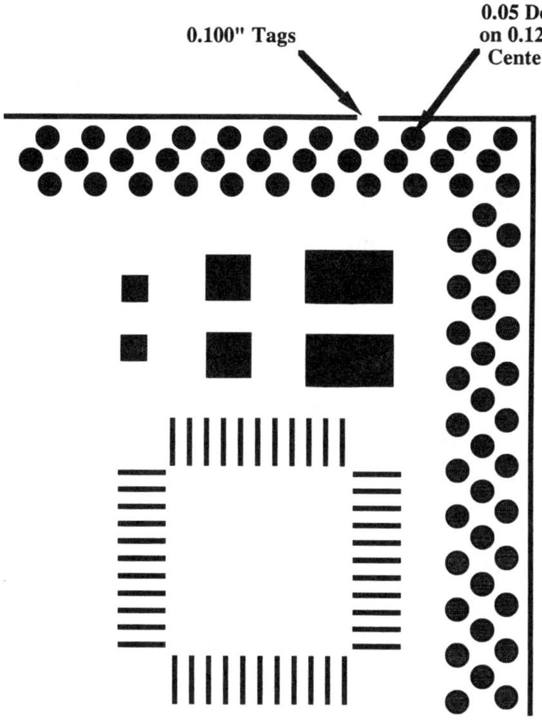

FIGURE 5.31. The border pattern etched into the edge of the stencil allows secure bonding of the stencil to a flexible membrane. The flexure that results allows the stencil to standoff the board during printing and improves the print quality.

When attaching the stencil and fabric to the frame, great care must be made that the plate is flat. Otherwise the plate develops waves which cause buckling at the end of the print stroke of the squeegee. This buckling lifts the stencil off the board, breaking the gasket seal, and allows paste bleed.

The maximum wave should be 0.3 to 0.6 mm anywhere on the plate. In order to determine wave height, the frame mounted stencil is placed on a flat block (such as granite) and the heights of any waves or buckles that appear are measured from the block's surface.

The stencil plate is bonded to the fabric using epoxy glue. This is usually a two part thermoset epoxy. The choice of epoxy must be compatible with the cleaning agents, flux type, spray pressure, and temperature that the stencil will experience during cleaning.

For automated printing, the squeegee should never traverse the epoxy bonding fabric to plate. However, this is a common occurrence during manual printing.

Therefore, the epoxy bond should be strong enough to endure the shear force of the squeegee under manual pressure. The attachment of the fabric to the frame uses a different adhesive. The frame adhesive is usually a lower temperature thermoset adhesive to prevent thermal expansion of the stencil, as any expansion will alter stencil tension.

The fabric should also be covered with an emulsion. The purpose of the emulsion is to seal the mesh and prevent it from passing solder paste. The emulsion, like the adhesives, must be compatible with the flux and cleaning systems used.

The Frame

The stencil frame is usually extruded, rather than cast, aluminum. Cast frames are less flexible, with poorer surface characteristics than extruded ones, and are usually machined flat. Mounting a cast frame to a printer does not affect the stencil's flatness unless the mounting clamps are tightened beyond the recommended torque limit. Extruded aluminum is more flexible and seldom machined flat. When clamped to the printer the frame tends to assume a contour, and warpage if the clamping torque are unequal.

Syringing Solder Paste

Stencils add time and expense to the prototyping of electronic assemblies but are minor cost factors in production. Many users do not want to wait one to three days for a stencil just to produce a few first articles, prototypes, or even a small volume production run. For these users syringing the paste may be a viable alternative. However, if the assembled units are to be used for reliability evaluation and customer acceptance, then the use of a stencil is encouraged over syringing of paste.

Applying solder for a small quantity of assemblies could be accomplished using a very fine tip soldering iron, thin solder wire, and hand soldering each lead. However, very few people have the patience to do this consistently for more than a few hundred fine pitch leads (about one component package).

Syringing solder paste, manually placing the components, and reflowing in a controlled oven is an effective process for small quantities. It is not a perfect process, as some solder shorts and opens will occur. But with proper care and a little easy touchup, fixing these faults may cost a little less than a stencil and may save the time delay.

To syringe solder paste on surface mount lands, simply deposit a blob of paste that approximately fills 2/3 the area of the land.

For fine pitch lands the syringe nozzle needs to be very small in diameter. For the solder paste to flow freely it should be about 85% metal by weight and 400,000 to 500,000 centipose. Instead of depositing individual solder paste dots, syringe a thin straight line of paste.

If the placement time to put the good solderable components on the paste does not exceed the tack time of the paste, then the solder, when melted, will flow off of the solder mask, breaking most solder shorts. The amount of solder syringed onto the land and the soldermask is usually more than enough to provide a reliable compliant solder joint. Some solder masks will trap solder balls, but these are easily removed with a pointed stick during inspection and touchup.

Stencil Inspection and Acceptance Criteria

Before acceptance and periodically during use, the stencil should be tested for quality. The common incoming (or acceptance) tests include:

1. Stencil flatness and tension when clamped
2. Proper fabric bonding
3. Clean image definition size, and sidewall etch
4. Correct materials
5. Proper orientation, position and direction

During use, the stencil should be carefully handled, cleaned, and stored. Periodically and after every cleaning, it should be checked for:

1. Clogged openings and general cleanliness
2. Grooves, scratches, and cuts
3. Missing portions and rips
4. Proper tension

The sidewall and image definitions tests can be performed using a microscope with a graduated eyepiece. The image size and sidewall variations should not exceed the user's and the stencil maker's mutually agreed limits.

One easy but subjective way to test flatness is to mount the stencil on the printer. Being sure to clamp it using equal and proper torque. Then examine the stencil for uniform, flat contact with a test board (be sure it is flat). And examine for and measure the height of any waves in the plate. Tension can be measured during the flatness test by raising the stencil to the proper off contact distance. Attach the squeegee and set the squeegee pressure to the proper force. Pour solder paste onto the stencil and squeegee a test print. Examine the paste pillars for proper height and definition. Be sure there is a minimal amount of deformities and paste bleed.

References

1. Ayllon, P. 1991. The Stencil Printing Process for the Attachment of TSOP Components in a High volume Production Environment. Proceedings of Nepcon West '91, Feb. 1991, Anaheim, CA.

2. Coleman, W. et. al 1990. The Design and Manufacture of Metal Mask Stencils. Meeting the Challenge of Fine Pitch Technology. In *Surface Mount Technology*, May, 1990.
3. Elias, I., 1990. Techniques to Improve Automated Screen Printing. In *Surface Mount Technology*, Sept., 1990.
4. Erdmann, G. 1991. Improved Solder Paste Stenciling Technique. In *Circuits Assembly*, Feb., 1991.
5. Hwang, J. 1989. Solder Paste in Electronics Packaging. New York: Van Nostrand Reinhold.
6. Jillings, R. 1991. Private correspondance
7. Jillings, R. 1991. Screens for Solder Paste Printing. Santa Clara, CA: Screen Manufacturing Technologies. Inc.
8. Prasad, R. 1989. Surface Mount Technology: Principles and Practice. New York: Van Nostrand Reinhold.
9. Silma, T. et al 1991. Selecting a Solder Paste. In *Electronic Packaging and Production*, March, 1991.
10. Socolowski, N. 1989. Electronic Materials Handbook. Vol. 1. Materials. Park, OH: ASM International.
11. Socolowski, N. 1990. Solder Paste for Fine Pitch Components. In *Electronic Packaging and Production*, Nov. 1990.
12. Socolowski, N. 1991. Private correspondence.
13. Thwaites, C.J. 1982 Soft-soldering Handbook. Greenford, Middlesex, England: International Tin Research Instite.
14. Wassink, R.J. 1989. Soldering in Electronics, 2nd edition. Ayr, Scotland, Electrochemical Publishing.
15. —1991. Requirements and Test Methods for Solder Paste, QQ-S-Paste. Lincolnwood, IL: IPC.
16. —1992. Solder Flux, IPC-SF-818. Lincolnwood, IL: IPC.

6

Package Placement

6.0 PLACEMENT

For surface mount size packages (lead pitch of 0.8 to 1.3 mm) the placement of the parts needs to be precise. For fine pitch sized packages (lead pitches of 0.15 to 0.65 mm) placement needs to be ultraprecise.

The ultraprecision placement required is due to several factors and their tolerance distribution. These factors (see Figure 6.1) are:

- The land area and location for each lead
- The land width to lead width ratio
- The package lead area and location
- The package size and tolerance
- The misalignment allowance
- The ability to self-center during reflow

The land area, especially the land width, is a function of the package lead width, lead pitch, and the minimum space between the lands. The recommended equation for calculating the land width is:

$$\text{Land width} = P - TS_{min}$$

where:

P = the package lead pitch
TS_{min} = the minimum trace space

The minimum trace space is the distance that the fabricator is capable of supplying *consistently*. If a user defines consistency from a supplier to mean 99.7%

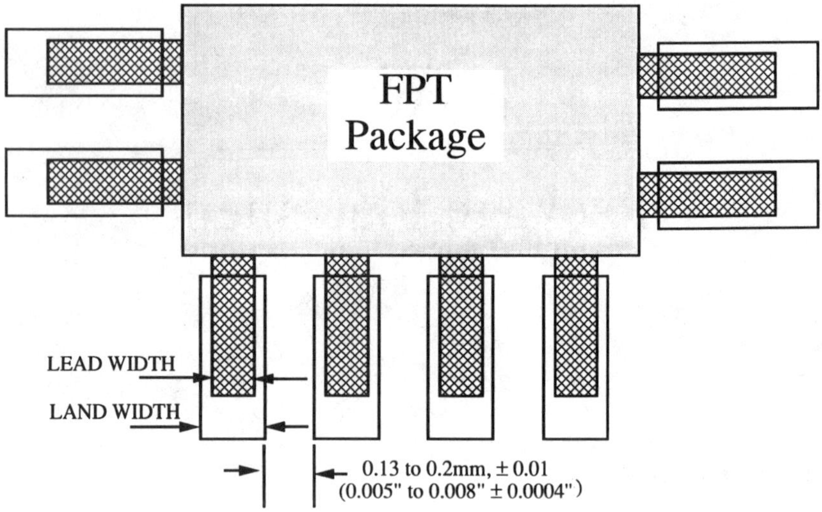

FIGURE 6.1. The key placement factors for accurately placing fine pitch packages include adequate land width, properly aligned leads, and the placement systems capabilities.

of the time (± 3 sigma), then that is the value of TS_{min}. If the supplier expects consistency 99.9997% (6 sigma) of the time, then that is the value of TSmin, which most assuredly would be larger than the ± 3 sigma TS_{min} limit.

Table 6.1 and Figure 6.2 shows the land width and land to lead overlap for a minimum land to land space of 0.15 mm (0.006″) and 0.18 mm (0.008″). Notice that the land width is smaller than the typical lead width when the lead pitch is 0.25 mm or less for TS_{min} = 0.15 mm and 0.3 mm for TS_{min} = 0.18 mm. The ± 3 sigma distribution for the land width using 1 oz. copper is typically ± 0.01 mm (± 0.0004″).

The ± 3 sigma distribution for lead width and location is generally ± 0.08 mm (± 0.003″). This variation in lead size and location is due in part to the supplier's lead forming process which can be a stamping or a chemical etch process. It is also due to the handling the package receives from the supplier, through the user's preassembly operations, and during placement. This variation also includes the variation in package size. The commonly accepted surface mount workmanship criteria allows a 50% gull wing lead to land misalignment for low reliability applications, 25% for industrial applications, and 10% for high reliability. Figure 6.3 shows maximum offset of the lead to the land for the three different misalignment limits. This figure assumes that the land width is the same as the lead width.

Because of the reduced solder joint area of the fine pitch devices and the

TABLE 6.1 This Table Shows the Calculations of the Land Width and Land to Lead Overlap when the Minimum Land to Land Spacings are Limited (usually by Either the Fabricator or by Specification) to 0.15mm and 0.18m.

Pitch	Lead Width	Trace space	Maximum Land Width	Land to lead overlap
TSmin = 0.5 mm (0.006")				
0.65	0.3	0.15	0.50	0.02
0.5	0.25	0.15	0.35	0.10
0.4	0.18	0.15	0.25	0.07
0.3	0.13	0.15	0.15	0.02
0.25	0.13	0.15	0.10	-0.03
0.2	0.13	0.15	0.05	-0.08
0.15	0.1	0.15	0.00	-0.10
TSmin = 0.18 mm (0.008")				
0.65	0.3	0.18	0.47	0.17
0.5	0.25	0.18	0.32	0.07
0.4	0.18	0.18	0.22	0.04
0.3	0.13	0.18	0.12	-0.01
0.25	0.13	0.18	0.07	-0.06
0.2	0.13	0.18	0.02	-0.11
0.15	0.1	0.18	0.03	-0.13

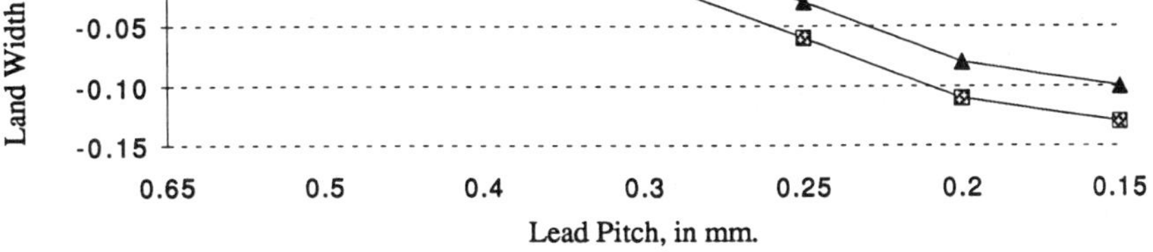

FIGURE 6.2. This figure is a graph of the calculations in Table 6.1. Note that the land width is narrower than the lead width in some lead pitch situations.

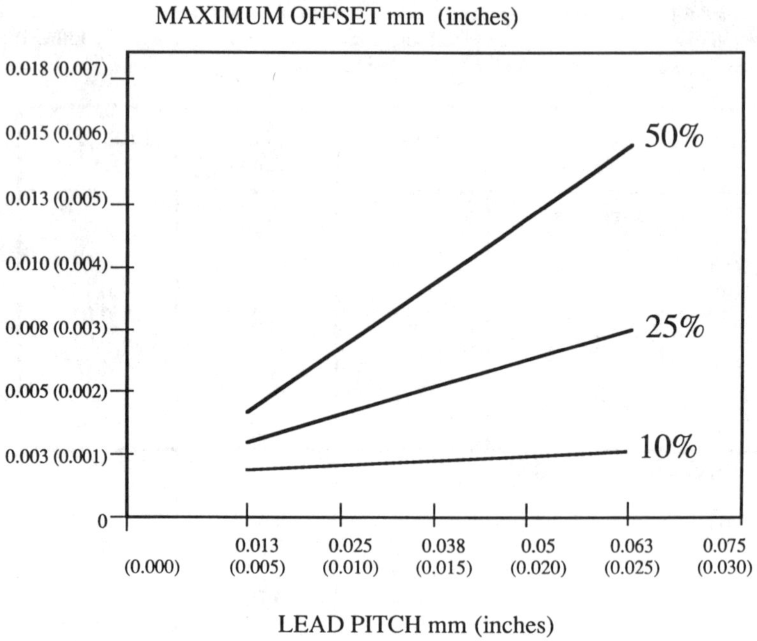

FIGURE 6.3. This graph shows the maximum misalignment distance allowed for 50%, 25% and 10% alignment criteria. This graph assumes that the land width is the same as the lead width. (Source: Sierra Research and Technology, Inc.)

potential for electrical leakage, many experts (including this author) recommend the maximum misalignment for fine pitch packages be 10% maximum for all electronic applications. The consequence of the reduced solder joint area is reduced reliability. The potential for electrical leakage occurs when a lead is close enough to an adjacent land that electrical charge can flow from the lead to the land and vice versa. The charge flow may be capacitive, inductive, or resistive, but the root cause is the proximity of the lead to the land.

Many of the very small SMT chip resistors, capacitors, diodes and IC packages are able to shift and self-center during reflow; this caused early assemblers to feel there is a lower criticality of placement. This shift phenomena is a result of the wetting action of the solder, its surface tension, and gravity. When the solder wets the lead or termination, there is a strong force opposite to the wetting force. Since the solder has a high surface tension, the wetting force needs to be relatively large to change the surface shape of the solder. If a lead or termination is off the land area by a small amount, and if the component mass is small, then the component will move with the opposing force of the wetting solder until the forces are balanced. This occurs when the component is either centered on the lands or the wetting has progressed as far as it can. The fine pitch and quad leaded SMT packages seldom exhibit this self-centering. This is because of the package mass and solder wetting speed. These larger IC packages usually have a large mass compared to the solder wetting force. Also the IC leads are very small and wet very quickly. The high mass and short wetting time counter the wetting force. As a result, the placement of quad leaded and high mass packages must be consistent and accurate. The leads of these packages solder to where they are placed, unless the assembly is mechanically vibrated during reflow. But mechanical vibration is not a universally accepted process, especially on high reliability assemblies.

Current placement machines use higher precision mechanisms and vision systems to significantly improve the placement accuracy then earlier generation systems. And if the board is designed and fabricated using the suggested guidelines in this text, the board location accuracy is significantly improved over prior years. Table 6.2 compares early non-package related placement accuracies with current placement and board fabrication accuracies.

The current placement accuracy of ± 0.09 mm (± 0.002″) is adequate for the 0.65 lead pitches, even with the 10% maximum misalignment criterion. This is shown by comparing the allowable placement window calculated in Table 6.3 with the current accuracy in Table 6.2.

The allowable placement window values in Table 6.3 are calculated by:

Placement Window = 10% of the lead width + ½ the land to lead overlap.

TABLE 6.2 Placement System Accuracies have Improved Since the Early Generation Systems of the 1980s. This Table Shows the Current Accuracies Compared with the Past.

	Early and Current Placement Tolerances Early (Circa 1985-6)	 Current (Circa 1992)
x, y, axis error	±0.15	±0.05
Rotational error	±0.05	±0.03
Land Location error	±0.10	±0.01
Total (Non-Package) Placement Tolerances ±mm ± Thousands of an inches	 0.3 12	 0.09 4

TABLE 6.3 The Allowable Placement Window is a Calculation of the Placement Tolerance that Remains After Allowances for Lead Misalignment (± 10% and the Land to Lead Overlap are Factored.

Pitch	10% of the Lead Width	1/2 of the Land to Lead Overlap (@ TSmin = 0.18 mm)	Allowable Window mm	inches
0.65	0.03	0.085	0.115	0.00453
0.5	0.025	0.035	0.06	0.00236
0.4	0.018	0.02	0.038	0.0015
0.3	0.013	-0.005	0.008	0.00031
0.25	0.013	-0.03	-0.017	-0.0007
0.2	0.013	-0.055	-0.042	-0.0017
0.15	0.01	-0.065	-0.055	-0.0022

This calculation does not include any error allowance for lead width and location variation, which as previously mentioned can be assumed to be as much as ± 0.08 mm (± 0.003″). When this amount is subtracted from the allowable placement window the resulting value is the required placement accuracy versus lead pitch required of a fine pitch placement machine. These values (shown in Table 6.4) assume ± 3 sigma (99.7%) quality consistency. Users striving to achieve 6 sigma (99.9997%) quality need much greater precision placement systems.

The required placement accuracy shown in Table 6.4 indicates that ± 0.04 mm is needed to properly place 0.65 mm pitch packages 99.7% of the time. Since the current typical system accuracy is ± 0.09 mm, then it can be readily expected that a significant portion of the packages placed by the typical system will require rework.

This rework can only be minimized by:

1. Improving the placement system accuracy to less than the ± 0.09 mm tolerance
2. Improving the package lead location and size tolerance to tighter than ± 0.08 mm

These tables may lead the reader to conclude that fine pitch placement generates significant quantities of misaligned packages. However, these tables assume normal gaussian distributions of the size and location variations. They

TABLE 6.4 The Required Placement Accuracy, Assuming a Package Lead Location Variation of ± 0.08mm (± 0.003″) Allows no Margin for Error.

Pitch, mm	Required Placement Accuracy	
	mm	inches
0.65	0.04	0.001378
0.5	-0.02	-0.00079
0.4	-0.04	-0.00165
0.3	-0.07	-0.00283
0.25	-0.10	-0.00382
0.2	-0.12	-0.0048
0.15	-0.14	-0.00531

do not account for the offsetting situations such as undersized lead widths being placed by the placement system operating at its misalignment extreme. Dealing with these scenarios is overly complex. Also, these tables do not account for the various placement options available to the user. The estimate of current placement tolerances (Table 6.2) is the tolerance distribution of several machines, some of which are capable of better tolerances than the ± 0.09 mm tolerance indicated.

6.1 PLACEMENT OPTIONS

There are three package placement options for fine pitch packages:

1. Manual-aided placement
2. Sequential automated placement
3. Combined placement and reflow

Manual-Aided Placement

It may seem farfetched in light of the prior discussion of required placement accuracy, but fine pitch packages can be placed manually using a vacuum quill if an alignment aide is used. The best alignment aide is a vision camera and monitor. This enables the operator to fine tune the package position over the paste covered land areas. Several inexpensive workstations are available that are ideal for small volumes of assemblies. Most of these stations have a board holder that secures the board in place, a vacuum quill to pick up and transport the package, and an adjustable pressure control for placement force. To align the package, the operator views the lead to land position through the monitor attached to the TV camera positioned above the package. The package or board is repositioned until the leads are aligned to a best fit. This sequence is illustrated in Figure 6.4.

When the leads are in alignment, the package is placed into the paste at a predefined pressure. This pressure is enough to assure positive contact with the paste, but without smearing the paste. Alignment tolerance, independent of package lead variations can be within ± 0.03 mm (± 0.001″).

Depending on package size, an operator and workstation can place 15 to 240 packages per hour. Workstation costs range from $20,000 to $90,000 and are available from several international sources.

Sequential Automated Placement

A sequential placement system is a programmable placement machine that will pick a component from a bank holding many different parts and place the component on the defined printed circuit board location.

The holding bank can store from 2 to 200 different part types depending on the storage medium. Typically, the parts are stored in tubes, trays, or tapes on reels. Tubes are the most desirable for low to medium volumes since they are small and

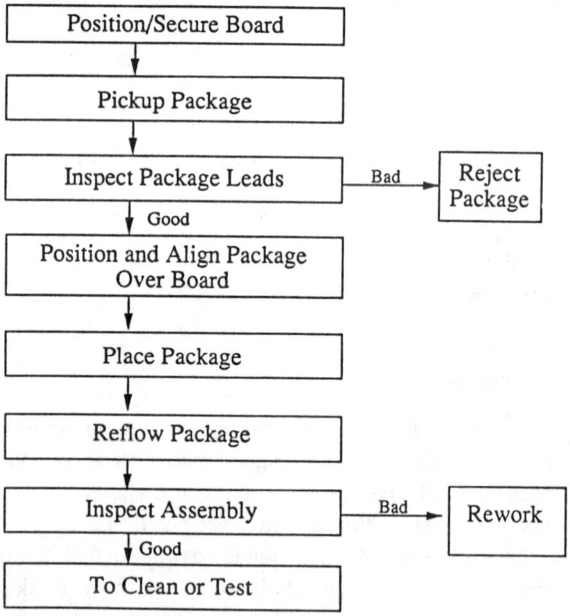

FIGURE 6.4. The sequence of events necessary for accurate placement of fine pitch packages.

easy to handle. Tubes also provide an easy way to present the part to the pickup quill. One end of the tube is simply elevated about 45° above the horizontal and the parts ski down into a cavity that roughly centers the part prior to being picked by the quill.

Tape on reel is the preferred method for handling parts in volume. Tapes are convenient and easy to store, and they occupy minimal space on the placement machine. The tape should be antistatic plastic with an easily removed cellophane covering. The tape is advanced and the cellophane covering is peeled from the top of the cavity holding the part. The vacuum quill then removes the part for placement. Trays require more complex automation than tubes and tapes and occupy more physical space. For this reason, they are the least desirable shipping method from a user standpoint. However, trays have become the standard handling method by the semiconductor suppliers of the QFP packages (see Section 2.6). They find that trays are the best way to ship the quad leaded packages without distorting lead planarity or causing damage. The

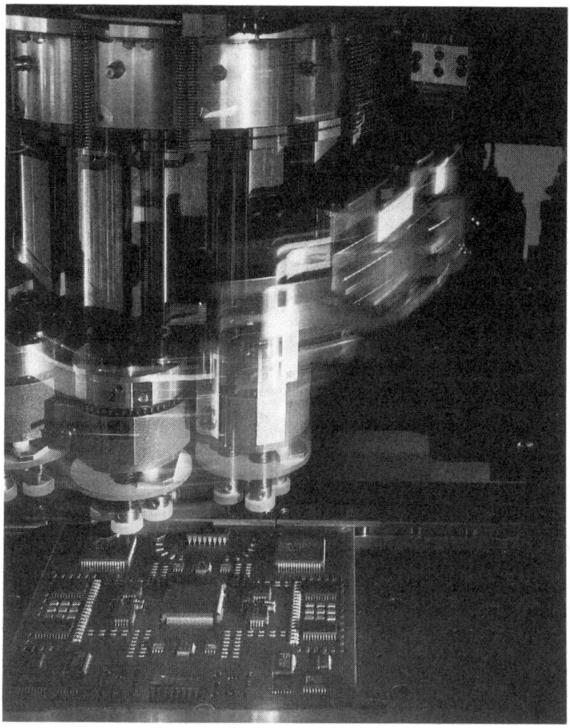

FIGURE 6.5. Photo of a sequential placement system. (Photo courtesy of Universal Instruments)

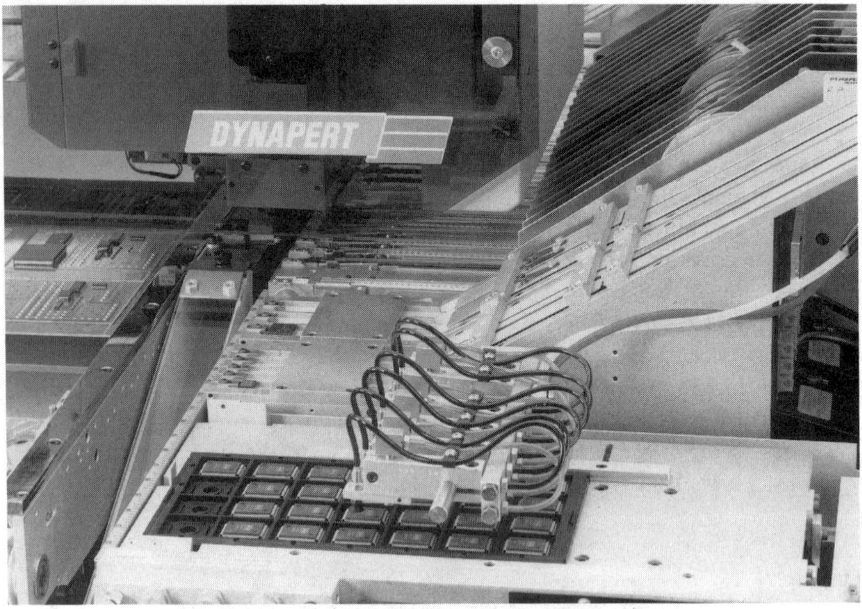

FIGURE 6.6. Photo of a parts holding back. The parts are advanced to a holding position so the pickup quill can pick from the same location. (Photo courtesy of Dynapert)

trays are also convenient for baking the parts and sealing in dry-bags prior to shipment so excess moisture is removed and remains out of the package.

Selecting a Sequential Placement System. When selecting a fine pitch sequential placement system the first important specification is placement accuracy. This applies to cartesian coordinate accuracy in the x and y direction and the angular coordinate of Θ. The amount of accuracy needed will vary with lead pitch, package size, and user workmanship criteria, as previously discussed. (see Table 6.4).

To assure sufficient accuracy for fine pitch placement, the system needs video camera assistance. After the board to be assembled is secured in place either mechanically or with a vacuum, the board viewing camera locates a fiducial mark on the board, in most systems such a mark is located near each FPT package to be placed. The camera compares the viewed location of the fiducial mark with its preprogrammed correct location. If the actual board location differs from the correct location, then the machine moves the board or the part until the correct location is achieved.

The accuracy of component placement is a function of the vision system, the

FIGURE 6.7. Close-up photo of a QFP package about to be placed. (Photo courtesy of Universal Instruments)

board, and package lead locations. The ability of the machine to adjust for correct location is determined by:

1. The position accuracy and resolution of the cameras looking at the board and the package
2. The lighting quality on the board and the package
3. The size accuracy and distance between the board fiducial marks used as location references for the package
4. The resolution of the linear motors that position the board and the package to the programmed correct position

Camera Accuracy and Resolution. Fine pitch placement systems will have one or two vision cameras (see Figure 6.8). A one camera system usually looks upward from a fixed position at the location of the package. The camera resolves

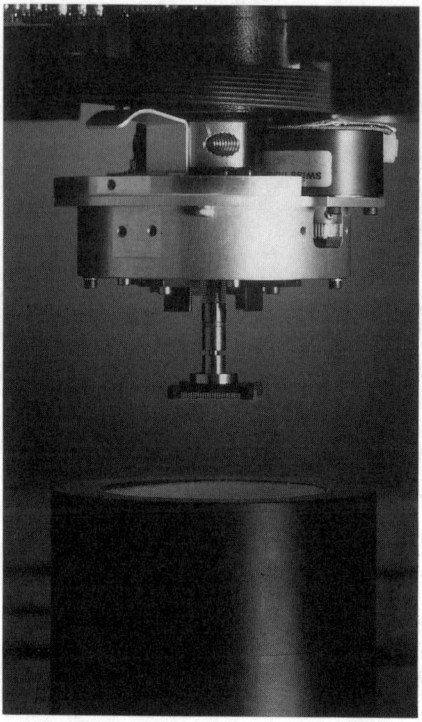

FIGURE 6.8a. Photo of an inspection camera scanning a QFP package for alignment prior to placement. (Photo courtesy of Quad Systems)

the package image using converted digital data in either a binary or gray scale image format. Both of these formats are very lighting sensitive. Gray scale format is also geometry specific and may cause rejection of packages that are much larger or smaller then the nominal size programmed in the machine's database.

The one-camera system assumes the board is located in its proper position using tooling holes or fixtures. After the package position is corrected for x, y and Θ mismatches the package is placed on the board.

Some early versions of automated fine pitch placement systems suffered in accuracy because of the need to precisely locate the camera position. Frequent calibration of the camera was needed. Current generation systems use relative camera locations. These use mathematical algorithms that normalize the location of the camera to a stable reference point. This method guarantees that the factory rated accuracy is maintained with time, temperature, and normal wear and tear.

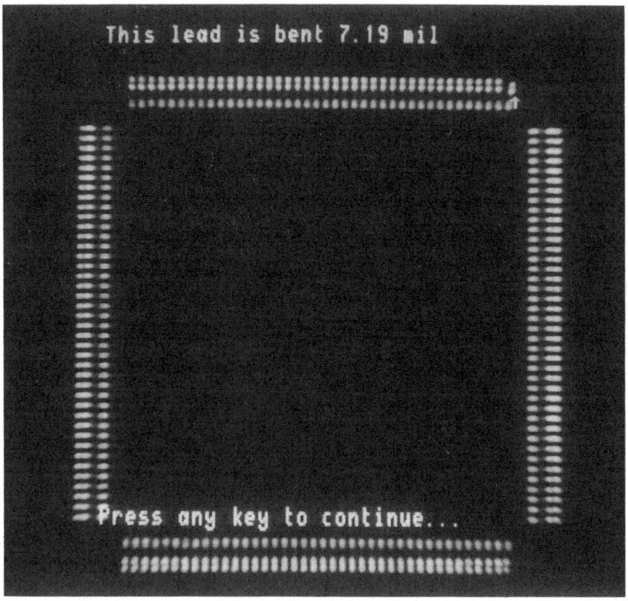

FIGURE 6.8b. A display of the process control monitors readout of misaligned leads. (Photo courtesy of Quad Systems)

When the camera views the bottom of the component it needs to look at a stable reference geometry to assure accurate package positioning. Most fine pitch packages are of the QFP variety that have large square or rectangular molded bodies with leads extending out all four sides. Many systems use a small field of view to look at the opposite corners of the molded portion of the package. By magnifying the image to concentrate on the corners, the system overcomes the inaccuracy of a low pixel resolution camera. When the distance between the two corners is determined, the centroid of the package can be calculated. However, this approach requires the camera or the part to be moved to view both corners of the package. This movement may introduce inaccuracies, which must be compensated for in the system algorithm.

The best approach is to view the entire body of the package as one object image. It is common to use the edges of the package leads even though the lead positions and lengths vary more in size then the molded package body.

The minimum resolution should be 512 pixels. If the system is programmable and capable of inspecting leads, then a minimum of 7 to 10 pixels per lead are required to be able to detect missing or bent leads reliably. The vision detection system should use a best-fit algorithm to optimize placement of packages with bent or skewed leads. The algorithm should be programmable so the placement system rejects any packages whose leads are bent or skewed beyond the users allowable misalignment criteria (the 10% maximum lead misalignment recommended earlier).

This may be difficult with some systems since their vision algorithms work on a pixel intensity gradient. This means that the package edges are defined as the point when a pair of pixels have an opposite intensity, i.e., one pixel is black and the next is white. If the lighting on the package is poor and the camera is a binary format then the leads appear as a dark area, confusing the camera as to where the correct corner location occurs. Systems with gray scale format eliminate this problem, but still require quality package lighting.

When selecting a gray scale vision system with the intent of using the entire package body as the viewing object, users need to verify that the system is accurate over the entire range of body sizes to be used. The distortion of the camera lens may require corner detection programming for large packages and full body detection for small packages.

Use of Fiducial Marks. The two camera system adds a second camera that looks at the location of the fiducial marks next to the board location for the package to be placed. The shape, size, and distance between the fiducial marks critically affects the accuracy. Most systems need two fiducial marks per package, located on opposite corners and close to the package. Examples of typical fiducial marks and their placement are shown in Figure 6.9.

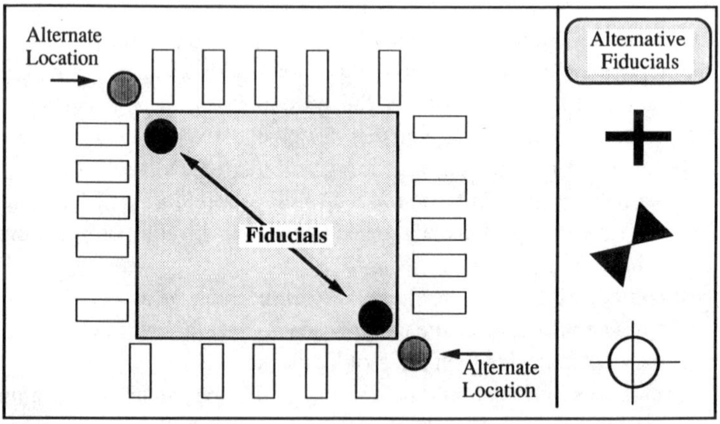

FIGURE 6.9. Localized fiducial marks, beside or within the land footprint enable a vision system to enhance the placement accuracy for fine pitch placement.

The choice of which fiducial shape to use is dictated by the placement machine manufacturer. This is unfortunate, since it constrains the assembly of the board design to specific machines (and owner's of such machines). It may also limit the resolution and hence the accuracy of placement. Crosshair type fiducial marks, for example, require a large number of pixels in order to calculate the centroid location. Fiducial size and image distortion will greatly affect the accuracy.

Fiducial mark size is a function of design and the board fabricators control in etching and solder leveling. If the fiducial mark is visible by the vision system under soldermask or if the board traces and lands are bare copper, then one fiducial size variation source is eliminated, that being the tin-lead coating. However, some systems prefer solder covered fiducials, since the solder reduces the amount of reflection.

The critical parameter is the centroid location of the fiducials. Therefore, the design of the fiducial should be one that enables the viewing system to calculate the correct centroid even when there is a size variation due to over- or underetching of the copper. Generally circular marks such as shown in Figure 6.10 are excellent choices and have a small size variation of typically ± 0.01 mm (±0.0004″).

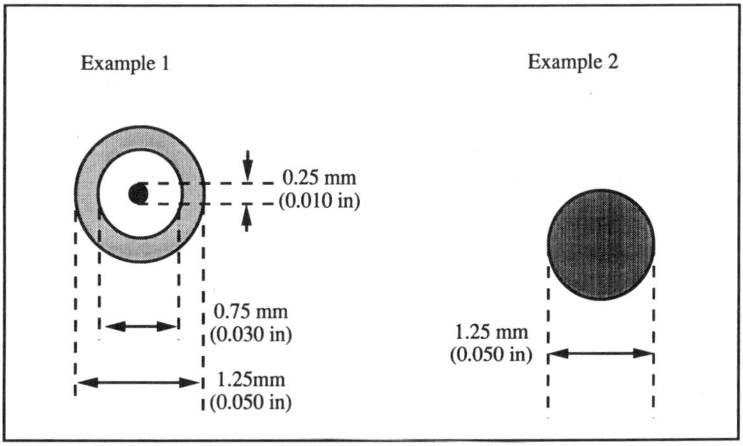

FIGURE 6.10. The circular fiducial mark is probably the best choice given the ability of a vision system to resolve its location accurately.

Place and Reflow

Another option for package placement that is proving accurate and economical, especially when there are only a few fine pitch packages on a board, is the combination place and reflow workstation. Often these stations are marketed as rework stations, but they are more easily justified in the place and reflow role.

A combination place and reflow workstation picks and places the package either automatically or manually. The manual placement option requires an operator to locate and center the package to be placed and then to locate where it is to be placed. The automatic option uses a camera and programmed coordinates as with the automated placement systems just described. In either case, an operator can intercept the package before it is placed in the paste and correct it's position to avoid misplacement and rework. Some systems also are capable of inspecting the package leads to assure that they will fit in the allowable placement window.

When the package is aligned and placed, the reflow unit built into the workstation can be engaged. This allows the selective heating and solder joint formation of the leads on that package. This avoids the potential problem of the package shifting during transport from the placement machine to the reflow oven. However, the main feature is that an operator is in control of two critical processes simultaneously, placement and reflow. Any problems such as shorts, opens, or misalignment receive immediate attention before significant rework is created.

The reflow heat source on the workstation may be hot air, IR, laser, or conduction. Hot air sources require special nozzles for each different package size. The IR and laser units are usually programmed sources; the user programs the lead coordinates and the heat source (IR or laser) is directed on those coordinates. The heat source may heat and reflow one lead at a time or all leads simultaneously.

The conduction heat method is commonly called *hot bar bonding*. It is described in more detail in Chapter 7. Basically, hot bar heating is done by passing electrical current through a metal bar. The bar is designed to heat uniformly along the edges that will contact the package leads. The bars are pressed at a controlled force onto the rows of leads on each side of the package. The hot bar force needs to be controlled separately from the placement force since the two forces will be different. After contact, the bars are heated to slightly above the solder reflow temperature, then they are cooled. After cooling to a preprogrammed temperature, they are retracted from the package.

6.2 ADDITIONAL PLACEMENT SYSTEM FEATURES

Certain additional features should be incorporated into the placement system. Most of these are process control enhancements to assure the optimum yield. The most important additional features are programmable pressure, lead planarity test, component electrical tests, and defect collection capability.

Placement Pressure

The amount of pressure applied to the package during placement is another parameter similar to placement accuracy in contributing to optimum yields. Often the package leads are slightly nonplanar, i.e., they do not all sit in a level plane. To solder properly, the leads need to be in contact with the solder paste. A good way to achieve this is to apply pressure to the package when it is placed. The pressure should be great enough to assure that the leads are pushed into the paste enough to contact the flux, but not great enough to smear the paste or bend the leads excessively.

Defect Collection

The placement system that is equipped with any form of problem testing (mechanical or electrical) should also have provisions to tally the problems that are found so yield and trend analysis can be performed. A simple data transfer to another computer each time a part is examined may be sufficient to make the data collected useful. The other computer can analyze the data, prepare Pareto and control charts, and send alarms when the process varies beyond the control limits.

Lead Planarity Testing

If lead planarity is a problem (and in all probability it will be), the defect collection capability can be used to test the lead planarity. A maximum limit of 0.05 mm (0.002″) lead difference is desired for packages with lead pitches from 0.4 to 0.65 mm, and 0.025 mm (0.001″) for 0.2 to 0.3 mm pitches.

Some systems use the component viewing camera and the variation in gray scale as the means to check lead position and planarity prior to placement.

Lead Excise and Form

The guardring and TAB packages require excising from their holding tape and lead shaping before they can be placed. The best point in the operation to perform the excise and form is immediately before placement, as the excise step will expose the base metal of the toe of the lead. If exposed to atmosphere for too long prior to placement and reflow, the toe metal oxidizes and becomes a solderability problem. Most experts agree that the solder on the toe contributes very little to the solder joint strength, but they are concerned about lead corrosion and cosmetic appearance.

Excise and form is performed with a die punch specially constructed to guillotine the leads at the proper length and to shape the leads in the correct form. An example of an excise and form tool is shown in Figure 6.11. Excise and form punches can be purchased for manual operation before placement. Automated punches are available for direct use on the placement system. These are mounted after the package tape feeder.

Mechanical wear is the largest problem with excise and form tools. Over time, the punch and die change shape and the ability to produce a clean, burr free lead cut is reduced. Also, the lead width may increase and deform as the punch edges loose sharpness and registration. Aside from these problems, preplacement excise and form provides the best cure for the lead planarity problems of the other non-guard ring package types.

Component Testing or Programming

Often, the fine pitch packages are the most expensive components of the final product. If they were housed in through-hole packages such as a DIP or PGA they would probably be mounted in sockets rather than directly soldered. The expense

FIGURE 6.11. Photo of lead excise and form punches. (Photo courtesy of Universal Instruments)

and lower reliability of using sockets can be eliminated by the option of pretesting or preprogramming the critical fine pitch packages before they are placed. This can be done at a location other than the placement machine, but in some machines there is a built-in nest where the user can program a stop prior to placement of the package on the board. This nest is usually a probe fixture that mates to the leads of the package. Through this probe, the IC can be tested, programmed, and retested automatically. This can prove very economical for certain custom ASICs that have electrical failure rates of 50,000 to 250,000 ppm per part.

6.3 THROUGHPUT REQUIREMENTS

The role of a placement machine is to position fine pitch packages on the printed circuit board accurately and consistently. If the system is a combination place and reflow system, then the soldering needs to be consistent also.

When selecting a system, most users are quick to focus on the placement rate of the system. This number is supplied by manufacturers in a variety of ways. Some suppliers specify the placement rate in packages per hour, noting the size of the package and distance between the pickup point and placement point. Other suppliers only specify the placement rate. Other suppliers specify cycle time without specifying the number of cycles needed to pick and place a given package.

Regardless, the user needs to derate the factory estimated throughput time by the amount of packages need rework due to misplacement, the number of packages that the quill may drop, and the system downtime. The rework quantity can be estimated from the spread in system placement accuracy versus the users allowable placement window.

The number of packages that may be dropped by the quill is not easy to estimate unless several test runs are made on the system. Some packages have sufficient variation in the contour of the top surface and enough mass that they fall off the quill. The rapid acceleration and deceleration of the quill on a high speed placement system may cause more package loss than a slower system would. When a fine pitch package is dropped its leads are usually distorted, sometimes enough to be rendered unusable scrap.

System downtime is a function of system robustness and ease of service. If certain control circuits in the system are not properly cooled or equipped for continuous operation then system shutdown occurs. Sometimes altering the cooling method is sufficient to overcome the problem. But if system elements need to be replaced, and replacement requires hours or days of waiting, then system throughput degrades substantially. Also, if it is difficult to access those elements of the system needing repair or periodic maintenance then throughput is degraded.

The wise user will thoroughly evaluate accuracy, serviceability, and downtime history before purchasing a system.

References

1. Bell, J. 1991. Private correspondence.
2. DeCarlo, J. 1991. A New Technology for Fine Pitch Placement Utilizing Active Lead to Pad Matching and Integral Reflow. Paper read at Nepcon West '91, Feb., 1991, Anaheim, CA.
3. Deley, M., et. al. 1990. Putting Fine Pitch Errors on a Budget. In *Circuits Manufacturing*, Jan., 1990.
4. Duffey, B. 1989. Applying machine Vision to Fine-Pitch Pick-and-Place. In *Circuits Manufacturing*. Dec., 1989.
5. Hyman, H. 1991. Fine Pitch Placement with Integral Reflow; the Logical Cost Effective Approach. Paper read at Nepcon East '91, June 1991, Boston, MA.
6. Peck, D. 1991. Private correspondence.

7

Solder Reflow

7.0 REFLOW DEFINED

Solder reflow is the term used to describe the process of heating the assembly until the solder paste melts and wets, forming a solder joint with the copper land and the package lead.

The success of this process is dependent on several variables:

1. The heat transfer rate from a heating source
2. The heat absorption rate and uniformity by the assembly
3. The solderability of the metals to be joined
4. The melting point of the solder
5. The heat tolerance of the materials on the assembly
6. The flux chemistry and the cleanliness of the metals to be joined

Each of these variables needs to be controlled in varying degrees if a consistently reliable solder joint is to be formed. Of these, the most process critical are the first four.

7.1 THE DYNAMICS OF THE REFLOW PROCESS

The primary purpose of the reflow process is to melt the small solder particles in the solder paste, wet the surfaces to be connected, and then solidify into a strong metallurgical bond. During the process many local actions occur that may help or hinder the global action. The general reflow time vs temperature profile is shown in Figure 7.1.

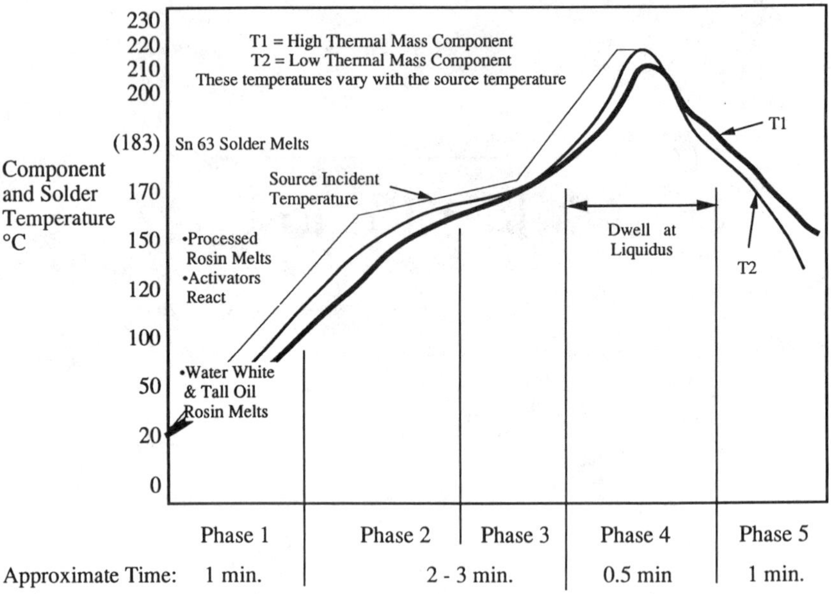

FIGURE 7.1. The typical time and temperature profile experienced by the assembly during solder reflow. (Source: Vitronics)

Phase 1—Solvent Evaporation

The first action upon the assembly's entrance into a reflow oven is the evaporation of paste solvents. Additionally, the solder particles compress, allowing trapped gas and bubbles to escape. If heat transfer is too rapid, then a typical problem is a "mini-Mt. Vesuvius" effect. That is, the escaping gases cause small eruptions and component movement if the assembly is heated too fast.

Phase 2—Oxide Reduction

During this phase the flux begins cleaning the oxides off the metal surfaces that are in contact. The increasing temperature during reflow causes the flux activity to increase significantly. The primary functions of flux are to remove oxides, to protect the surfaces from reoxidation, to assist in heat transfer, and to help remove the reaction products so a uniform and strong metal connection is made. A suitable flux completes these duties by a chemical reduction process that removes the main contaminants (oxides, hydroxides, and sulfides). The source of these contaminants is usually the reaction of the base metals with air.

The amount of time needed during phase 2 is dependent on the thickness of the oxide on the metal surfaces and the flux strength. The oxide thickness increases with time and exposure to oxidizing atmospheres, such as humidity. The approximate oxide growth rates for common tin-lead alloys is shown in Figure 7.2.

The time to reduce a given thickness of oxide from tin-lead is shown in Figure 7.3. This graph is very approximate and is intended to roughly show the approximate time versus flux strength needed to remove an oxide layer. For example it would require an M (medium strength) flux, which is the designation for the prior RMA flux, approximately 50 to 70 seconds to remove 1000 Å of oxide at a flux temperature of 125°C.

A typical problem related to fine pitch lead soldering in this phase is runout of the solder paste, resulting in bridging. Runout occurs if the oxide reduction goes on too long, or if there is too much paste, or the paste has too much flux compared to the metal content.

Phase 3—Solder Melt

As the fluxing action proceeds, the solder particles in the paste start to melt, replacing the flux, and wetting to the clean metal surfaces. This wetting action results in wicking of the solder onto any clean metal surface. The speed of this action depends upon the local temperatures. Any temperature nonuniformity can cause comparable nonuniform wetting and wicking action, with the wicking action favoring the higher temperature region.

Phase 4—Complete Solder Melt

Soon all the solder particles have melted and a liquid solder volume is formed. The surface tension of this volume of molten solder impedes further wicking action.

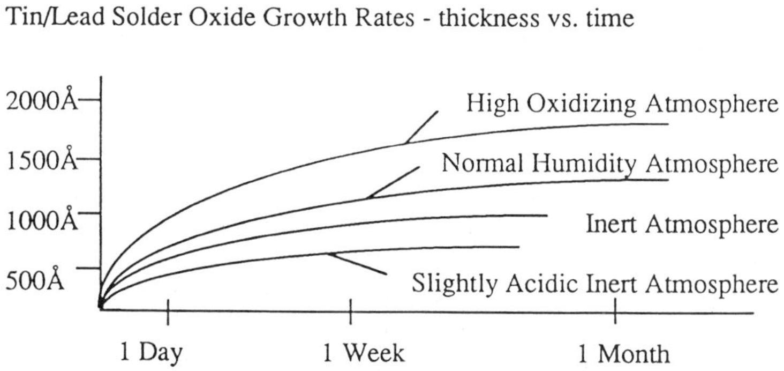

FIGURE 7.2. Tin-lead develops oxide layers of various thicknesses and densities depending on atmospheric conditions. The relative oxide thickness as a function of time and atmosphere are graphed in this figure. (Source: International Tin Research Institute)

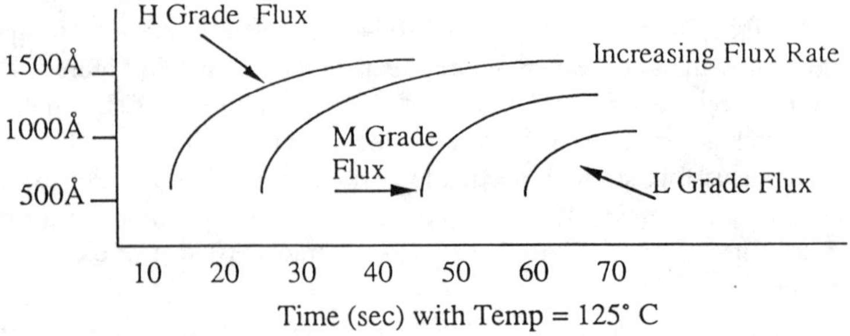

FIGURE 7.3. Flux is used to chemically reduce the oxide layers. To remove a given oxide thickness, the various activity level fluxes must be aided with the time and elected temperatures shown. (Source: International Tin Research Institute)

The wetting action tends to spread the solder out, maximizing surface area. The global action of the surface tension tends to hold the solder in place and bridge the gap between a component lead and the land on the board, and to form the fillet upon cooldown. The volume of solder and the physical geometry of the metal surfaces (lead, land, via, etc. shape and spacings) determine the resultant shape, fillet formation, and solder joint strength.

Timing is critical in this phase. Excess dwell at liquidus can cause component damage; low mass components may float off the land areas because of incomplete wetting if dwell time at liquidus is insufficient, or if the molten solder volume is subjected to large mechanical shocks and bumps. The solder melt may be incomplete. Partial wetting may occur, or excessively grainy solder joints may be formed.

Depending on the length of phases 3 and 4, the buoyancy of low mass components may cause them to float to the most convenient or inconvenient location. This is usually determined by the design of the land, the temperature, and the map difference between the lead and the package.

Phase 5—Cool Down

Package cracking in chip capacitors is usually due to trapped expanding gas in the capacitor and the thermal expansion mismatch of the ceramic and fiberglass. Cracking in fine pitch integrated circuit packages is sometimes due to the expansion of trapped moisture. In other cases, it has been traced to molding defects in the thermoset plastic packaging the integrated circuit.

The cooldown rate is also important, though not as critical as the heatup phases. With a moderately fast cooldown the solder solidifies into a smaller grain structure, resulting in a slightly stronger joint. With a slow cool down, there is more time for continuation of Cu-Sn intermetallic and Pb-rich pool formation. This results in a higher ratio of Pb at the metallurgical interface and a weaker joint. Also, a slower cool down increases the possibility of component misalignment when the pcb exits the solder machine. A typical cooldown rate is 10°C/sec.

Phases 1 through 4—Temperature vs. Time Rate

To minimize component package cracking, especially in chip capacitors, TSOPs, QFPs, and PCCs, the suggested maximum thermal rate of change 4°C/sec. Many users prefer to control the thermal heating rate to 2 to 3°C/sec.

Prebake

Prior to reflow and even before use in the assembly, it is advisable that many of the FPT packages be prebaked to remove trapped moisture. Package cracking due to trapped moisture which turns to steam has been reported by several sources. The most susceptible packages are the TSOPs, large QFPs($<$ 2.0 cm in any dimensions), and PQFPs ($<$ 2.0 cm in any dimension).

While the amount of moisture is the main contributing factor, the package thickness (specifically the package thickness between the IC chip mounting tab, the die flag, and the package edge) and the internal IC size also are major determinants of cracking, as shown in Figure 7.4. The amount of trapped moisture needed to create a crack when the package is subjected to the typical reflow process

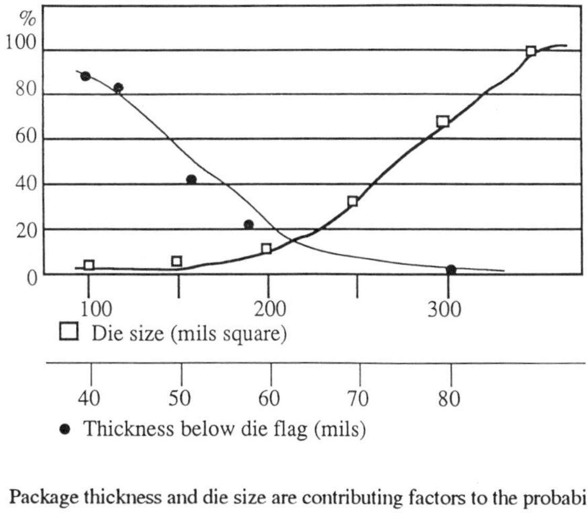

FIGURE 7.4. Package thickness and die size are contributing factors to the probability that a plastic IC package may crack during too rapid exposure to reflow temperature. (Source: Signetics-Philips)

is between 0.1 and 0.25 weight % of moisture. If the moisture content exceeds this, a separate prebake is recommended as follows:

Temp.	= 125°C (± 5°C)
Duration	= 24 Hours
Oven Humidity	=< 50% RH

Most suppliers prebake sensitive packages and ship them in dry bags with clearly marked warning labels. They also supply instructions on user precautions, such as how long packages can be out of the dry bags prior to reflow.

Prebaking the board before assembly will help improve solderability and prevent board warpage. Prebaking helps remove or bring to the surface trapped process solutions (such as in the solder mask) that can decrease solderability. Board warpage is reduced by removing the natural moisture and organic volatiles. The boards should be cleaned and dried before and after bake and before assembly.

7.2 THE HEAT SOURCE

Several reflow systems are commercially available which uses different heat sources. These heating sources include:

1. Infrared (including "pure" IR and IR mixed with convection)
2. Vapor phase
3. Convection—hot gas
4. Conduction
5. Laser
6. Thermosonic

Some of these sources may be combined such that the actual heat source incident on the assembly differs from the prime source. For example, several commercial reflow ovens embed infrared heat emitters into metal or ceramic platens. When the platens are heated from the power source, they reemit heat in the form of both convection and IR; these, in turn heat the assembly until the solder melts. This heat source mixing is done to overcome some of the limitations of the prime source.

The first three heat sources, (IR, vapor phase, and convection) are considered mass reflow sources, i.e., the entire assembly is transported through an oven using one of these sources for heat. The last three (conduction, laser, and sonic) are used selectively to heat certain areas of the assembly.

Infrared

True infrared (IR), heat transfer is done by emitting light from the infrared region of the spectrum. IR emission is that having wavelength of 0.72 microns to 1000 microns and is just beyond the visible light region. The IR region is shown with respect to the other light regions in Figure 7.5.

In the wavelength range of 0.72-2.5 microns the radiation is called near or short wave IR. Between 2.5 and 5.0 microns it is referred to as medium IR; wavelengths longer than 5.0 microns are long wave IR.

The IR emitter temperature determines the emitted wavelength of IR radiation; the higher the temperature the shorter the emitted wavelength. The temperature of the assembly is a function of the absorbitivity of the assembly. The absorptivity is IR wavelength and material mass dependent if the assembly surface is nonlinear and has multicolored bodies. Unfortunately, typical electronic assemblies have very nonlinear surfaces, varying masses, and color variations ranging from off-white to black. This variation results in nonlinear IR absorption and wide temperature variations.

In general, the rate at which heat can be transferred by IR radiation from the source to the assembly is proportional to the difference in the absolute temperatures of the assembly and source raised to the fourth power, i.e.:

$$q = \sigma A F\epsilon F_a (T_S^4 - T_A^4) \tag{7.1}$$

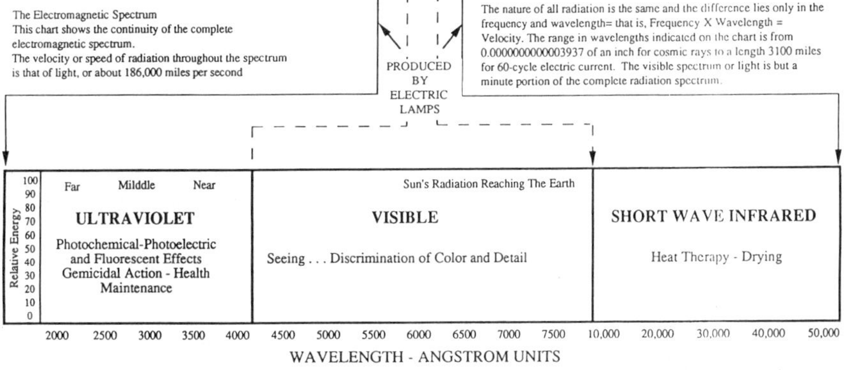

FIGURE 7.5. Infrared light is an excellent heat source. IR light occurs just beyond the visible light spectrum. (Source: General Electric Company)

where:

q = The average radiative heat transfer from the source to the assembly
$F\epsilon$ = The emissivity factor for the source and product
F_a = The configuration or geometric view factor of the assembly
σ = The Stefan-Boltzman constant
A = Area being heated
T_A = Temperature of the area being heated
T_S = Temperature of the source

Because of the $(T_S^4 - T_A^4)$ term, IR heat provides very rapid heating when the temperature differences are great. As the assembly temperature approaches the source temperature, the heat transfer rate diminishes. And $q = 0$ when $T_A = T_S$. In practice, T_A should never equal T_S. The source temperature is always greater than the melting temperature of the solder. The source temperature is set higher to compensate for any reflected radiation and to assure rapid initial heat transfer rates.

When IR radiation is incident on the assembly, some of the radiation is absorbed, reflected, or transmitted. The amount of IR radiation incident on the package as a percentage of the total emitted energy by the source is given by the geometric factor, F_a, shown in Equation (7.1).

The energy balance for the assembly can be expressed as:

$$\Phi_{\text{incident}} = \Phi_{\text{absorbed}} + \Phi_{\text{reflected}} + \Phi_{\text{transmitted}} \tag{7.2}$$

The ratio of absorbed energy to incident energy, Φ absorbed/Φ incident, is known as absorptivity, or α. The ratio of reflected energy to incident energy, Φ reflected/Φ incident, is known as reflectivity, or ρ. The transmitted ratio is called transmissivity, or τ.

These ratios, relative to the incident energy add to unity:

$$\alpha + \rho + \tau = 1 \tag{7.3}$$

In the practical use of IR for solder reflow, the emitter temperatures are low enough that the value of transmissivity is very low, therefore:

$$\alpha + \rho \approx 1 \tag{7.4}$$

If the assembly was of uniform mass and completely black then:

$$\alpha = \int \infty\, \alpha (\lambda)\, d\lambda = \int \infty\, \epsilon (\lambda)\, d\lambda = \epsilon \tag{7.5}$$

where ϵ equals the emissivity, which is the ratio of the radiant emittance of a given source to that of a black body. Radiant emittance is the power radiated per unit area.

As stated in Equation (7.5), if the assembly has uniform mass and completely black, then the absorptivity of the assembly would equal the emissivity of the source.

Unfortunately, this is never the case. The assembly usually contains large mass integrated circuit and connector packages which are near black and as shown in Table 7.1, have an absorptivity of about 98% (and a reflectivity of 2%). Solder paste, on the other hand, has an absorptivity of approximately 69% in its paste form but transitions to a lower absorptivity of about 30% at liquidus.

The absorption and reflection values are averaged over the IR wavelength spectrum. Some materials, such as fiberglass, have little variation in α and ρ over the medium and long region of the spectrum, but vary greatly at the near (or short) wavelengths (see Figure 7.6). Solder has a wide variation, particularly as it transitions from dull paste to shiny liquid form.

It has been common belief that the widely varying values of absorptivity and reflectivity in an assembly accounted for the phenomenon common with

TABLE 7.1 Materials Vary in their Absorption and Reflection Properties to Light. This Table Shows those Properties Relative to IR Light for Common Electronic Assembly Materials.

Material	$\bar{\alpha}$	$\bar{\rho}$
Aluminum		
polished	.39	.61
sandblasted	.42	.58
heavily oxidized	.42	.58
anodized	.15	.85
Fiberglass	.85	.15
Gold Plate on stainless steel and polished	.30	.70
Tin - Lead solder in paste	.69	.31
at liquidus	.30	.70
Tin-Lead Indium solder in paste	.73	.27
form at liquidus	.35	.65
Black Epoxy Novolac IC package mold compound	.98	.02

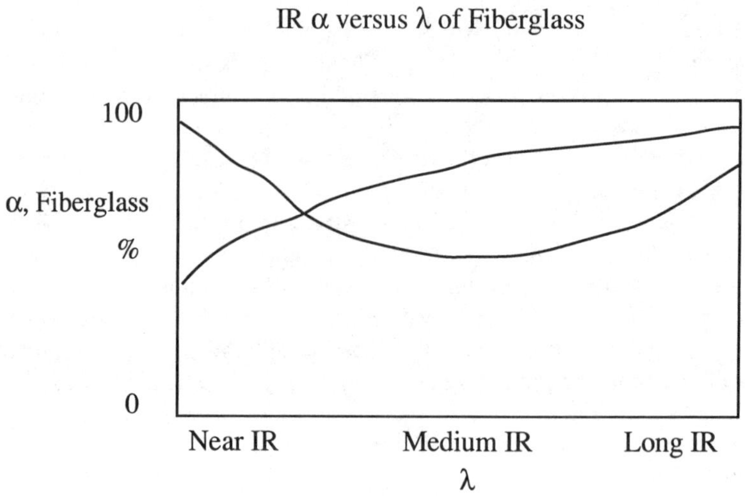

FIGURE 7.6. The absorptivity of fiberglass as a function of IR frequency.

IR soldering that some solder joints melt readily and others do not. However, studies show that the variation in mass of the components to be soldered is the primary cause of this temperature variation. The higher mass packages, being thicker and denser, slow the transmission of heat to the attached joints because of their high heat absorption rates. This situation is discussed further in Section 7.3.

Commercially available infrared reflow systems may not use true IR heating as the incident heat source. They may use a mixture of IR and convection heating. To reduce possible confusion, the Surface Mount Equipment Manufacturers' Association (SMEMA), has created a classification system. Class I systems are predominantly true IR systems, Class II systems are a mixture of IR and convection heat transfer from a circulating heated gas, and Class III are systems where greater than 80% of the incident heat is due to convection.

Vapor Phase

When most fluids are heated to their boiling point, they exist in two phases, vapor and liquid. In the vapor phase, these exist as a a boiling fluid and a saturated vapor at the temperature of the boiling fluid. If an assembly at a lower temperature is placed in the saturated vapor, the vapor immediately condenses on the assembly and transfers heat extremely rapidly. The rate of heat transfer is:

$$q = h\,A_A\,(T_S - T_A) \tag{7.12}$$

where

q = the heat transfer to the colder object, the assembly A
h = the vapor heat transfer coefficient (varies with time)
T_S = the saturated vapor temperature (constant temperature)
T_A = the assembly temperature (varies with time)
A_A = the assembly area

The characteristic of vapor phase heating is rapid initial heat transfer and rapid rise of the assembly's surface temperature, followed by a slowed asymptotic rise of the entire assembly temperature to approach the vapor temperature. The parameter h (fluid conductivity) is high for the saturated vapor used in electronic vapor phase systems, so heat transfer is very fast. This is an important benefit of vapor phase heating; the assembly temperature can not exceed the source temperature. Selecting a fluid that has a boiling temperature slightly above the melting point of solder enables the assembly and all of its constituents to be exposed to a known maximum and presumed safe temperature. This also provides the material suppliers a fixed maximum temperature target when specifying new component materials. Another benefit of vapor phase is that the vapor is inert, which allows the assembly to be soldered in a nonoxidizing atmosphere. Unfortunately, two things diminish the benefits of vapor phase as a reflow heat source. The first is the rapid initial rise in the assembly surface temperature. The second is the high cost of the inert fluids commonly used in the vapor phase process.

The initial temperature rise may be as rapid as 10—20°C/sec. This rapid rise can cause chip capacitors to crack due to rapid expansion of gases trapped and the differing thermal coeficients in the capacitors. Chip capacitor manufacturers specify a heating rate of 4°C/sec as the maximum tolerable by the capacitor's materials.

This rapid heating also causes the solder paste and the leads with the lowest mass to heat more rapidly. When the solder melts and the adjoining lead is hotter than the land on the board, the solder flows up the lead and off the land. This problem is called *wicking* and is prevalent on J leaded PCC packages when vapor phase heating is used.

The common fluids used for vapor phase are initially perchlorinated fluids. Through an expensive complex process, the chlorine ions are replaced with fluorine ions, resulting in a perfluorinated liquid, that is also expensive. At the time of printing, typical prices were between $700—$900 U.S. per gallon.

Convection

The process of heat transfer from a hot gas flowing over a cooler surface is called *convection heating.* If the gas circulation is caused by the differing densities of the masses being heated then the term *free convection* is applied. If gas circulation is done by mechanical means such as a fan or pump then it is termed *forced* convection.

In the convection heat transfer process, heat is transferred from a solid source to a gas by conduction to the gas particles adjacent to the source surface. This thermal energy will increase the internal particle heat and energy and set them into an accelerated motion. As the particles move through the gaseous fluid they encounter lower temperature particles, and upon encounter transfer part of their energy to the cooler particles. Typical gases in electronic assembly are air, nitrogen, or nitrogen/hydrogen.

In general, the heat transfer coefficient for free convection is low, resulting in slow heat transfer. Reflow systems with free convection will often supplant IR heating which, as noted in Equation (7.1), has a much faster heat transfer rate. The heat transfer coefficient for forced convection is usually higher than the combined free convection and infrared coefficient. SMEMA classifies the convection systems as class III and the mixed convection and IR systems as class II.

In free convection, the gas particles near the source are set into the highest turbulent motion since they have the highest energy. This gas region is called the *turbulent core* in Figure 7.7, which shows the gas flow regions in a typical convection oven. Below this layer the gas flow is less turbulent even though heat transfer continues. This region is called the *buffer layer*. Above the flat surface the gas flow is rather smooth and laminar. Since the laminar flow exposes the cooler surface to less gas fluid, the heat transfer to the surface proceeds at a relatively slow rate. However, as shown in the right side of Figure 7.7, if the surface is significantly

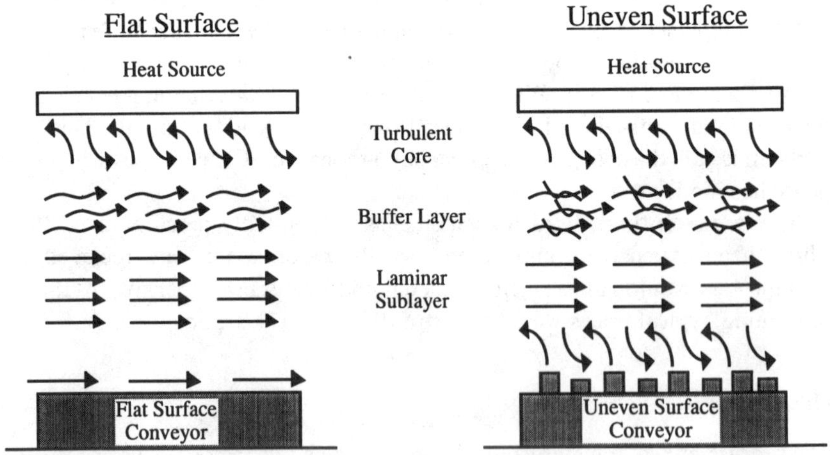

FIGURE 7.7. A diagram representing heat transfer via convention through the "free" transfer of heat and the movement of heated gas.

uneven it creates a new turbulent layer, forcing more particle flow and greater thermal energy exchange in a given time.

For a flat surface, the rate of heat transfer (using either free or forced convection) from the gas, G, to the flat surface, S, can be expressed as:

$$q = h_c A_S (T_G - T_s) \tag{7.13}$$

where:

q = the heat transfer rate from the gas to the surface
h_c = thermal convection conductance based on fluid dynamics
A_s = surface (the assembly) area
T_G = the gas temperature near the surface
T_s = the surface (assembly) temperature.

In the case of convection to a flat surface, the value of h_c is low, and significantly higher for the uneven surface. One way to increase the transfer rate of either case is to greatly increase the source temperature, which increases the T_G value. In practice, this must be done with care since different assemblies will have different h_c values. An assembly with a low h_c value will safely reflow in an oven with a very high source temperature, and an assembly with a high h_c value will burn in the same time period.

By forcing gas circulation in the oven, the gas will be turbulent. The number of gas particles available to transfer the heat increases and the h_c value is increased. And the heat rate does not vary significantly with surface texture. Therefore, a forced gas flow enables a faster heat transfer at a lower source temperature.

Free convection heating using normal atmospheric air is the least efficient in terms of heat transfer rate of the heat sources used in electronic mass solder reflow. However, the slower heat transfer rate of forced convection heating closely matches the maximum safe heat transfer rate specified by component manufacturers. Convection heating is probably the current optimum source for reflow when the entire assembly is exposed to the heat source during mass reflow. It fits all user's budgets, and equipment options range from commercial toaster ovens for lab usage to highly controlled multizoned ovens for serious production.

Figure 7.8 shows the heat flux comparisons for free and forced convection heating, and other common heat sources. This figure illustrates the amount of surface heat flux on the area being heated in Watts/cm^2 when the source temperature is at the indicated ΔT (in °C) above the surface temperature.

Conduction

Conduction and laser heat sources are generally selective as opposed to mass heating techniques. (Conveyor equipped hot plates are one version of a mass

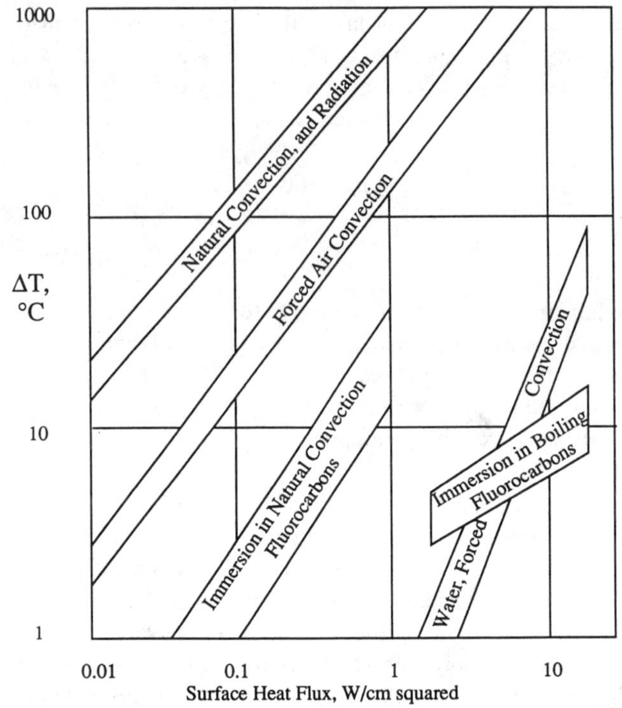

FIGURE 7.8. Comparison of heat flow through "free" convention vs "forced" convection heating.

conduction process that is used for reflow soldering on highly conductive substrates such as ceramic.) That means the heat is transferred to a select area, such as the package leads and solder area, instead of to the entire assembly. These sources offer the advantages of being combined with the placement and inspection operations and being embodied into one piece of equipment. Even though the per unit throughput is lower than mass reflow, when the defect rates and resulting rework time for the complex FPT packages are factored, the selective process may offer the highest yield and be the most economical method for some users.

For optimum results, the solder on the lands should have been reflowed prior to package placement when using the selective process. If the solder is in paste form the rapid heating causes solder splatter and solder balls.

In conduction heating, heat is transferred between two solids in contact with each other. (Laser heating will be discussed later in this chapter). The most common example in electronic assembly is heat transfer from the tip of a soldering iron to the work. In FPT assembly, the best examples are hot-bar and thermode bonding.

Hot-Bar and Thermode Bonding

Hot-bar bonding uses heated metal bars that contact all the leads on a package simultaneously. A quad flat pack requires four bars; a VSOIC only two. In thermode bonding the pre-heated thermodes are brought into contact with the leads and the solder after package placement. When the thermodes have been adjusted to assure lead planarity and uniform lead contact with the solder on the pads, the thermodes are heated. Heating to the necessary reflow temperature is very rapid, requiring only 10 to 20 seconds. The thermodes are then cooled before removal from the newly formed solder joints.

Other hot-bar bonding approaches combine hot-bar and hot-gas heating techniques. As the component is held by the placement quill above the lands, hot gas is used to heat the solder on the lands to liquidus. The package leads are then lowered into the molten solder and reflowed using the unit's hot-bar fixture. After reflow the bars are cooled while still in contact with the leads. The combination approach helps to overcome initial solder wetting and post-reflow joint vibration problems.

The way the bars are heated is important to assure uniform heating. The choice of bar metal and its the shape are also critical. The bars must resist heavy oxidation and be easily cleaned to assure optimum and uniform heat transfer.

The equation expressing the heat flow due to conduction heating is:

$$q = \frac{kA}{e}(T_B - T_A) \qquad (7.14)$$

q = heat flow through the material being heated
k = the average thermal conductivity at the heater and material interface
A = The contact area
T_B = temperature of the bar at the interface
T_A = temperature of the assembly at the interface
e = thickness of the region to be heated

The heating of the bars can be done electrically using I^2R, by radiation heating, or by forced convection heating. I^2R heating uses the natural resistance of a conductive metal to electrical current. This internal resistance causes the bar to generate heat. The typical heating profile of an I^2R heated bar, a thermode, is shown in Figure 7.9. To assure uniform thermode resistance and heating requires skillful bar design. The basic thermode bar design is shown in Figure 7.10 which illustrates forced convection heating of a hot bar.

Note in Figure 7.9 that the thermode temperature is higher (230 to 250°C) then the required temperature for reflow of 63/37 tin-lead solder (T_m = 185°C). The higher temperature is due in part to the contact resistance across the interface between the thermode and the leads. It is also due to the fact that the hot bar area

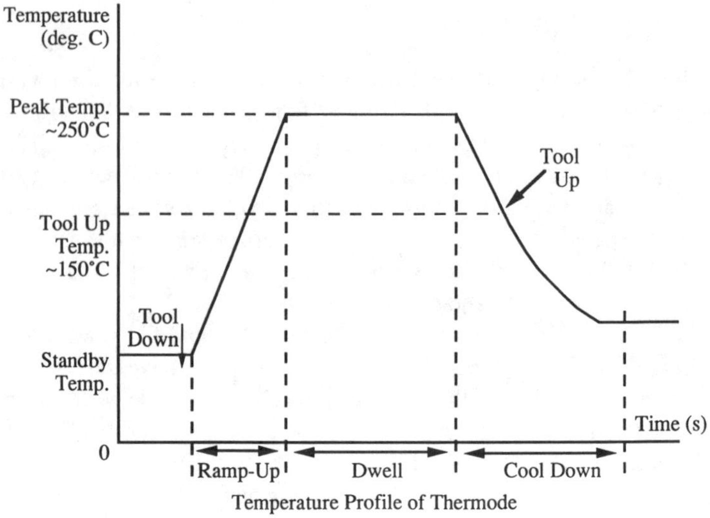

FIGURE 7.9. The typical heating profile for reflow soldering with a hot-bar unit is shown in this figure. (Source: National Semiconductor)

in contact with the leads is small. Additional factors are the surface finish, the heat flow from the leads to the solder, the flatness, and material hardness.

The higher thermode temperature invites rapid and heavy oxide growth on the thermode, which degrades its conductivity even more. This problem represents a poorly controlled process. Another way to increase the interface conductivity is to

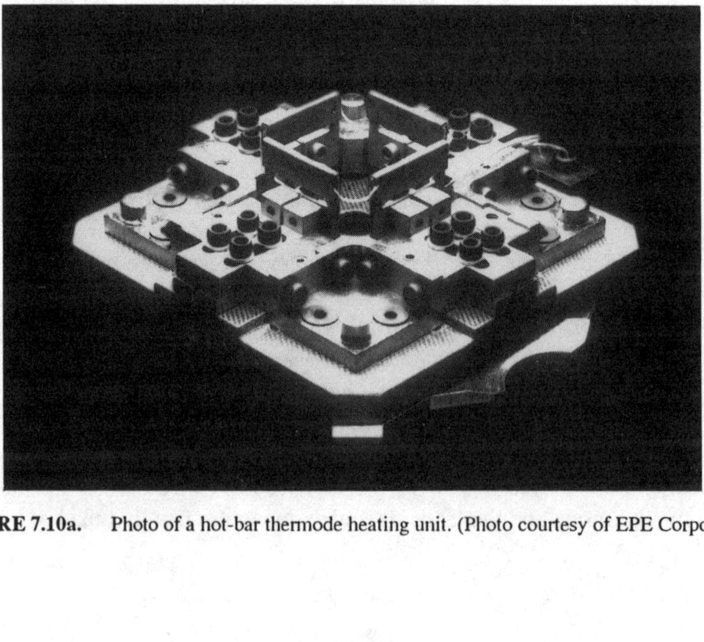

FIGURE 7.10a. Photo of a hot-bar thermode heating unit. (Photo courtesy of EPE Corporation)

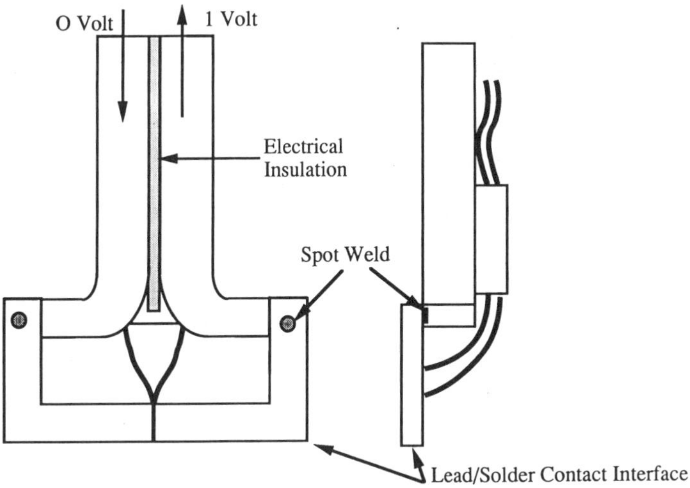

FIGURE 7.10b. Drawing of a hot bar thermode heating unit.

apply greater force on the bar. The relationship of conductance to force for an aluminum, copper, and stainless bar is shown in Figure 7.11.

The use of force as a process parameter is still poorly controlled. An improvement on thermode technology may be to heat a larger bar mass, which effectively increases the heating area, A, in Equation (7.8). The indirect use of a radiant IR or a hot gas (such as nitrogen) as the hot bar heating source

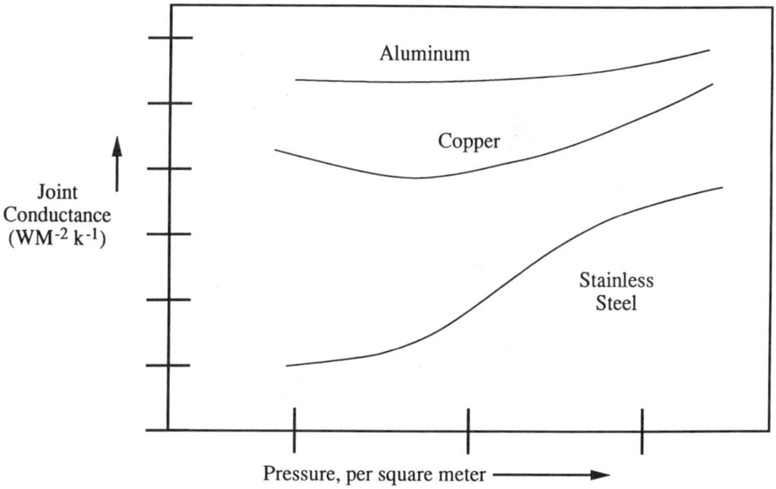

FIGURE 7.11. A graph showing heat conductance as a function of force on the thermode.

appears to have many advantages here. The IR or heated nitrogen heats a large metal block of shaped copper or aluminum, as shown in Figure 7.12. The metal readily conducts the heat to the leads and solder due to its high conductivity. Since this improves the efficiency of the heat transfer, the IR or hot gas temperature does not need to be much higher than the reflow temperature. The lower temperature and the use of an inert gas (nitrogen) reduces the amount of oxidation that occurs on the bar.

Lasers

Laser heat sources operate on the same principle as infrared radiation soldering, except that the heated area is only the area exposed to the laser light. The common lasers used for soldering are neodymin-doped yttrium aluminum garnet (Nd:YAG) and carbon dioxide (CO_2). Nd:YAG laser operates in the near infrared region of 1064 nanometers, which is readily absorbed by tin-lead, copper, nickel, and other conductive materials. CO_2 lasers operate at 10,000 nanometers, which is more readily absorbed by insulative materials such as fiberglass and plastics.

Lasers emit ph coherent radiation, polarized in a single orientation. This orientation property allows minimal divergence and energy reduction over great distances. Utilizing light sources in the infrared region, lasers can emit a beam that is focusable down to the size of small leads of an FPT package. When the laser beam (see Figure 7.13) is focused on a lead sitting on solder, heat is generated. Following the heat transfer shown in Equation (7.1), if the source temperature exceeds the solder melting temperature, then solder reflow will occur.

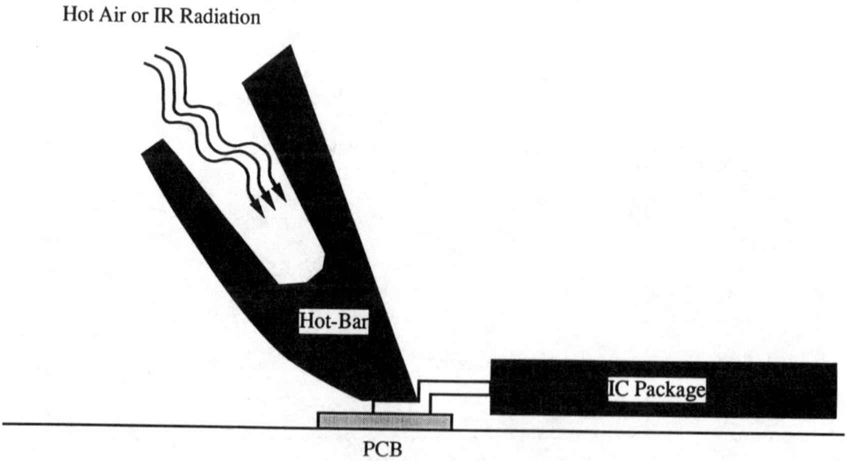

FIGURE 7.12. Improvements in thermode design have overcome earlier problems. (Figure courtesy of Sierra Research and Technologies, Inc.)

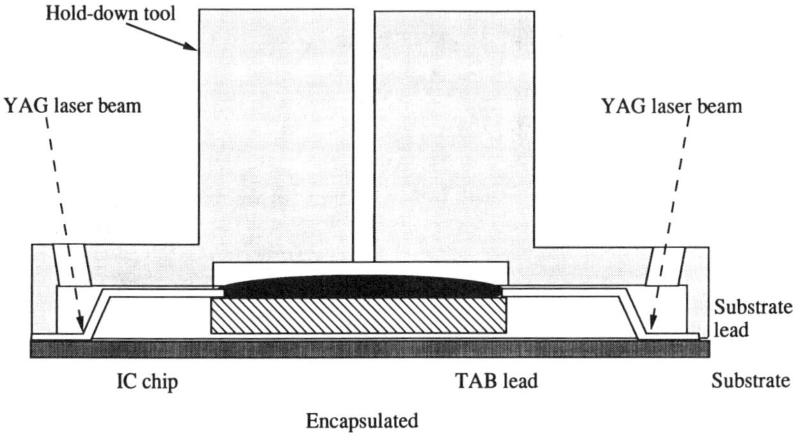

FIGURE 7.13. Laser beams can be used effectively to form solder joints. This figure diagrams the use of a YAG laser for this application.

Generating sufficient source temperature using a laser is not very efficient. The most efficient lasers are only able to convert 2 to 15% of the input electrical energy into output light energy. However, once generated and focused that amount of energy is more than enough to heat and reflow the solder. Sometimes the energy is too much; there have been cases where lifted TAB leads sitting off the solder by more than 0.05 mm have been vaporized by the laser.

One promising use of lasers for reflow soldering, with provisions of avoiding lead vaporization, is the transmission of the laser beam along a packet of fiber optics. The fibers are bundled together at the source so they transmit the same energy, but the individual fibers are unbundled and positioned over the individual leads at the package. As shown in Figure 7.14, this allows the use of a fixture that holds the laser reflow fibers and that will automatically place and reflow the package and solder joints. The fiber optics can be set into a high temperature clear plastic or quartz block to allow "through-the-block" viewing. The block holds the individual fibers perpendicular to the surface of the lead and pushes the leads flush to the solder surface. The solder crown on the land areas should be as flat as possible to prevent the leads from sideslipping off the lands during soldering. This is also true for the other selective reflow heat sources described.

The potential advantages of laser reflow are precise placement of the reflow heat in very dense areas, and the fast heating times. A fiber optic of 0.1 or 0.15 mm diameter is sufficient to transmit enough thermal energy to reflow a typical FPT lead joint. Soldering, if the laser is operated in a continuous wave mode with a shuttered transmission gate, is possible in only 200 to 800 microseconds. This rapid heating cycle is too short to allow significant and

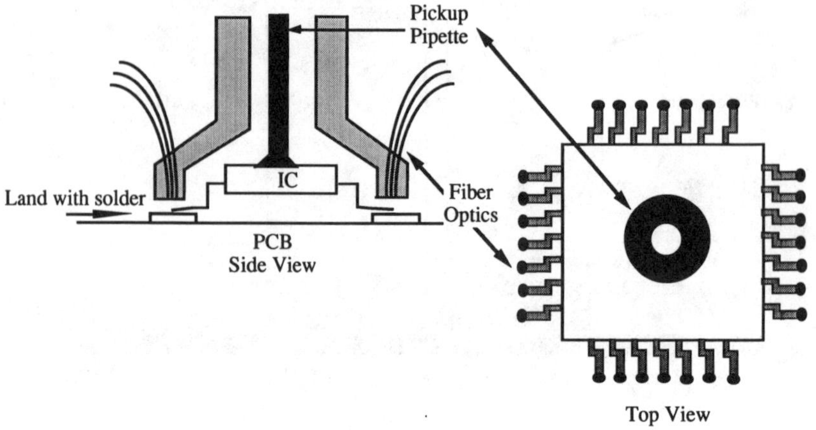

FIGURE 7.14. By transmitting the laser beam down a cluster of fiber optic strands that have been positioned over the package leads, a laser can simultaneously solder several leads at a time. (Source: Sony Corporation)

potentially damaging amounts of heat to transfer from the lead into the IC package. The commercially available laser systems at the time of this printing are first and second generation systems. These have some problems of thermal control, higher expense, and low thruput. Over time and usage this heat source method may gain in popularity.

Thermosonic

A mature attachment technique for inner lead attachment is sonic wire bonding. For decades this has been the method used to bond gold and aluminum conductor wires to the silicon chip. This method, with augmentation, is now being used for fine pitch outer lead attachment. This is particularly the case for TAB style fine pitch packages where the very fine pitch leads are more easily welded than soldered to the land.

The thermosonic energy is transmitted to the lead through a thermosonic horn that is illustrated in Figure 7.15. For a variety of technical and economic reasons the thermosonic bonding tools are configured for single lead versus multilead attachment. The thermosonic energy is heat plus standing wave sonic vibrations which are transmitted along the acoustic actuating horn. These waves cause the bonding tip to oscillate the lead and to create a solid-state weld between the bottom of the lead and the top of the land. Typical weld times are 10 to 30 milliseconds, enabling a thermosonic bonding machine to attach 10 to 20 leads of a fine pitch package per second. The extra time is used by the up and down travel of the horn and for lead to lead travel.

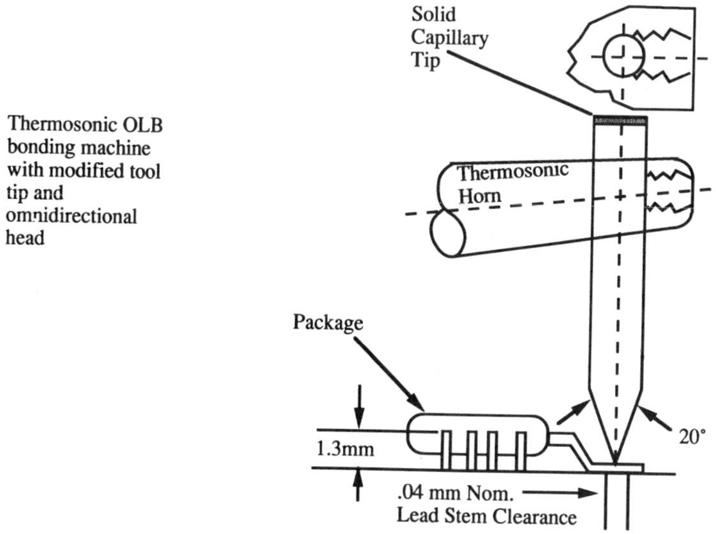

FIGURE 7.15. A thermosonically activated bonding tip is an effective tool for bonding very fine leads to the land site. (Figure courtesy of Tribotech, Inc.)

The disadvantages of thermosonic attachment compared to the solder reflow methods are:

1. The need for increased cleanliness
2. The difficulty of rework

However, to its advantage, thermosonic bonding uses no flux or third metal, such as solder, to join the lead to the land. But oxides and contamination must be pre-removed from the metal areas to be joined or they will be trapped in the joint. These impurities will reduce the bond strength. Thermosonic bonds can withstand over 100 G's of force when properly formed.

Since the bond is a weld and very strong, attempts to perform rework usually result in broken leads or pulled lands. If the land area is designed to be longer than the lead, then a new lead can, potentially, be attached in a previously unused area.

The bonding tip design can help optimize yield and thruput. Trapezoidal trench shaped tips as shown in Figure 7.16 help improve the lead holding and positioning. Waffle patterns, or grooves embedded in the tip at 90° to each other, enable the tip to be omnidirectional, improving throughput and consistency of the bonds. The grooved waffle pattern is shown in Figure 7.17.

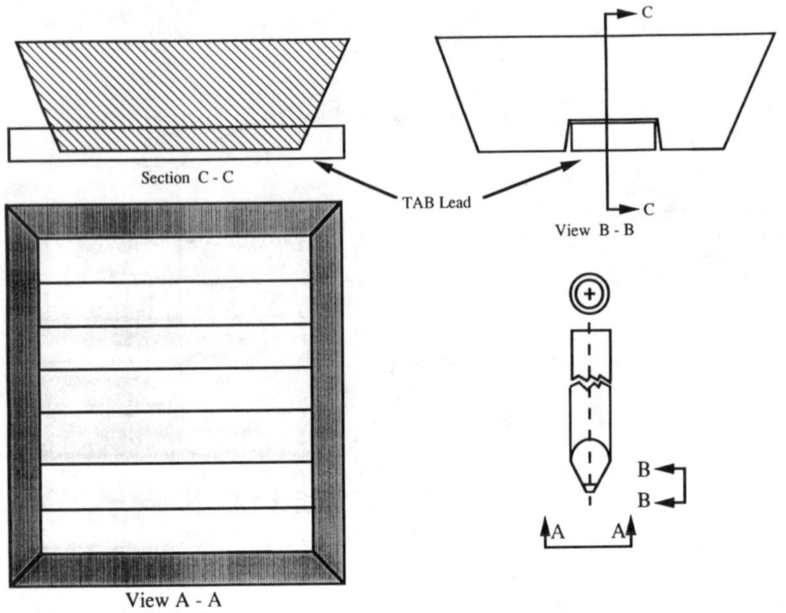

FIGURE 7.16. A close-up of a bonding tip with a trapezoidal trench to help improve lead holding and positioning. (Courtesy of Tribotech, Inc.)

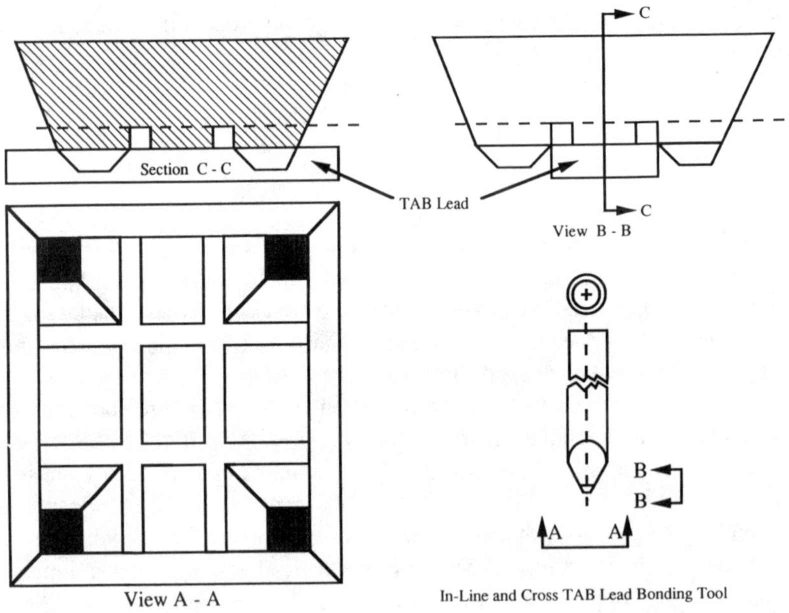

FIGURE 7.17. A close-up of the waffle pattern used on some bonding tips. This pattern allows omni-directional bonding. (Courtesy of Tribotech, Inc.)

7.3 SELECTING A REFLOW METHOD

Table 7.2 shows a summary of the attributes of the mass reflow lead attachment methods. Table 7.3 shows the selective reflow attributes. Both of these tables list the general heating rate equation, the advantages and the disadvantages for each of the methods discussed in Section 7.2.

The user's selection of the reflow method should be based on several factors. The most important of these are:

1. The ability to produce consistent high quality, high reliability joints
2. The ability to meet economic and throughput requirements and,
3. Environmental safety.

To produce a quality, reliable joint and overall assembly, the method should meet the following general criteria:

1. The source temperature, T_S should not be much greater than the melting point of solder, T_M, e.g. $T_S - T_M < 100°C$

TABLE 7.2 A Summary of the Heat Characteristics of those Sources Capable of Mass Reflow.

	Infrared	Vapor Phase	Convection
Heating Rate	$q = \sigma A F_e F_a (T_s^4 - T_a^4)$	$q = hA_a\ (T_s - T_a)$	$q = \bar{h}_c A_a (T_g - T_s)$
Advantages	Rapid heat transfer Rapid heat recovery wide range of temperatures available	Rapid uniform heating on a wide variation of assembly thermal masses. Temperature is constrained to a known maximum Rapid heat recovery.	Low equipment and low production use cost. All objects heat uniformly. The slow heat transfer minimizes component cracking. The heat transfer provides adequate flux preheat.
Disadvantages	Different surface features and body colors cause nonlinear heating. Source temperatures higher than the Tm of solder Source temperature difficult to monitor Each assembly requires a unique thermal profile	Heat flow is too rapid, damaging some components and materials	Slow heat transfer. Slow recovery rate. Equipment may be large.
Relative Cost of equipment: in production use:	Medium to expensive Medium to high depending on profile time	Medium Low to medium	Medium Low

TABLE 7.3 A Summary of the Heat Characteristics of those Sources Commonly Used for Individual Lead or Package Reflow.

	Conduction	Laser	Thermosonic
Heating Rate	$q = \frac{kA_a}{L}(T_s - T_a)$	$q = \sigma A F_e F_a (T_s^4 - T_a^4)$	$q = \frac{kA_a}{L}(T_s - T_a)$
Advantages	Individual comps. can be soldered, such as expensive FPT packages. Can be combined with the placement, rework, and inspection operation to save process steps.	Same as conduction.	Same as conduction except one lead at a time is attached. Good for very fine pitches. (< 0.2mm pitch)
Disadvantages	Not for mass assembly. Oxidation and force change the heat transfer rate and de-control the process	Not for mass assembly	Slow Rework is difficult
Relative Cost of equipment: in production use:	Low Low in low volume High in high volume	Medium Low in low volume High in High volume	Low to medium Medium to high depending on labor costs.

2. The heat transfer rate must be controllable to less than the maximum safe rate of the materials, typically less than 4°C/sec.
3. The heat absorption rate should be uniform

These criteria are for the attachment method. Other criteria that must be met are discussed in Chapter 10.

The summary of reflow attributes is divided into mass and selective methods. Mass methods compared to selective methods are usually the most desirable as they:

- Use the least labor
- Are the most economical
- Have the potential to produce the most consistent joints

For fine pitch, the mass reflow methods (infrared, vapor phase, convection) are successfully used for 0.5 and 0.65 mm pitches in large volumes.

With careful board design, paste selection and process control, mass techniques can be practical in volume for 0.3 and 0.4 mm lead pitches. Below 0.3 mm selective reflow techniques (laser, conduction, or thermosonic) are the most practical.

Of the mass reflow methods, the convection method using forced hot gas (air or nitrogen) is very popular. This is due to its low cost and slower heat transfer rate. The heat transfer rate is very close to the 4°C/sec limitation from the sensitive component suppliers. The other two methods, infrared and vapor phase, can be successful, but they require greater control and the user's knowledge of the heat absorption properties of the materials in the assembly.

Up to this point very little has been said about the ability or desire of the assembly to absorb and use the heat. Heat absorption is dependent on the thermal resistance of the material being heated. A review of heat absorption basics may allow users to use the infrared and vapor phase methods or may enhance their implementation of another emerging reflow process using any of the methods discussed in Section 7.2.

Heat Absorption Basics

Heat absorption is defined as the ability of an object to transfer heat from a source and to conduct that heat across itself until the object and the source reach thermal equilibrium. The classical expression of heat absorption, as described by J. B. J. Fourier in 1822, is:

$$q_z = -k\,A\,\frac{dT}{dz} \tag{7.9}$$

where:

q_z = the rate of thermal conduction of the z direction
k = thermal conductivity of the material absorbing the heat
A = Heat transfer area normal to the z direction
$\frac{dT}{dz}$ = the temperature gradient in the z direction

The thermal conductivity, k, of the material being heated indicates the quantity of heat that will flow across a unit area, A, if the temperature gradient, dT/dz is unity. The units for k are Watt per meter per °Kelvin (1 W/m °K = 0.578 Btu/hr ft °F).

The thermal conductivities for the materials used in electronic assemblies vary widely. This is best illustrated in Figure 7.18 and Table 7.4. Metallic materials with high conductivities such as copper, tin, and aluminum are called *thermal conductors*, while low conductivity materials such as fiberglass, plastic, and steam are called *insulators*. Heat absorption by the materials in an electronic assembly is via conduction. Heat is transferred from a region of higher temperature (usually the assembly's surface) to one of lower temperature. The heat is transferred by direct

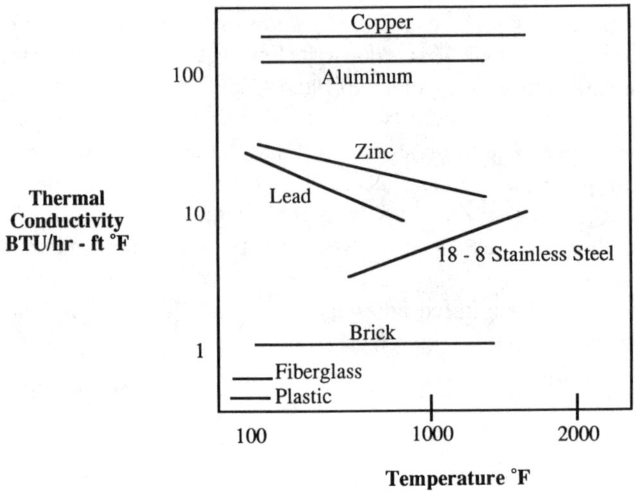

FIGURE 7.18. The thermal conductivities of various materials as a function of temperature.

molecular transference. The temperature of the molecules is proportional to their mean kinetic energy. The kinetic energy is called the molecule's *internal energy* and is a function of its velocity and relative position to other molecules. The faster the molecules are moving and the closer they are to other molecules, the higher the temperature and the internal energy.

TABLE 7.4 A List of the Thermal Conductivities for Some of the Common Materials Used in Electronic Assemblies.

	BTU/hr - ft °F	W/M - °K
Pure Metals	30 to 240	52 to 410
Alloys	8 to 70	14 to 120
Insulating Materials	0.02 to 0.12	0.03 to 0.21
Gases at Atmospheric Pressure	0.004 to 0.10	0.007 to 0.17

A cross-sectional view of a typical electronic surface mount and fine pitch assembly is shown in Figure 7.19. In this drawing, the thermal conductivities have been indicated for some of the major materials.

When this section of the assembly is mass heated up to the T_M of solder (~ 200°C) neglecting the heat source for the moment, the copper heats to thermal equilibrium very quickly, the plastic and the fiberglass slowly. Since the copper is hotter than the two insulating materials, it becomes an additional heat source conducting heat into the board and the plastic package.

The difference of thermal conductivities (copper ≈ 400 W/m °K, plastic ≈ 0.2 W/m °K, ceramic ≈ W/m °K) would imply that the plastic bodies of the integrated circuits and the printed circuit board would be the slowest heat absorbers. However, it is these areas that usually show early overheating damage. In fact, the corners of the board usually show the earliest damage. The reason the corners show earliest damage can be understood by imagining the heat areas as small cubic volumes. The cubic volumes in the center of the board are heated on only one surface, the top. The corners of the board are heated on three surfaces, the top and two sides. This is illustrated in Figure 7.20. The corner cubic volume surface receives three times more thermal energy than the bulk board cubic volume.

If the source temperature is significantly higher than the charring (or burning) temperature of the circuit board, then the board's corners will char in about one-third the time it takes for the bulk of the board to discolor. Vapor phase reflow has the advantage of limiting the source temperature to just above the solder melting temperature. However, the rapid heat transfer from the saturated vapor to a fast absorbing material such as ceramic causes cracking of chip capacitors and other ceramic components.

Convection reflow seldom discolors the board since, in normal usage: (1) the source temperature is set slightly above the solder melting point, and, (2) the board is removed from the oven before it has enough time to exceed the safe thermal energy limit. This limit is approached much more slowly in the convection oven because of the slower heat transfer rates.

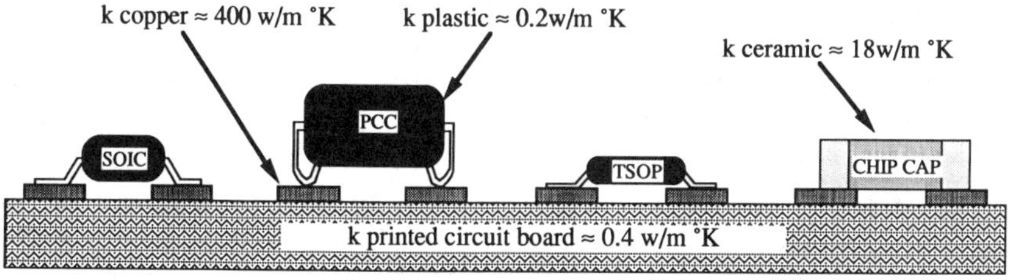

FIGURE 7.19. A cross-sectional view of an assembly showing the thermal conductivities for the various elements.

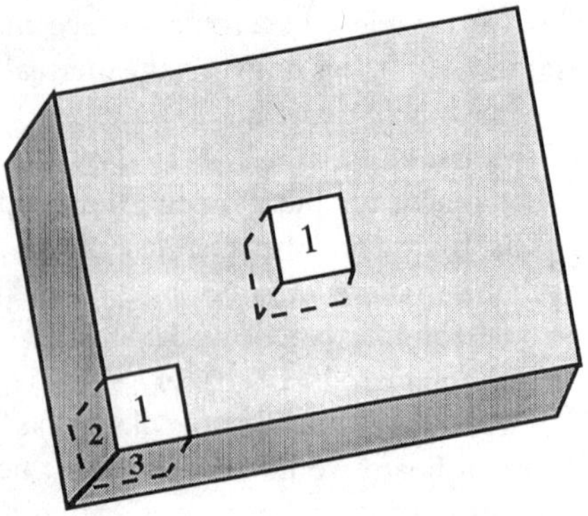

FIGURE 7.20. The corner cubic volume receives 3 times more thermal energy than an embedded cubic volume. As a result, there's a higher probability that the corners will char and burn when the volume is heated with IR energy.

Infared can be operated at temperatures close to solder melting temperature, but the higher heat transfer rate of infrared heat ($q = K\ (T_S^4 - T_A^4)$) compared to convective heat ($q = k\ (T_S - T_G)$) means that thermal energy is transferred to the sensitive board corners to the power of four faster. This is not to say that infrared can not be used successfully for fine pitch solder reflow. However, certain procedures must be followed to achieve a capable process:

1. Spread all large, slowly heat absorbing bodies out as uniformly as possible across the board. Leave as much space as possible between them to allow heat to flow as rapidly as possible to the joint areas. If the large *heat insulators*, such as J-lead integrated circuit packages, are too close, their thermal mass prevents rapid heat transfer to the joint areas below the package.
2. Leave copper foil on the board edges and around cutouts. The copper acts as a heat spreader to remove the heat from the sensitive corners. Placing the board assembly in a fixture is another way to accomplish the same goal. Be sure that the fixture material will withstand repeated exposure to the thermal energy.
3. Use an oven thermal profile like that as shown in Figure 7.21. This profile consists of four parts. The first part is very rapid temperature rise of 4 to 7°C/sec to the flux activation region of about 150°C. The second part is a *soak* at the 150°C temperature for more than a minute. During this soak, the temperature differentials across the board disappear and the flux has sufficient time to generate very solderable surfaces. The third part is the reflow phase, with heat transferring at a rate of 2 to 3°C per second. The final part is cooldown.

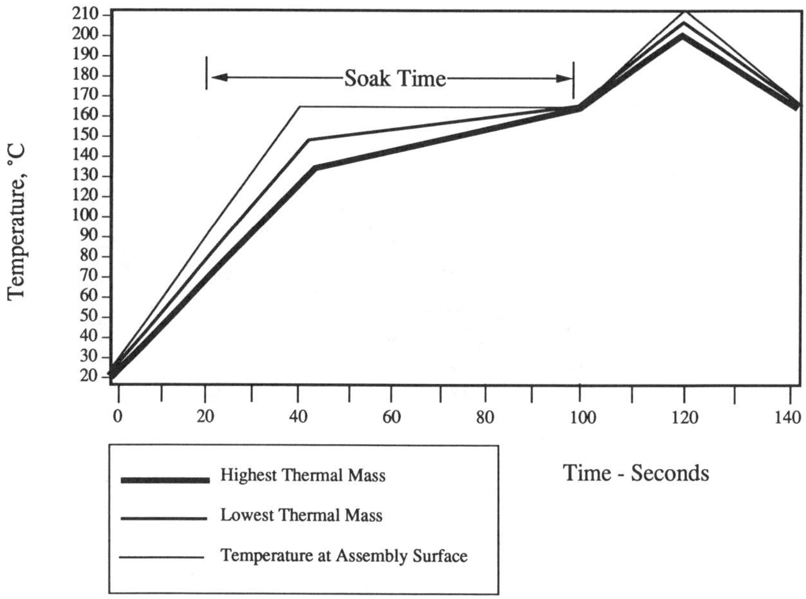

FIGURE 7.21. The recommended IR time temperature profile for SN 63 type solder. This profile allows sufficient time for all thermal masses on an assembly to equalize prior to entering the highest temperature of the reflow zone. (Source: Vitronics)

Combined together, these three precautions allow the advantages of the rapid heat transfer of infrared but without excess thermal energy absorption and resultant damage. This profile takes advantage of the rapid heat transfer character of infrared for the initial and reflow phases.

7.4 CHANGING COMPONENT LOCATION DURING REFLOW

In Chapter 6, much of the text dwelled on the importance of accurate placement before reflow. It was also stated that small, low-mass components would self-center as a result of the solder wetting force and vibration during reflow. The fine pitch packages generally are too high in mass to self-center with the aid of the solder wetting force. However, mechanical vibration while the solder is molten may allow the high mass packages to vibrate into proper alignment. One study showed that solder rework to reposition improperly aligned components reduced significantly when a mechanical vibratory system was attached to the reflow oven's belt. The reduction in defect rates is shown in Figure 7.22.

Users interested in exploring this attachment to the reflow oven need to be aware of the potential reduction in solder joint reliability that may occur. If the

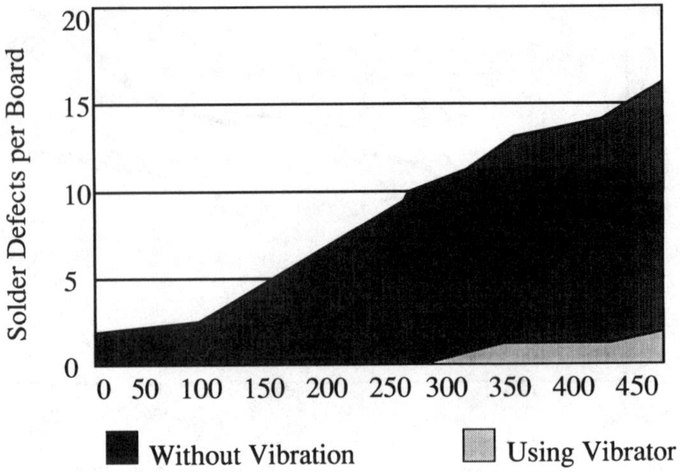

FIGURE 7.22. The reduction in rework of solder defects as a result of vibrating the assembly while the solder is liquidus. Care must be taken to minimize the magnitude of the vibration during the cooling phase. (Source: Gartner Group Inc.)

solder joint experiences excessive vibration during cooling, it will develop excessive grain and weak grain boundary bonding. These joints can fatigue and crack early, with thermal and mechanical cycling. To avoid this, users should avoid allowing the joints to be vibrated during the cooling phase.

In general, it is better for users to implement a placement system that is capable of placing components within the allowable window consistently, and to avoid the use of mechanical vibration.

References

1. Arslancan, A.N. and Flattery, D.K. 1990. Infrared Reflow for SMT: Thermal and Yield Considerations. Proceedings of Nepcon West '90, Feb., 1990, Anaheim, CA.
2. Carlson, J. 1990. Recent Advances in Single Point TAB Bonding Tools. Paper read at ASTM F1.16 TAB meeting, Oct. 1990, Costa Mesa, CA.
3. Cox, N. 1990. Cost Optimization of Near I.R. Reflow Soldering. Proceedings of Nepcon West '90, Feb. 1990, Anaheim, CA.
4. Dow, S. 1990. Forced Convection for SMT Reflow Soldering. In *Hybrid Circuit Technology*, Oct., 1990.
5. Giedt, W. 1957. Principles of Engineering Heat Transfer. New York: Van Nostrand Reinhold.
6. Graham, G. 1990. Achieving Perfect SMT Component Location After Reflow. In *Surface Mount Technology*, Sept. 1990.
7. Hampshire, B. Private correspondence.
8. Hammond, R. 1990. Laser Ultrasonic Tape Automated Bonding. In *Surface Mount Technology*, Sept. 1990.

9. Holland, J. et. al. 1970. Heat Transfer. New York: Elsevier Publishing.
10. Keith, F. 1973. Principles of Heat Transfer. New York: In Text Educational Publishers.
11. Mah, H. O. 1990. Mounting Sub 5 mil Pitch TAB Components by Solder Reflow. Proceedings of SMART VI, Jan., 1990, Lake Buena Vista, FL.
12. Newbury, P. 1990. TAB Technology's Emerging Debate. In *Circuits Assembly*, Oct., 1990.
13. Peck, D. 1991. Private correspondence.
14. Rowland, R. 1991. Private correspondence.
15. Sato, Y. 1991. Fine-Pitch Solder Joint. Read at IBM SMT '91 Seminar, March, 1991, Austin, TX.
16. Spigarelli, D. 1991. Private correspondence.
17. Steiner, T.O. and Suhl, D. 1987. Investigation of Large PLCC Package Cracking During Reflow Exposure. IEEE Transactions on Components, Hybrids and Manufacturing Technology. CHMT-10 (2) pp 209-216.
18. Thwaites, C.J. 1982. Soft-soldering Handbook. Greenford, Middlesex, England: International Tin Research Institute.
19. Zarrow, P. 1990. IR Reflow Soldering Systems and Steps. In *Circuits Manufacturing*, Feb. 1990.
20. Zarrow, P. 1990. Process Concerns in Infrared Reflow Soldering. Proceeding of Surface Mount '90, Aug., 1990, Boston, MA.
21. Zarrow, P. 1991. Fine-Pitch Soldering. In *Circuits Assembly*, Sept. 1991.
22. Zarrow, P. 1991. Private correspondence.

8

Post Reflow Cleaning

8.0 WHY CLEAN

Cleaning electronic assemblies has caused heated debates for the past several years. Cleaning is done primarily for cosmetic and reliability reasons. Suppliers want products to look good. Customers want highly reliable products. A clean assembly implies, in many minds, a high reliability assembly.

That thinking needs to be revisited. The emissions of the cleaning agents commonly used for assemblies, solvent vapor waste and water, are causing environmental problems. These include depletion of the ozone layer due to CFCs in cleaning solvents and pollution of the ground water with untreated aqueous effluent. These are costly to control and fix.

Some studies and users challenge the common wisdom that uncleaned boards are less reliable. These sources indicate that leaving certain fluxes on the board do no harm and find that the elimination of the cleaning steps is environmentally and economically preferred.

If the flux is mild, has low solids, and a halide-free content as do most of commercially marketed no-clean formulations, then no cleaning is possibly more reliable than cleaning the assembly. Most of the no-clean fluxes leave a rosin or inert substance that acts to encapsulate any contaminates and corrosives. If a cleaning operation is used it removes the inert substance, which exposes the board to the contaminants and corrosives if they are not adequately removed from the board. There are several excellent references on cleaning assemblies. Some of these are listed in the references at the end of the chapter. These references thoroughly cover the cleaning agents, equipment and problems available. For readers interested in cleaning, those references are recommended. This text recommends using a solder paste containing a no-clean flux and eliminating the cleaning of assemblies.

8.1 SUCCESSFUL NO-CLEAN

To adopt a no-clean process one needs to implement several steps. The primary steps are:

1. Select a no-clean or low-solids flux that meets the user's and customer's process and reliability needs.
2. Design surface insulation resistance (or SIR) test patterns (at least two) onto all boards, especially under the larger FPT packages. Use the SIR test pattern and SIR test (per IPC-SF-818 One-day SIR test procedure) as a process monitor for the inertness of the residues left on the board.
3. Carefully select and qualify a flux, a solder reflow profile, and paste environmental controls so the possibility of forming solder balls that will be left on the uncleaned board is minimized.
4. If bed-of-nails test fixtures are used, choose a flux that does not corrode the probes. Also test the boards as soon as possible after reflow; this will minimize probe and test problems. The least number of probes should be used (see Chapter 11) to further minimize the potential of test problems.
5. Routinely clean the test fixture probes. Consider using stainless steel probes.
6. Apply a label to shipping bags notifying customers that the assemblies are not cleaned as a policy to protect the environment, and that the reliability is as good as cleaned boards.
7. Always use the most solderable boards and components, and use suppliers and storage procedures that maintain the optimum solderability.

Selecting a No-Clean Flux

Many solder paste suppliers now offer no-clean fluxes. These are basically halide-free fluxes made of rosin or resin that have removed or minimized the corrosive elements incorporated in present cleaning-required fluxes.

The choice of a compatible no-clean flux involves experimentation with the flux candidates to determine compatibility with the board's materials, especially the soldermask. It also involves determining the proper solder reflow profile that works for the paste and minimizes the potential for solder balls and probe problems. If solderability problems occur using a normal atmosphere reflow oven consider converting to a nitrogen atmosphere, as inert nitrogen aids solderability when many of the commercial no-clean flux bearing solder pastes are used.

Design Surface Insulation Resistance Test Patterns Onto Boards

The surface insulation resistance test pattern is generally free of soldermask and is measured to determine the surface insulation resistance, or SIR, between the test pattern's traces. The test pattern can assume an infinite number of shapes. Figure 8.1 shows the normal patterns for use under IC packages and chip components and assumes there are no via holes. If the space between the land areas is generally left free of soldermask, then the test pattern traces should be free of soldermask. The trace width and space of the test pattern are the minimum used elsewhere on the board, typically 0.18 mm (0.007″). Figure 8.2 shows the more practical design of comb patterns under IC packages. These designs work around the via holes normally required to route traces away from the package interconnections.

Bed-of-Nails Testing

Besides assumed customer objections, the next potential problem is contamination of the beds-of-nails test probes used in in-circuit test fixtures. Bed-of-nails are clusters of small probe points oriented in a fixture so the appropriate points on the board make contact with the nails when the board is mated to the fixture. These probes need to make good physical contact with the test points to provide electrical contact.

In practice, the electrical contact of probe points with cleaned boards is a significant problem that is not increased by no-clean flux or flux residue. The use of sharp multicrown probes and an adequate mechanical pull down fixture to form a good vacuum/mechanical bond generally helps reduce the contact problem. However, the choice of a no-clean flux should include testing for compatibility with the probes. Some users are switching to stainless steel probes as one way of assuring compatibility.

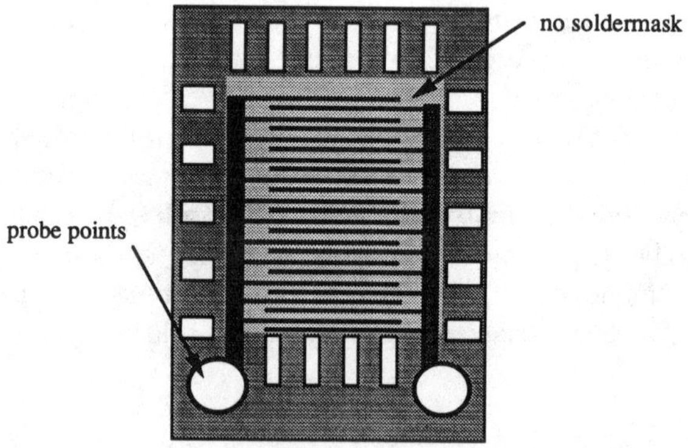

FIGURE 8.1. The surface insulation resistance, or SIR test pattern that is used on test coupons or open board areas.

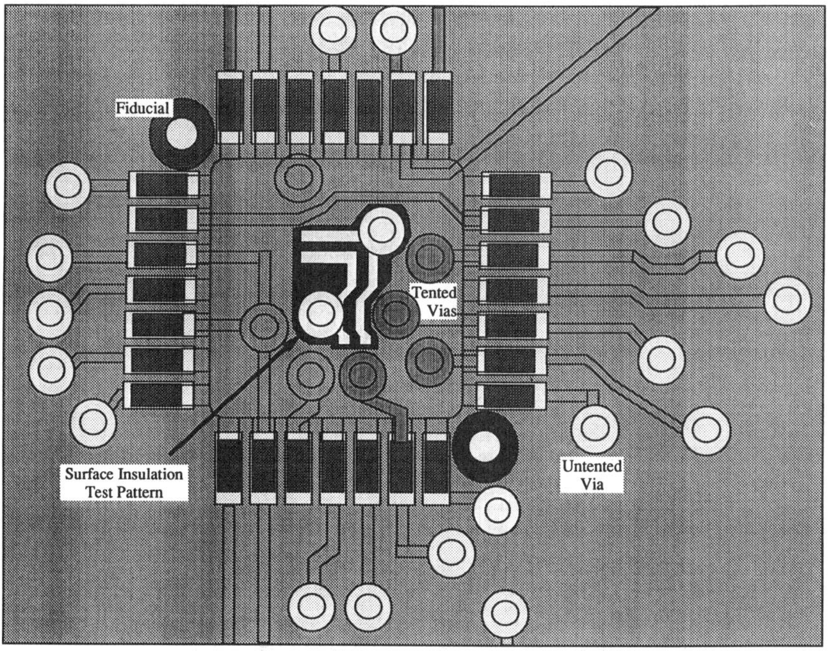

FIGURE 8.2. A possible SIR pattern that can be placed under packages to assure cleanliness. Be sure to remove the soldermask from the parallel trace areas.

In the longer term, as electronic assemblies become denser, the use of bed-of-nails testing will become more difficult whether the board is cleaned of flux or not. Users should begin taking steps now to minimize bed-of-nails testing in order to prevent problems later. See Chapter 11 for testing strategies.

Testing For Cleanliness

There are several tests for cleanliness, including ionic, visual, and surface insulation. These tests are covered in great detail in the IPC-SF-818, "General Requirements for Electronic Fluxes," as well as other references. Of these tests, the surface insulation test (or SIR) is highly recommended. The SIR test measures the conductivity between the traces due to contaminants that will dissolve in moisture. If the contaminant goes into solution and conducts appreciable electrical current then there is a strong possibility of metallic corrosion and electromigration. If there is low electrical conductivity then there is generally little contamination to cause problems. The best SIR test is to design several (at least two or three) patterns onto a board and under IC packages. If the spaces between the IC land patterns are left free of soldermask, then be sure to remove the soldermask over the test pattern traces. To conduct the SIR test, assemble the packages on the board. Then subject the assembly to the temperature and humidity test conditions shown in Figure 8.3. The humidity source should be deionized water and the moisture chamber should be free of any conductive contamination sources.

Diagram No. ________

Test Conditions and Measurement Timing

Figure Number	Event
1	Test Begins.
2	Two hours later. Initial measurements taken. All subsequent test timing referenced to this point.
3	Initial measurements complete. Chamber ramp-up started. Reverse bias of 50 volts DC applied to all specimens.
4	24 hours after point #2. Measurement cycle begins.
5	24 hour measurement cycle complete. All specimens biased.
6	96 hours after point #2. Measurement cycle begins.
7	96 hour measurement cycle complete. All specimens biased.
8	BENCHMARK Measurement. 168 hours after point #2.
9	Benchmark measurement complete. All specimens biased. Chamber ramp-down initiated.
10	Chamber stabilized at 25C/50% RH.
11	Two hours after point #10. Final measurement sequence started.
12	Final measurement sequence complete. Turn off power supplies. Open chamber. Remove test specimens.

T0: Initial ambient measurement
T1: 24 hour measurement
T2: 96 hour measurement
T3: 168 hour measurement (BENCHMARK)
T4: Final ambient measurment

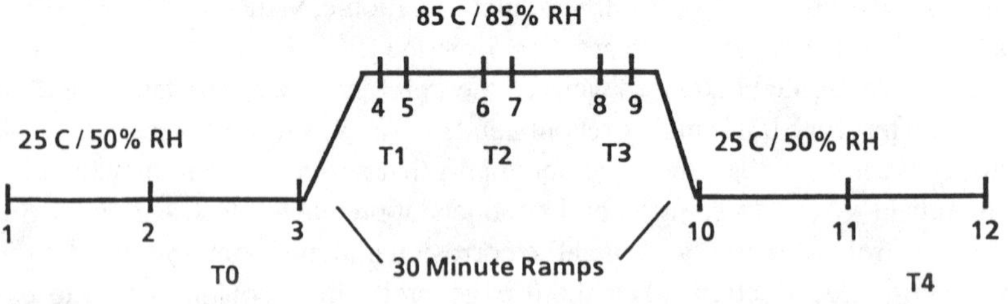

FIGURE 8.3. The SIR test conditions and measurement timing. (Source: IPC-TR-580)

TABLE 8.1 The Minimum Desired SIR Test Resistance Values. (Source: IPC-CH-65)

Application	Minimum Resistance (Ohm)
Consumer (Class I)	1×10^8
Industrial (Class II)	1×10^8
High Reliability (Class III)	5×10^8

After the exposure period measure the conductivity of the SIR patterns under the indicated bias conditions (100 Volt DC voltage bias) shown in Figure 8.3 and record the readings. The conductivity is measured on a high impedance resistance meter or equivalent. These meters operate by forcing a high voltage (for example, the recommended 100 V); and measuring the amount of current. The meter displays the resultant resistance ($R = V \times I$). The measured resistance should be higher than the minimum resistance limits shown in Table 8.1.

As an extension of this test, one may then dry the board and remove the IC packages that sit above the test pattern sites. After removal, inspect the comb patterns for any evidence of corrosion, electromigration and dendrite growth.

Shipping and Storing No-Clean Assemblies

After reflow soldering, the assemblies should be handled in dust-free containers. If possible these should be individual antistatic bags. After final acceptance and chassis assembly, the final product should be covered with a protective bag. If the customer will see the board assembly then it is helpful to place a label on the board or its protective bag that notifies the customer that the board uses a *no-clean process*, is uncleaned, and why. If the customer knows that the board is reliable with the flux or flux residue, there will be minimal or no objection to the no-clean approach as an environmental protection measure.

References

1. Denyon, W., and Wargotz, W., and Turbini, L. 1990. Guidelines for Cleaning of Printed Boards and Assemblies, IPC-CH-65. Lincolnwood, IL: IPC.
2. Marcoux, P. 1992. PCA Design Guidelines Manual. Sunnyvale, CA: PPM Associates.
3. Morris, J. and Conway, J. No-Clean Reflow Process Implementation. In *Circuits Assembly*, Aug., 1991.
4. —1991. Cleaning and Cleanliness Test Program, IPC-TR-580. Lincolnwood, IL: IPC.

9

Inspection, Rework, and Repair

9.0 INSPECTION

The inspection, rework, and repair of improperly assembled or electrically defective FPT packages is difficult. The high lead counts and fine lead pitches make visual inspection a challenge. The common method of desoldering a fine pitch part is to use hot air or a hot bar to remelt the solder so the package can be removed or repositioned on the board. Remelting the solder uniformily is tough and may introduce the problem of lifted lands or bent leads.

For years, sage advice has been offered to all manufacturers to "do it right the first time." The challenges of fine pitch inspection rework and repair may finally force this advice to become reality.

Surface Inspection

When assembled boards are inspected, the inspector is looking for a wide variety of defects. These defects result from processes drifting out of specification, from poor materials, or from random anomalies. Inspection generally means examining the solder joints, component positions and markings, board quality, and hardware attachment.

Surface mount solder joints are generally considered good if they appear to be smooth with a satiny luster and wetted to the metals to be joined. If the land area is larger than the lead area, then the wetting should form a concave fillet up the sides and toe of the lead. This is shown in Figure 9.1.

Fine pitch lands are generally only as wide as the lead, therefore no appreciable fillet is expected on the sides of the lead. Also, if the toe of the lead is one that is excised from a guard ring or TAB tape, then it is possible that the toe will not wet with solder because the exposed metal has oxidized. Therefore, a suggested visual workmanship specification for a fine pitch solder joint lead is

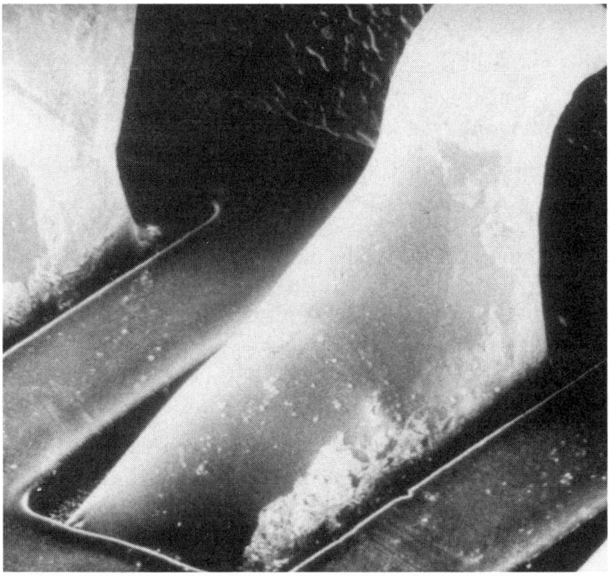

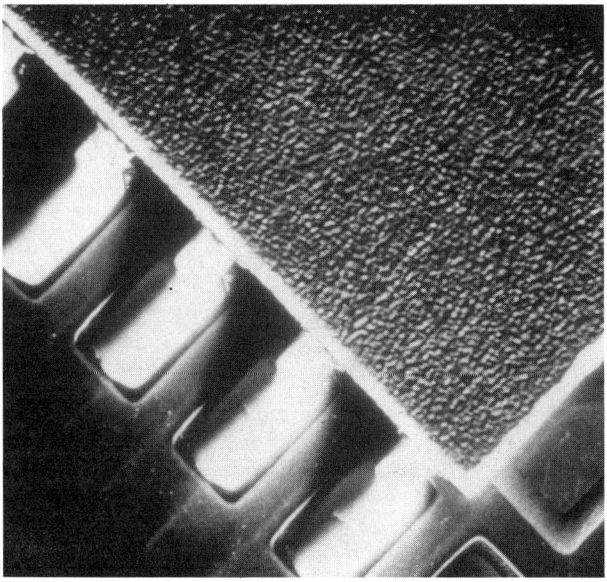

FIGURE 9.1. (a,b) A SEM photo of the surface appearance of properly soldered joints. (Photo courtesy of Dupont)

1. A smooth, satiny luster joint that is free of foreign material, unless it is an allowable no-clean flux
2. Evidence of solder wetting up the two sides of the leads and at least partial wetting on the top of the lead
3. An absence of excess solder that wets beyond the midpoint of the lead bends

Figures 9.2, 9.3, and 9.4 show examples of acceptable, minimally acceptable and rejectable examples of fine pitch solder joints when inspected visually.

Other visual surface workmanship criteria should include:

- No evidence of lead damage or cratering due to excess hot-bar or bond tip pressure
- No evidence of opens or solder bridges
- Uniform solder joint appearance for each package lead
- No appearance of package damage such as fine cracks near the package leads or surface
- A maximum lead to land misalignment of 10% of lead width, or an amount that reduces the lead to adjacent land spacings below the safe electrical distance.

FIGURE 9.2. An example of the preferred appearance of a fine pitch solder joint. (Photo courtesy of Omni Training Corporation)

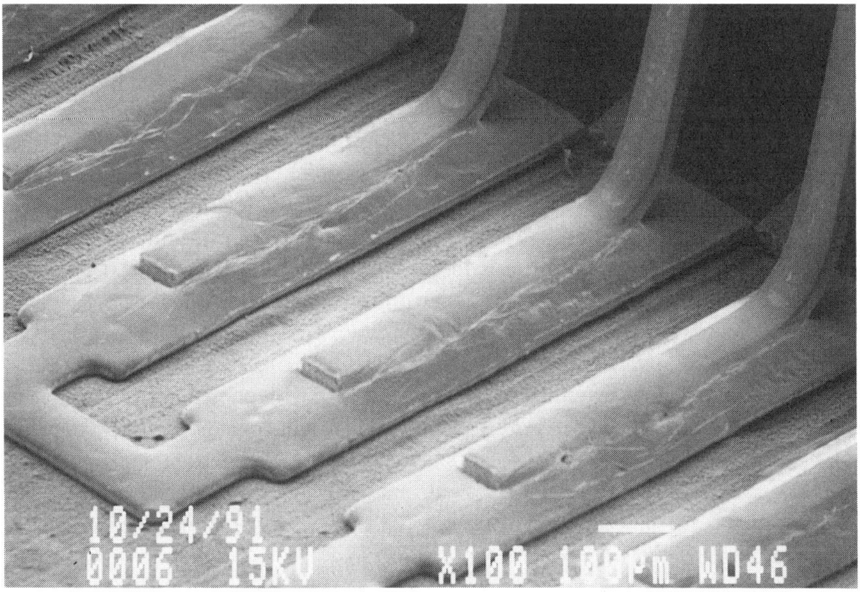

FIGURE 9.3. An example of a minimally acceptable FPT solder joints. In this photo, the toe of the lead side not completely wetted with solder. (Photo courtesy of MCC)

FIGURE 9.4. An example of a rejectable FPT solder joint, the lead is not wetted. (Photo courtesy of Omni Training Corp.)

These criteria should be obvious and acceptable to anyone with visual inspection experience. The misalignment criterion may need further explanation, as most SMT workmanship standards accept a larger lead to land misalignment. A maximum 25% misalignment for high reliability applications and 50% for consumer and commercial applications are common SMT standards. These are acceptable for SMT because it is accepted belief that the lead area in contact with the solder is sufficient to hold the lead to the land, and the lead to adjacent land distance is greater than the distance that would cause an electrical short or leakage.

The spacings for fine pitch leads are so close that a significant amount of misalignment may allow a lead to get close enough to the adjacent land that an electrical leakage path develops. The amount of leakage will be a function of the humidity, the ionic content, the voltage difference, and the rate of signal change, dV/dt, in the lead or adjacent land. If any of these factors is excessive, then electrical leakage in the form of crosstalk, spurious noise, or current will develop. Table 9.1 shows the minimum distance as a function of the voltage differences between the lead and the adjacent land. Since fine pitch leads are so close, the most practical alignment specification is to allow a maximum of 10% lead width misalignment, or as an alternative, the amount of misalignment that still provides the minimum allowable distance between the lead and adjacent land for proper electrical performance.

Subsurface Inspections

If one examines the subsurface below the package lead, one may find voids and nonwetting which act to reduce the solder joint area and possibly reduce the joint strength. To inspect the subsurface of the solder joint involves using x-ray inspection. The use of x-rays enables the viewing of subsurface voids, cracks, and other defects in the solder joints. The inspection area is exposed to x-rays from a focused source, which are then collected by a detector. The collected rays are viewed on monitors and analyzed by software. Any voids, cracks, and misalignment show up as different (usually darker) images than the solder.

TABLE 9.1 The Minimum Trace Space Dimensions Needed to Provide the Voltage Isolation Listed. (Source: IPC-D-275).

Minimum lead to land, and trace to trace space versus voltage differences	
Uncoated external copper/lead at < 3000 meters attitude	0.005mm/V
Uncoated external copper/lead at > 3000 meters attitude	0.0025mm/V
Conformal coated copper/lead	0.03mm/V

There are two basic approaches to x-ray: radiography and laminography. Radiography allows the x-rays to pass through the circuit board being inspected. Those areas of the circuit board that are thick and dense reflect more of the x-rays than thin, less dense areas. The x-rays that pass through are collected by a stationary detector and displayed on a monitor.

Laminography overcomes one shortcoming of radiography. If the circuit board has components on two sides and multiple trace layers, it is difficult to isolate and clearly see a small area on one of the layers. In laminography, the detector is rotated around the focal plane, which is the location plane of the circuit board. Isolating a specific feature on any of the multiple circuit board layers is done by varying the laminographic angle, i.e., the angle of the x-ray beam relative to a line normal to the plane of the object. This is illustrated in Figure 9.5.

When used in the radiographic approach, the laminographic angle is 0° and the detector is stationary. In laminography, the detector is rotated in sync with the rotating source, and the typical laminographic angle while rotating is 32°. Because

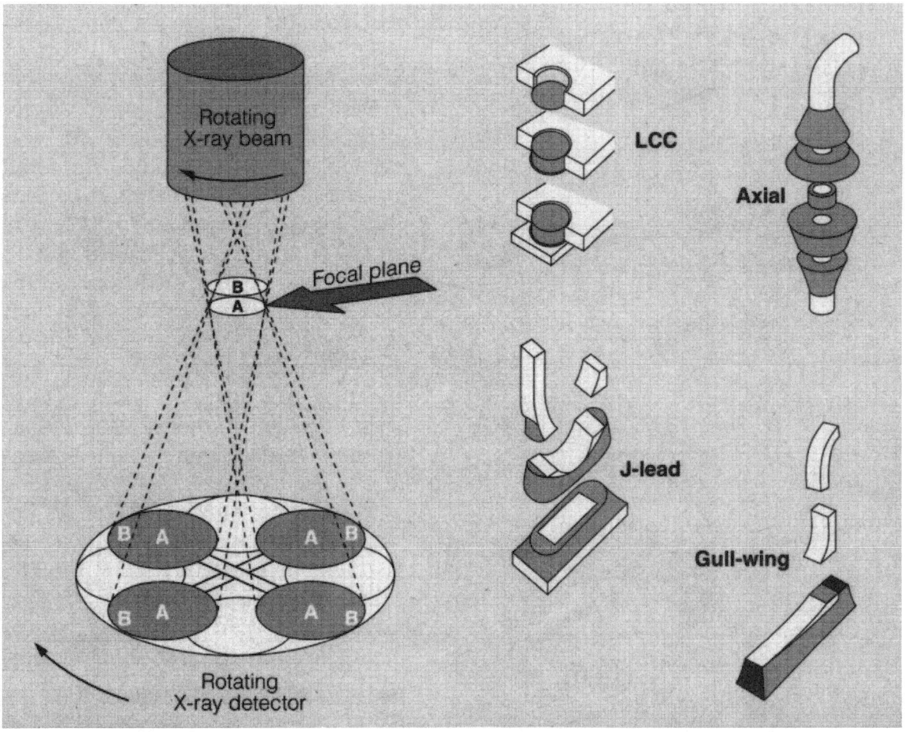

FIGURE 9.5. As solder joints continue to shrink, sub-surface inspection is warranted. X-ray techniques, such as the laminography approach diagramed here, may provide a viable solution. (Photo courtesy of Four-Pi Systems)

of the rotating source, the unit is able to resolve a feature or object on a specific layer. By changing the source angle slightly, a laminated representation of the feature can be composed and displayed.

The laminographic image allows detection of voids, lifted leads, misalignment, cracks, some foreign materials in the joint, and solder bridges. There is some confusion in the industry as to the acceptability of some of the things detected by the x-rays. This is not surprising given that there is still confusion and varied opinions on what is acceptable using surface visual inspection.

The acceptability of some of the x-ray detected flaws will be a function of the product's use environment and customer requirements. In general, lifted leads, solder bridges, obvious solder joint cracks are rejectable criteria in all applications. Solder voids, misalignment and evidence of foreign material may not degrade the reliability of the joint in less demanding use applications.

The appearance of solder voids (shown in Figures 9.6 and 9.7) may reduce the joint area to where it is marginally capable of securing the lead during the products exposure to mechanical or thermal shock and cycling, and vibration. Solder voids are possibly trapped gas or flux that wasn't able to escape to the solder surface before the solder joint cooled and set. They may also be sites of nonwetting between the solder and the lead or the land. Regardless, excess voids reduce the

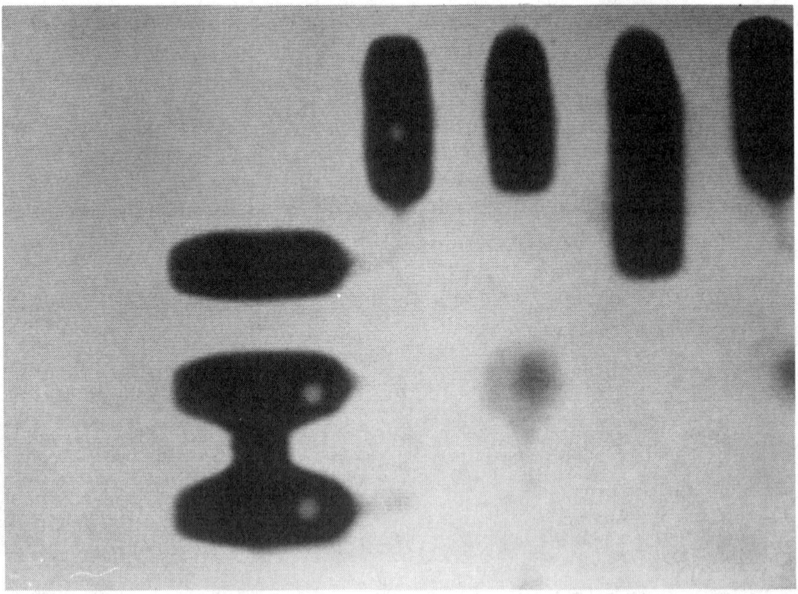

FIGURE 9.6. Voids in solder joints as detected by a x-ray inspection system. (Photo courtesy of Four-Pi Systems)

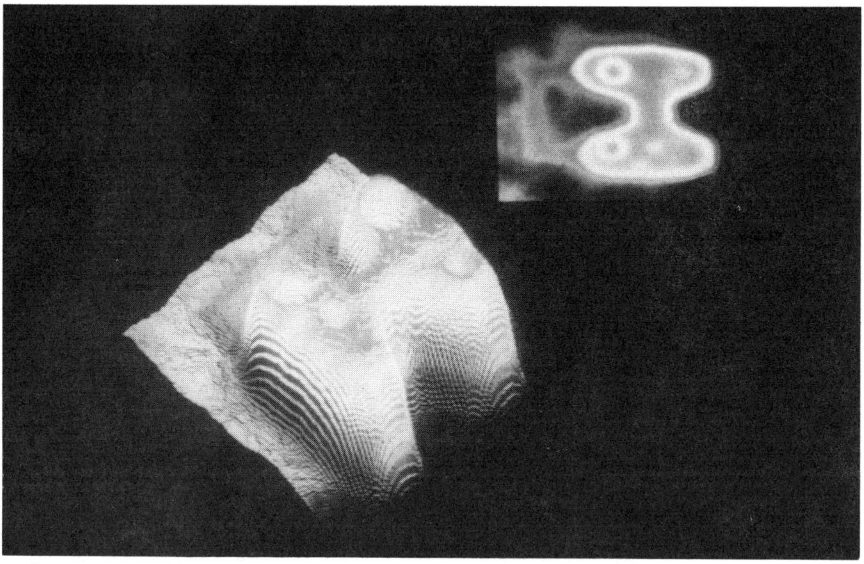

FIGURE 9.7. Voids in solder joints as detected by a x-ray inspection system. (Photo courtesy of Four-Pi Systems)

effective solder contact area and increase the probability of joint failure. Particularly if the use conditions are severe enough to initiate a crack or solder separation at a void site.

The specific amount of void area allowable in a solder joint may be estimated using the solder fatigue model equations discussed in Chapter 10. At present, many industry experts are specifying 25 and 50% as the maximum void area in the solder joint for high reliability and commercial applications, respectively. This criterion assumes all other conditions such as good wetting and 10% misalignment are satisfied.

9.1 REWORK AND REPAIR

Rework and repair are not process steps that should be easily tolerated in any factory. FPT rework and repair adds a new level of complexity to a process that is already more of an art form than a repeatable science.

Package Removal

A defective fine pitch package can be removed from the circuit board successfully by using a heat source that uniformly heats each lead quickly. Common heat sources are hot gas, hot bars, or IR light. Once each joint is molten, the package can be gently lifted from the board. There is a wide range of equipment to perform this task. This equipment includes hand tools such as hot air blowers with special nozzles, or IR emitters that can be aperture controlled over the heated area. On the more expensive and automated end of the spectrum are rework stations with automatic pickup, temperature, and time controllers.

The removal of a fine pitch package is successful only if it can be removed without lifting any of the pads to which it is attached, and without damaging adjacent components including the board. Without the proper skills and heat transfer, it is easy to lift several lands and traces off the board when the package is removed. The proper skills require knowing how much heat to apply and knowing when the solder is molten on each lead. The latter is tough since the lands are connected to differing copper masses. Some lands are dummy lands with no traces, and some are attached to massive ground planes. The dummy lands heat very fast, and the ground plane lands heat slowly. This can result in the dummy lands lifting from the board because of overheating, and the ground plane lands lifting because the solder on these is still semi-solid. This problem can occur as readily on expensive workstations as with the hand tools commonly used by a skilled operator, unless proper operator training and board design is employed.

There are a couple of design tips to reduce the potential for these problems. The first is to design thermal relief connections to the lands and grounds. Thermal relief is simply reducing the copper volume around the land and at the ground connection. This increases the thermal resistance and slows the heat transfer rate. The common method at the land is to use a thin trace to separate the land from a via or the ground plane. The common design to form a thermal relief pad in the ground plane is to make a cross-hatched pattern in the via pad and possibly in the ground plane. These methods are shown in Figure 9.8.

Another way to reduce the potential for lifted lands is to assure rapid and uniform heat transfer to the solder joints. Each of the heat sources used in the common rework tools offer effective heat transfer. However, if the solder is oxidized, it resists the heat transfer. The application of a small amount of flux will reduce this resistance and aid uniform heat transfer. The choice of flux is a function of the user's cleaning policies, but choices should be limited to those ranging from no-clean to Type L (the previous R or RMA classifications). The no-clean flux remains on the board while others are removed using water, alcohol, or (as a last resort) stronger solvents.

The other mentioned problems of adjacent component and board damage are functions of the heat source and component spacing. If other components are very close to the package to be removed, then their joints may reflow when exposed to the hot air or IR that is being directed onto the package to be removed. Many tools offer nozzles and apertures that restrict the heat to a small area.

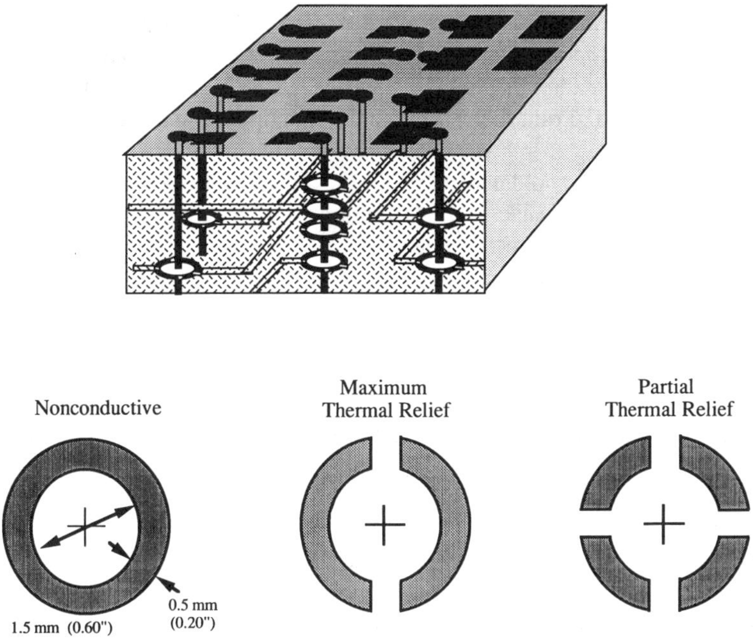

FIGURE 9.8. Side view of a pcb showing via thermal relief locations and examples of thermal relief techniques to prevent excessive heat sinking of certain leads during rework.

These restrictors should also minimize the heat to the package and maximize the heat to the leads. This precaution is taken to prevent damaging the fine pitch package when it is to be reused. This is the case when repositioning a misaligned package.

Board delamination is a possible problem due to excessive exposure to concentrated heat. Board delamination raises concerns of broken or intermittent trace and via connections. The best way to avoid board delamination is to select a tool that provides rapid heat transfer to a restricted area and to carefully teach the operators how to use the tool (even on the most automated station) and how to react when problems begin. In this author's opinion, the most effective tool for this rework step and all the others is to implement process controls and feedback monitors that reduce the possibility and amount of rework.

Package Replacement

To solder a new package onto the board it is a good practice to first prepare the board by removing the old solder and retinning the lands. Then place the new package using a workstation or manual tools. Using a workstation, it is possible for the leads to be simultaneously reflowed using the built-in heat source. If the operator uses hand tools, then it is best to use a very fine tip solder iron to tack down four or more leads while holding the package in place. After tacking, use thin solder wire (0.4 to 0.6 mm diameter) and the fine tip iron to solder the remaining leads.

As an alternative to solder wire, a thin line of solder paste can be syringed on each row of leads. The package is placed into the paste and reflowed using a hot air nozzle that directs hot air to the leads.

Land Replacement

As previously mentioned, lifted lands are a common problem. The consequences of lifted lands are minimized if the land is separated from any vias by a small trace. Not only does this provide thermal relief, but if the land is broken from the board, the via connections to the board's sublayers are protected since the land will break at the trace. The recommended land to via connection is shown in Figure 9.9.

If only a few lands need replacement, it is usually easiest to place and solder the new package on the remaining lands, and then to solder short jumper wires onto the top of the leads at the missing land sites. The other end of the jumper wire can be inserted and soldered in the via hole or onto a cleaned portion of the trace that connected the missing land. The jumper wire can be covered with an adhesive to provide extra protection and attachment strength. This method is illustrated in Figure 9.10. The jumper wire approach is more practical then attempting to add replacement lands.

If many lands are lifted, it may be most economical not to do any rework and to scrap the assembly instead.

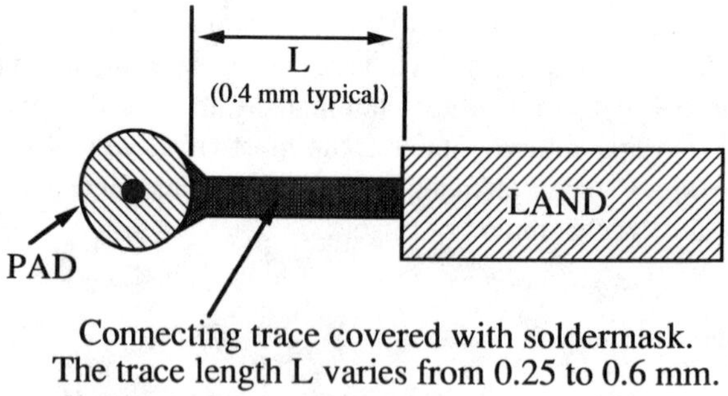

FIGURE 9.9. The recommended connection of a land to a via. The land and via should be separated so rework is easier if a land lifts during rework.

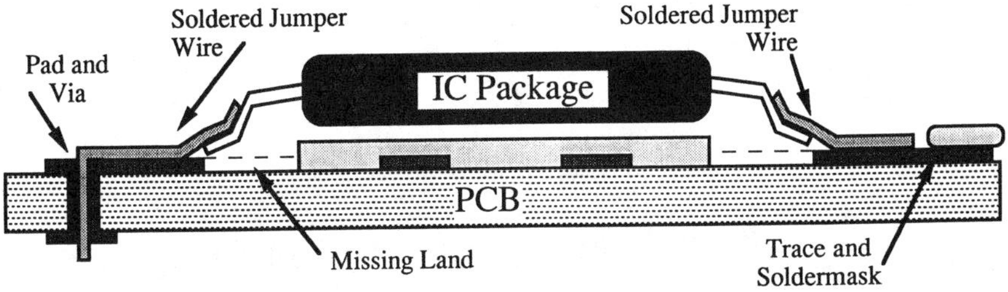

FIGURE 9.10. Missing lands can be replaced with jumper wires. Replacement of fine pitch lands is too tedious and costly.

References

1. Marcoux, P. 1991. Operator SMT Inspection, Rework and Repair Techniques. Sunnyvale, CA: PPM Associates and Dearborn, MI: Society of Manufacturing Engineers.

10

Design for Reliability Guidelines

10.0 SOLDER JOINT RELIABILITY

Solder joint reliability is a significant issue as the solder joint area becomes smaller. Achieving reliability over the expected service life in a product using fine pitch packages must be addressed at the design level. After all, the surface mounted fine pitch leads are the *only* means of mechanical support and electrical connection to the package.

As will be seen from the models in this section, designing for reliability is accomplished by:

1. Providing sufficient solder joint area
2. Selecting or designing a lead shape with high flexure
3. Keeping thermal differences low
4. Use the smallest package size possible
5. Minimize lead planarity differences

10.1 DEFINING SOLDER JOINT RELIABILITY

Solder joint reliability is usually thought of as "when will the joint break in the field". In fact, it is more. Solder joint reliability is "the ability to function as expected for an expected time, without exceeding expected failure levels."

Reliability in this sense is a different concept from quality. When a product lacks quality, it may experience low in-process yields, high infant mortality, and short term disasters. When a process lacks quality, it usually responds well to

TABLE 10.1 The Steps Necessary for Achieving and Maintaining Reliability and Quality for Electronic Products. (Source: Engelmaier Associates)

Achieving and Maintaining	
Reliability	**Quality**
• Proper Use Conditions	• Incoming Inspection
• Appropriate Service Life	• Process Verification
• Acceptable Failure Probabilities	• Statistical Process Control
• Knowledge of Material Properties	• Environmental Stress Screening
• Capable Processes	• Yield Improvement Programs
• Adequate Thermal Design	

immediate attention and fixes. And because of the high visibility that quality problems create, they usually demand immediate attention and repair.

When a product or process lacks reliability, this generally appears as long term problems and high warranty returns. These are usually discovered only after difficult failure diagnoses. Reliability problems usually start quietly and remain unnoticed for long periods. As a result, they do not receive immediate attention and solutions take a long time to implement. In some cases, the problems linger because the cause is blamed on an older process, and it is assumed that the current process is free of the reliability problem.

Achieving and maintaining reliability requires different steps than achieving and maintaining quality. This is illustrated in Table 10.1. As seen from this table, achieving reliability and quality requires a solid technical understanding of the materials and of process capabilities.

10.2 FAILURE PROBABILITIES

The integrated circuits inside the package have a finite probability of failure due to infant mortality. *Infant mortality* is a catch-all phrase for problems such as wire bond or circuit trace breakage, electrostatic discharge (ESD) failures, power surge, and other electrical failures. Since these problems occur early in the product's life they are referred to as *infant mortality* failures. The probability of these failures is displayed in reliability *bathtub* curves of electronic components with and without attachment to the board. The failures on the left of Figure 10.1 are those due to infant mortality. These failures are usually due to poor quality. The failures on the right are dominated by attachment problems, primarily solder joint wearout. Figure 10.1 shows the probability of solder joint wearout for leadless and leaded package attachments.

The leadless package is the leadless chip carrier, or LCC, a surface mount package still used in many high reliability products. Its only advantage is hermeticity, which is overshadowed in most applications by its disadvantages of higher solder failure rate and intrinsically high cost. The leaded failure rates are the combination of J and gull-wing leaded attachments.

The failure rates are very general and vary depending on the product's use environment and specific design parameters. The use environment, to be defined in greater detail in Section 10.5, is the thermal and mechanical environment the product will experience during normal usage. This includes the product's external environment such as the change in temperature from day to night. And it includes the change in temperature when the integrated circuit in the package is turned on and off.

The design parameters most impacting reliability, also to be described in more detail, include the package size, solder joint volume, lead flexure, and power on and off sequencing.

Figure 10.2 shows the cumulative failure probabilities. Failure rates are com-

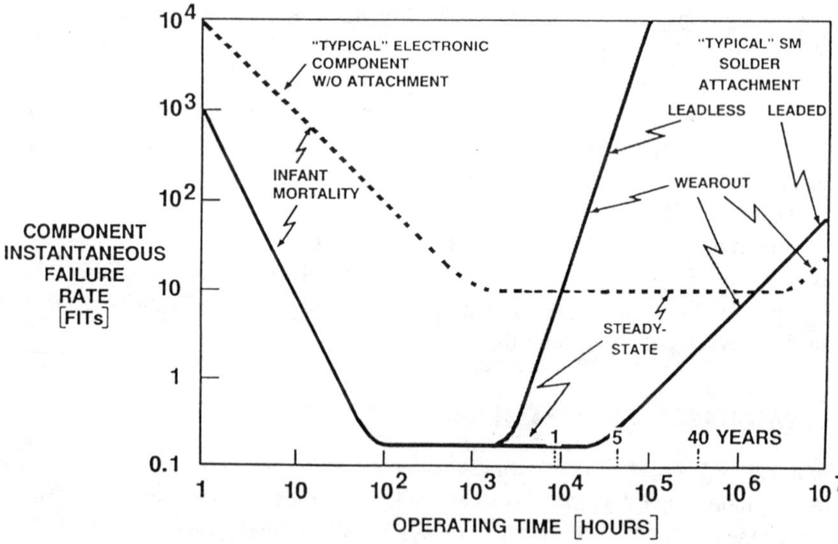

FIGURE 10.1. The reliability *Bath tub* curve, showing rapidly decreasing *infant mortality* failures in early product life and gradually increasing *wearout* failures in later life. (Figure courtesy of Engelmaier Associates)

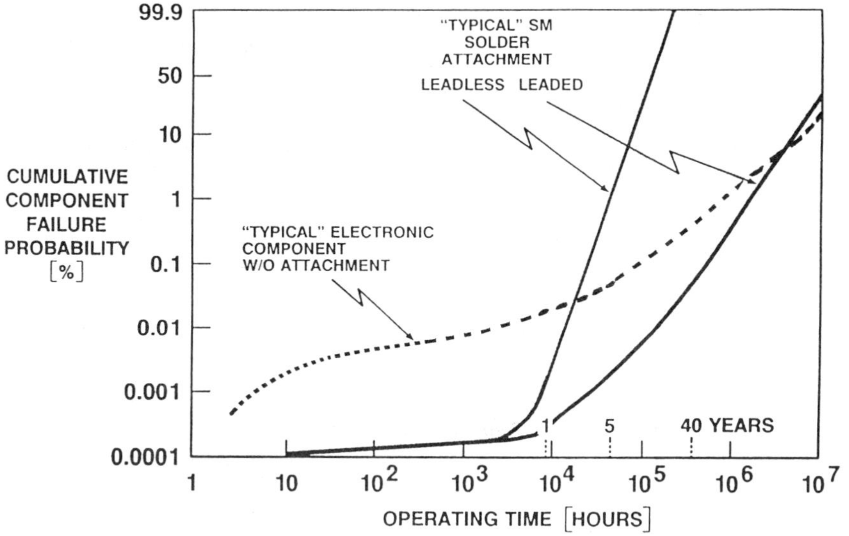

FIGURE 10.2. The possible magnitude of cumulative failure probabilities when the consequences of infant mortality and wearout are combined. (Figure courtesy of Engelmaier Associates)

monly expressed in failure in 10^9 device-hours. The conventional term for this is fits. In the early service of a product from 0 to 1000 hours the failures are component and quality related problems. These are due to previously mentioned causes. Beyond 1000 hours the failures are generally due to solder joint fatigue failures.

The curves in Figure 10.2 predict that failures in the first 1000 hours of a product are very low, usually less than 100 ppm. This is due to the traditionally high emphasis on quality measures such as extensive component and product testing, environmental stress screening, and quick reaction to quality problems.

Starting at approximately 5000 hours leadless solder joints in severe use environments begin to show a rapid rise in failures. The most severe use environments are under-the-hood automotive and avionics applications. Integrated circuit packages with flexible, compliant leads display a slower increase in failures starting at about 10,000 hours.

10.3 CAUSE OF SOLDER JOINT FATIGUE

Soldering is the joining of two metals using low melting temperature alloys that are soluble in the metals to be joined. Solder acts by wetting the base metal surfaces so as to fill the space between them completely. And solder forms a metallurgical bond to the wetted surfaces when it solidifies. The more time solder joints spend in liquid form, the thicker the intermetallic compound layers that will grow.

The formation of intermetallic compound layers, shown in Figures 10.3 and 10.4, which are solutions of the bonded surface metals and tin, is what is meant by solder wetting. Thus, intermetallic compounds are essential for the soldering process. However, excessive intermetallic compounds are brittle and deprive adjacent solder regions of tin. Thick, brittle intermetallic compound layers with equally thick lead-rich regions are detrimental for solder joint reliability. Figure 10.4 shows the typical acceptable amount of intermetallic area. And Figure 10.4 shows excessive area. In Figure 10.3 the lead rich regions show up as dark masses.

For electronics applications, solder needs to be electrically conductive and mechanically strong. Although tin-lead solder alloy is mechanically suitable for component attachment it must be understood that tin-lead solder alloy is a relatively soft metal. Tin-lead solder melts at 183°C and the shear strength of a typical joint ranges from 1700 to 7500 pounds per square inch. This depends on solder temperature and strain rate.

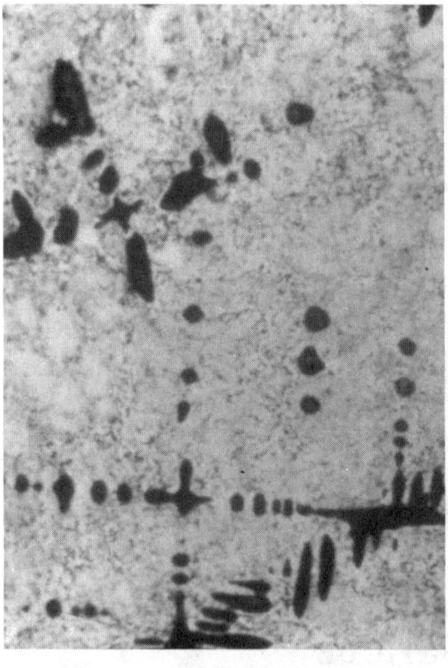

FIGURE 10.3. Photo of solder alloy displaying several lead-rich (dark areas) regions. These regions may be sites of fracture during severe stress. (Photo courtesy of Alpha Metals)

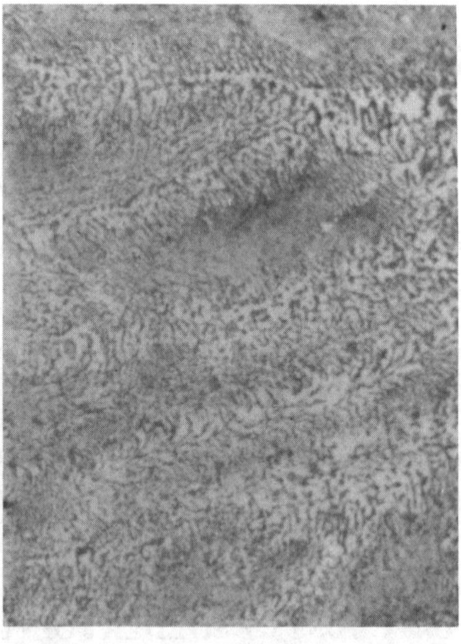

FIGURE 10.4. Photo of the desired appearance of a solder alloy, uniformly free of lead rich areas. (Photo courtesy of Alpha Metals)

Tin-lead is also very ductile. For example, it can be elongated to 75% of its length or more at room temperature. Ductility is an important property of the short life fatigue behavior of solder. The material's ductility is quite temperature dependent. As shown in Figure 10.5 the percentage of elongation increases expotentially as the temperature approaches the melting point.

After the solder solidifies, it is subject to stresses and strains. Cyclic stresses and strains are sources of fatigue. Fatigue damage accumulates with each fatigue cycle until the ability of the material to accommodate the fatigue damage is exhausted. If the fatigue loads are severe or if the fatigue loads occur for many cycles, than solder joint failure can occur prematurely.

For years the industry has debated the sources of solder fatigue and the way in which fatigue can be predicted and prevented. Studies over the past decade have shown that the greatest cause of solder joint failures is thermomechanical cyclic fatigue. The cyclic strains are generally caused by environmental temperature changes and/or power changes, such as turning the power to the device on and off.

When a PCB assembly, shown in Figure 10.6 as illustrated by two bonded metallic strips, undergoes temperature change it creates deformation. The amount of deformation is a function of the individual materials' coefficient of thermal

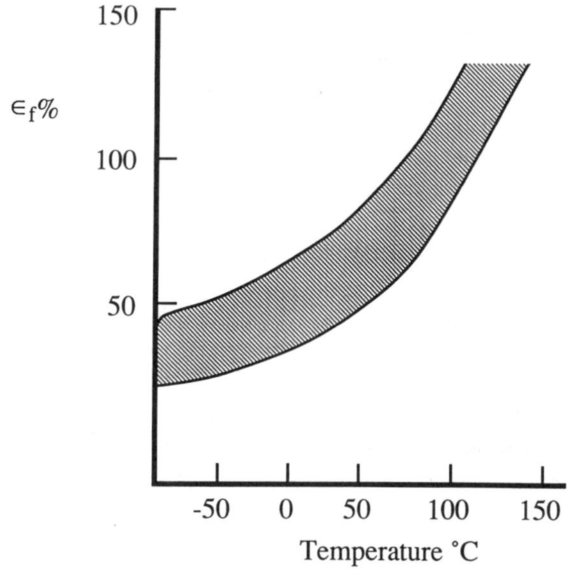

FIGURE 10.5. Solder is a highly ductile metal, it can be elongated up to 75% of its length at room temperature.

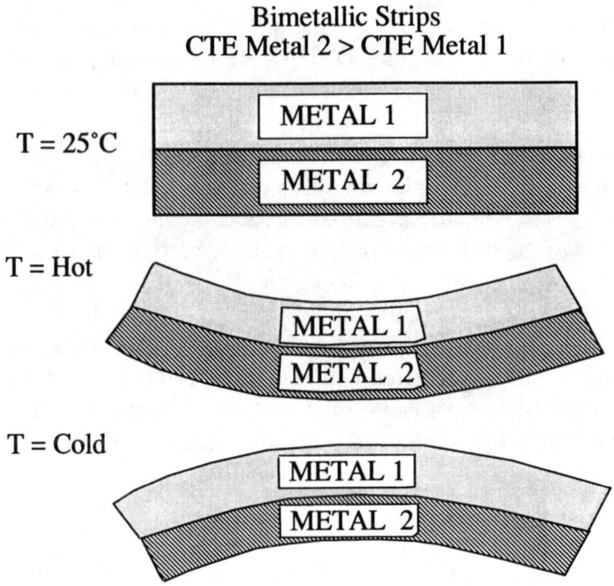

FIGURE 10.6. Two metal strips with different coefficients of thermal expansion. The CTE of metal 2 is greater than the CTE of metal 1.

expansion, (CTE). It is also determined by the geometry of the connected assembly.

At room temperature the bonded strips show no mismatch. If it is assumed that the lower strip has a higher coefficient of thermal expansion, then as the temperature is increased, the strips assume the appearance shown in Figure 10.6. The warpage is due to the differing CTEs.

If the view of CTE is expanded to examine a surface mounted solder joint, we see three elements that vary: the package, the solder, and the board. At room temperature there is generally no stress within the solder joint, although the lead stiffness and lack of planarity may introduce some tensile stress if the soldering is done with a heat bar.

When the temperature is raised, the different CTEs cause different thermal expansions. If the package illustrated in Figure 10.7 is made of ceramic, it has a low CTE of about 5 to 7 parts per million per degrees centigrade. If the board is fiberglass epoxy, such as FR-4, it has a CTE of about 14 to 18 ppm/°C. The solder has a CTE of about 25ppm/°C.

When the temperature is lowered the thermal expansion differences reverse and the assembly assumes the shape shown in Figure 10.5.

An equivalent source of thermal expansion is turning the power on and off to

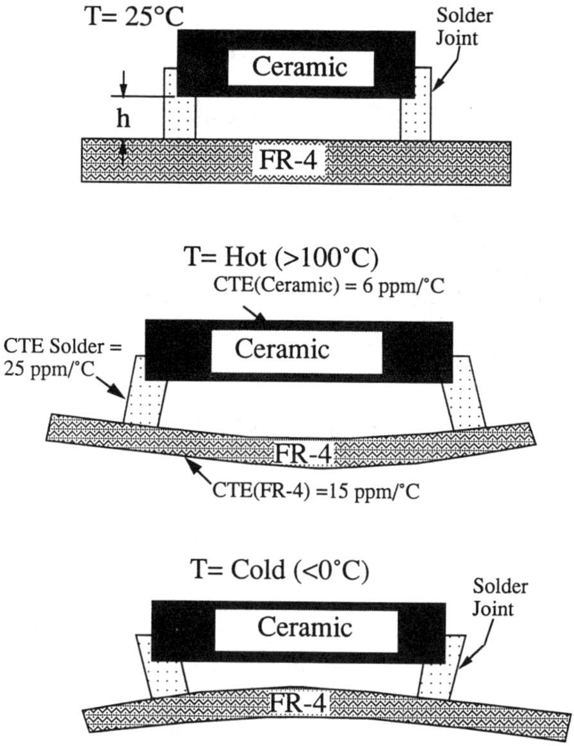

FIGURE 10.7. The effects of CTE mismatch on a solder joint at three temperature extremes. Because the ceramic package has a low CTE and the fiberglass a high CTE, the solder may experience an extreme amount of stress and possible fracture.

the integrated circuit. Turning the power to an IC on and off can introduce similar stresses due to the different temperatures of the package, the solder, and the board.

In the case of a rigidly attached ceramic package to an FR-4 board the solder joints must withstand a factor of 3 to 4 difference in CTE of the ceramic and the fiberglass. If the solder volume is insufficient and the solder itself is poorly formed then some or all of the joints will crack after a few or several thermal cycles.

The damage done during temperature induced cyclic fatigue can be demonstrated in the form of cyclic hysteresis loops in a stress-strain diagram. In Figure 10.8 is such a diagram for a typical engineering metal such as aluminum. During each fatigue cycle, the material undergoes the stress-strain loading described by the hysteresis loop. The hysteresis loop area is proportional to the visco-plastic strain energy. This area provides a measure of the fatigue damage.

Solder, however, does not behave like a typical engineering metal. It readily creeps and stress relaxes at normal use temperatures. And the higher the tempera-

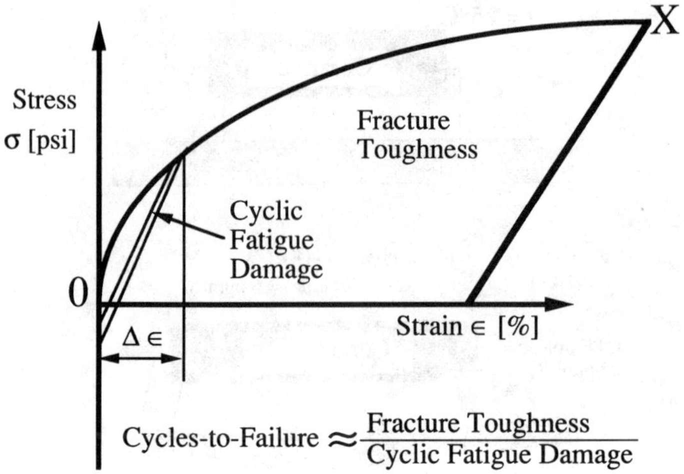

FIGURE 10.8. The cyclic hysteresis stress-strain diagram for a the metal, aluminum. (Source: Engelmaier Associates)

ture, the faster the stress relaxation. For example, at 80°C more than 90% of the stresses will have relaxed within 5 minutes. This is illustrated in Figure 10.9. Given this complex behavior of solder, the fatigue of solder joints is substantially different than for aluminum. Because of the stress relaxation, the cyclic fatigue hysteresis loop for solder is much larger for the same cyclically applied strains (see Figure 10.10). The solder joints will fail significantly earlier because of stress-relaxation induced fatigue damage, unless the applied strains are reduced by design practices that will be described later.

The response of the solder joint to cyclic displacement can characterized by hysteresis loops in a shear stress-strain diagram. The hysteresis loops shown in Figure 10.11 illustrate the difference between stiff leadless solder attachments and those with compliant leads. The large hysteresis loop is for a stiff leadless solder joint. It shows that twice during each cycle the solder joint is loaded to the maximum yield strength of solder followed by stress relaxation.

Compliant leads prevent the loading of the solder joint to the yield strength. So for the same strain range, shown as delta gamma, $\Delta\delta$, the hysteresis loop area is significantly smaller and the fatigue reliability is higher. If the leads are too stiff, then the leads do not reduce the solder joint stresses below the yield strength and little reliability benefit results from the leads. Compliant leads on fine pitch packages, especially the TAB styles, can become stiff when excessively soldered. The dashed lines in Figure 10.11 show the stresses during typical accelerated reliability tests. The solid lines show stresses that the typical product experiences in the field. Accelerated testing does not and can not allow the time for full stress

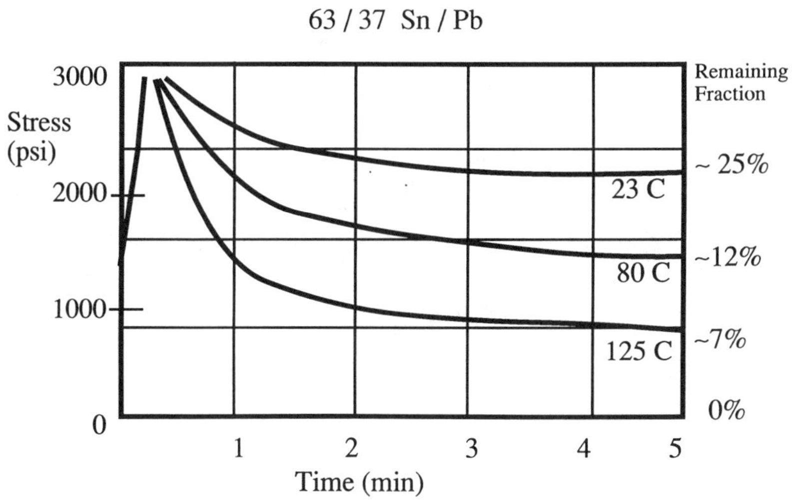

FIGURE 10.9. After exposure to stress, solder readily creeps and stress relaxes. The higher the solder temperature the faster the stress relaxation. (Source: Engelmaier Associates)

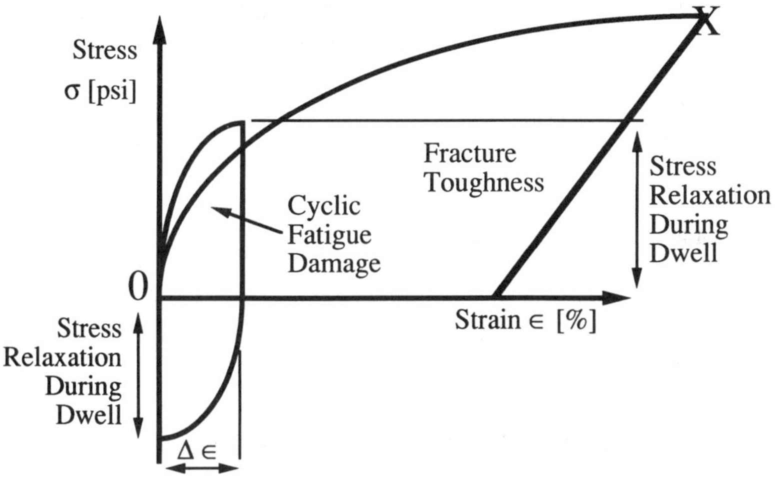

FIGURE 10.10. The cyclic hysteresis stress-strain diagram for SN63 tin-lead solder. (Source: Engelmaier Associates)

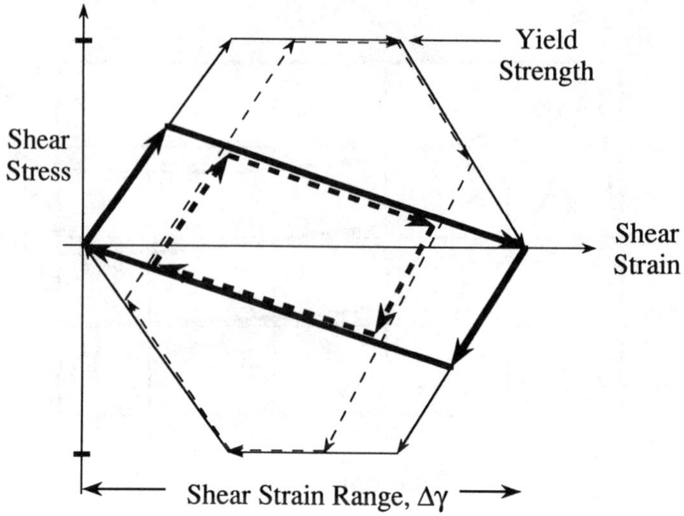

FIGURE 10.11. The stress-strain hysteresis loops for stiff leadless solder joints (the large outer loop) and compliant leaded joints (the small solder line inner loop). The dashed loop are the stress-strain magnitudes that occur during accelerated reliability testing. Note that accelerated testing (the dashed loops) don't load the joint with the same amount of stress and strain as normal usage (the solid loops). (Source: Engelmaier Associates)

relaxation. This is due to time constraints. Therefore, the fatigue damage per cycle in accelerated testing is less than the damage per cycle in the field. The number of cycles to failure in the field will, as a consequence, be less than the number of test cycles. The phrase *acceleration test* comes from the fact that many more cycles per unit time are possible for the accelerated tests.

Accelerated test conditions that are significantly different from actual operating conditions should be avoided. If more stringent conditions are used, there is a danger that a different damage mechanism will result. This mechanism may have no correlation with actual field reliability and will possibly create the impression that a reliable product is bad. Thermally shocking the joints with rapid temperature changes of more than 30 degrees centigrade per minute introduces cyclic transient warpages. These can cause tension fatigue and tension overstressing. Tension is much more damaging than shear stress and represents a damage mechanism different than the field operating mechanism on the product.

10.4 SOLDER JOINT PREDICTION MODELS

Several studies, such as those illustrated in Figures 10.12 through 10.17, have measured the solder joint failure rates as a function the variables listed for each figure. These variables include lead material, lead foot length, and lead shape.

From these studies, models have emerged that predict the number of cycles to failure of solder joints. The models, Equations (10.1) and (10.2), consist of two parts. The first part gives the mean cyclic life, and the second is the statistical failure distribution. The closeness of these prediction models is shown in Figure 10.18.

For Leadless 60/40 Sn/Pb Solder Attach

$$N_f(x\%) = \frac{1}{2}\left[\frac{F\, L_D \Delta\alpha\, \Delta T_e}{0.65 \quad h}\right]^{\frac{1}{c}} \left[\frac{\text{In}\,(1 - 0.01x)}{\text{In}\,(0.5)}\right]^{\frac{1}{\beta}} \tag{10.1}$$

For Leaded 60/40 Sn/Pb Solder Attach

$$N_f(x\%) = \frac{1}{2}\left[\frac{F\, K_D L_D \Delta\alpha\, \Delta T_e}{0.65 * 200 * Ah}\right]^{\frac{1}{c}} \left[\frac{\text{In}\,(1 - 0.01x)}{\text{In}\,(0.5)}\right]^{\frac{1}{\beta}} \tag{10.2}$$

IMPORTANT DESIGN FOR RELIABILITY PARAMETERS

A	=	Effective solder joint area ≡ 2/3 solder-wetted lead area projected to solder pad,
F	=	Design-specific input; F (leadless column) ≈ 1.5 to 1.0, F (leadless w/fillet) ≈ 1.2 to 0.7, F (leaded) ≈ 1
h	=	Solder Joint height, for leaded ≡ 1/2 solder paste stencil depth,
K_D	=	Diagonal flexural lead stiffness,
L_D	=	1/2 Maximum distance between component solder joints,
$N_f(x\%)$	=	Product service life in cycles of operation to x% failure probability
T_{SJ}	=	Mean cyclic solder joint temperature = 1/4 $[T_S + T_C - 2T_o]$
x	=	Cumulative failure probability after N cycles,
ΔT_e	=	Equivalent cyclic temperature range in powered operation = $\left\lvert \frac{\alpha_S \Delta T_S - \alpha_C \Delta T_c}{\Delta\alpha} \right\rvert$
$\Delta\alpha$	=	CTE - Mismatch, the difference in the substrate CTE, α_S, from the component CTE, α_c
β	=	Weibull probability slope _ 4 for leadless, _ 2 for leaded
C	=	$-0.442 - 6 \times 10^{-4}\, T_{SJ} + 1.74 \times 10^{-2}\, \text{In}\,(1 + \frac{360}{t_D})$
t_D	=	The half cycle dwell time (in minutes), average time available for stress relaxation at T_S, T_C and T_o.
T_S, T_C	=	Steady state operating temperature for the substrate (s) and component (c)
T_o	=	The temperature during off-half-cycle.

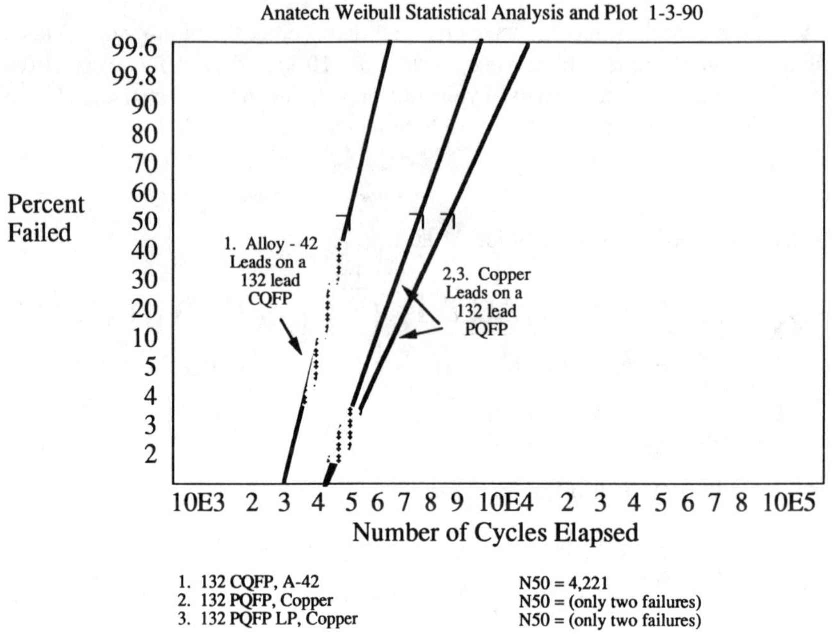

FIGURE 10.12. The failure rates of Alloy 42 leads and copper leads soldered with SN63 solder on FR-4 boards. The induced stress was a thermal cycle from −55°C to +125°C of a 30 min dwell, 30 min change sequence. (Source: Motorola)

Equation (10.1) predicts the number of cycles, $N_f(x\%)$, to a confidence level of x%, before a solder joint fails on a leadless solder attachment, such as on the LCC package. Equation (10.2) predicts the number of cycles to failure for a leaded solder attach such as a gull lead.

The first part of the models, which describes the cyclic fatigue damage, includes all the design parameters of primary importance to reliability. These are the component size, L_D; the difference in coefficients of thermal expansion between component and substrate $\Delta\alpha$; the cyclic thermal conditions, ΔT_e and c; and the solder joint height, h. For compliant leads, additionally included are the diagonal lead stiffness, K_D; and the solder joint area, A. For leaded attachments, the product of solder joint height and area represents a measure of solder volume.

The ductility exponent, c varies with temperature, T_{sj}, and T_s, T_c, T_o, and the time available for stress relaxation during the thermal cycles, t_D. The parameter F, represents the combined effects of second-order importance for reliability. Some of these effects can make sizeable contributions. These include the local expansion mismatch between the solder and the substrate materials,

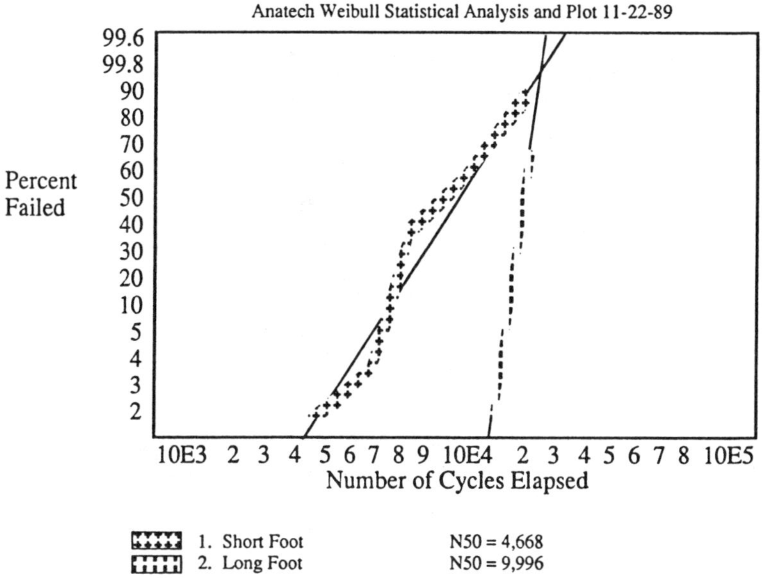

FIGURE 10.13. The failure rates of long foot lengths versus short foot lengths of 44 lead QFP packages that were soldered using SN63 solder to FR-4 boards. (Source: Motorola)

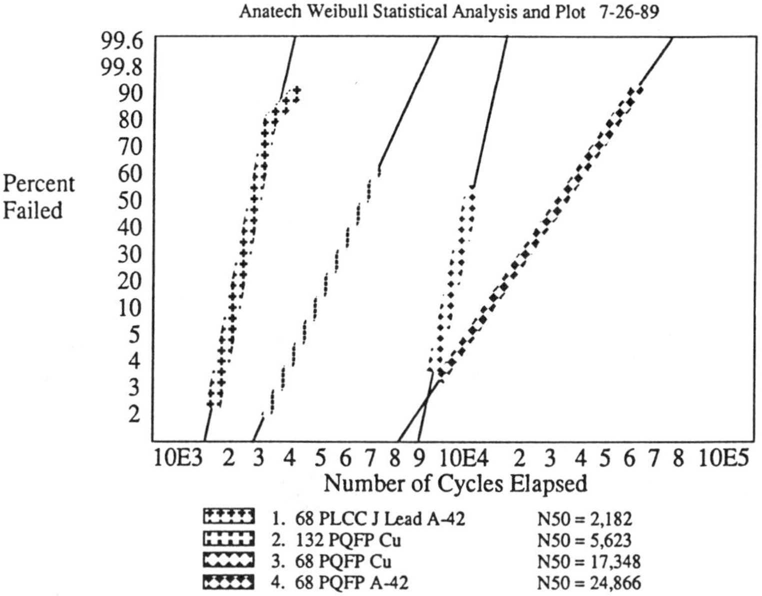

FIGURE 10.14. The measured failure rates of solder joints using a 68 lead PCC "J" lead package vs "gull" lead PQFP packages. The leads are copper and alloy 42 metal. (Source: Motorola)

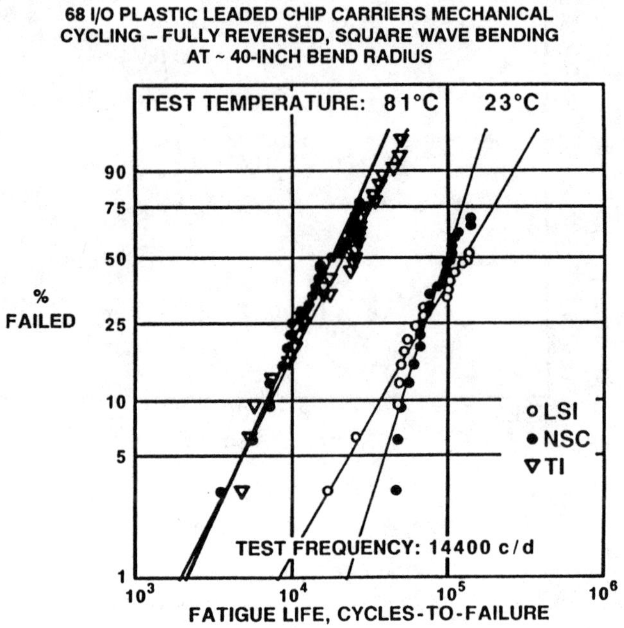

FIGURE 10.15. The failure rates of solder joints as measured by an IEEE sponsored study. Several different IC suppliers participated, supplying similar 68 pin "J" lead PCC packages. (Source: Engelmaier Associates)

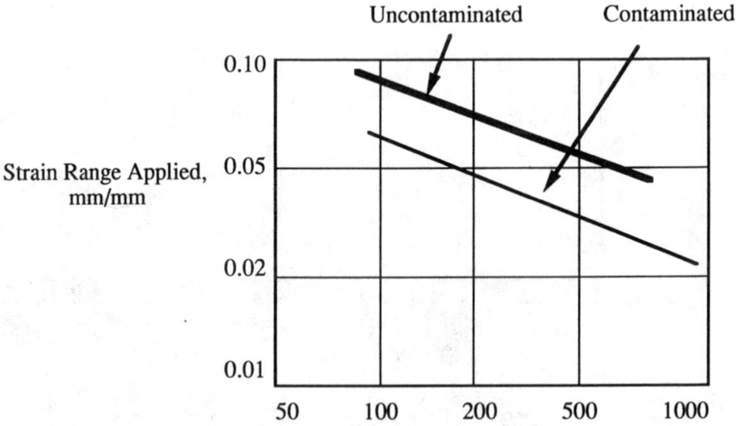

FIGURE 10.16. This figure shows the reduced strength and earlier failure probabilities contaminated vs uncontaminated solder joints. (Source: Screening Systems, Inc.)

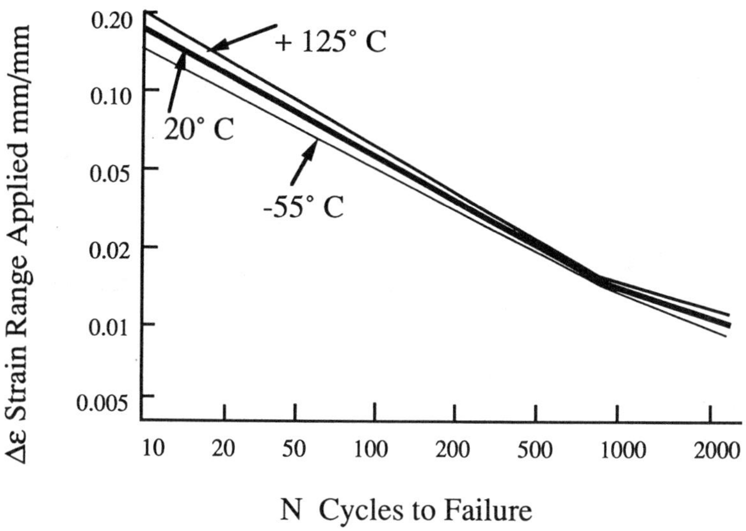

FIGURE 10.17. The predicted cycles to failure of a solder joint as a function of applied strain and joint temperature. (Source: J. Evans, NASA)

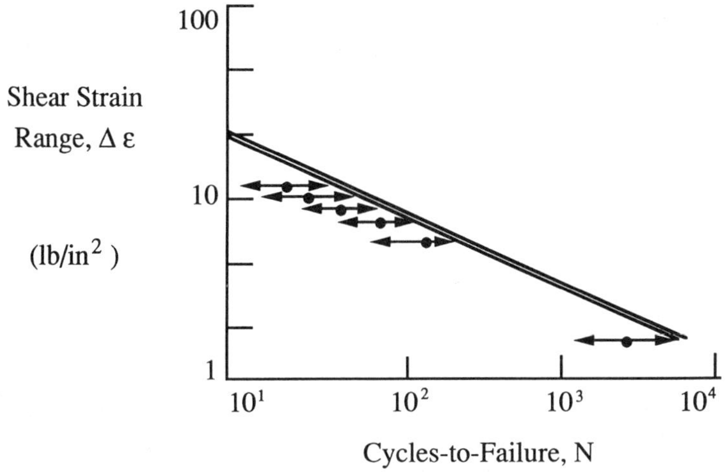

FIGURE 10.18. The calculated results from equations 10.1 and 10.2 yield a close approximation of the number of thermal cycles a solder joint will endure before failure. These equations correlate well to the measured test failures that were shown in the prior figures. (Source: Engelmaier Associates)

the nonuniform strain and stress distribution in a solder joint due to the joint geometry, and transient effects.

The ductility exponent, c, represents a measure of the completeness of the solder stress relaxation process during each cycle. Depending on the parameters influencing the ductility, the exponent can be in the range of about -0.4 to -0.5.

Therefore, the solder joint reliability depends inversely on the component size, L_D, the thermal expansion coefficient mismatch, $\Delta\alpha$, and the cyclic thermal excursion, ΔT_e.

This dependence is to approximately the power of 2 for leadless solder attachments, and to the power of 4 for compliant leaded attachments.

These equations are very useful for calculating the minimum solder volume (height and area) when a user knows the minimum number cycles and to what confidence level a solder joint must meet. This is found by transposing Equation (10.2) to solve for the solder joint volume term, Ah:

$$Ah = \frac{F\,Kd\,LD\,\Delta\alpha\,\Delta T_e}{(0.65 \times 200\,(2\,Nf\,(x\%))^c \left[\dfrac{\ln(1 - 0.01x)}{\ln(0.5)}\right]^{\frac{c}{\beta}}} \qquad (10.3)$$

These equations are also useful for predicting the probability of joint failure on misaligned, dewetted, and void containing solder joints. The predicted results should be checked with actual studies to verify accuracy.

Several field conditions impact solder joint reliability. External temperature changes can create cyclic steadystate expansion mismatches and warpage. Internal temperature changes can also create expansion mismatches and warpage. All of the expansion mismatches cause shear strains.

10.5 ELECTRONIC PRODUCT USE CONDITIONS

Use categories have been created for electronic applications according to the severity of the use environments. The environmental information shown in Table 10.2 represents realistic, worst-case conditions. Some of these environments are represented by more than one condition depending on time of year and geographic location.

It should be noted that the worst-case temperature swing, ΔT, is not always the difference of T_{min} and T_{max} since they may occur during different times of the year

TABLE 10.2 Electronic Products are Used in a Wide Variety of Environments. For Convenience, Nine use Environments have been Defined. This Table Lists the Realistic Worst-case Thermal Cycle Conditions Joints will Experience in the Specified Application environment. (Source: IPC)

USE CATEGORY	WORST USE ENVIRONMENT					Service Life, SL (years)	Acceptable Failure Probability % @ SL
	Tmin (°C)	*Tmax* (°C)	ΔT (1) (°C)	t_D (hrs)	Cycles/ year		
1 CONSUMER	0	+60	35	12	365	1-3	~1
2 COMPUTERS	+15	+60	20	2	1460	~5	~0.1
3 TELECOMM	-40	+85	35	12	365	7-20	~0.01
4 COMMERCIAL AIRCRAFT	-55	+95	20	2	3000	~10	~0.001
5 INDUSTRIAL & AUTOMOTIVE -PASSENGER COMPARTMENT	-55	+95	20 & 40 & 60 & 80	12 12 12 12	185 100 60 20	~10	~0.1
6 MILITARY GROUND & SHIP	-55	+95	40 & 60	12 12	100 265	~5	~0.1
7 SPACE leo geo	-40	+85	35	1 12	8760 365	5-20	~0.0001
8 MILITARY AVIONICS a b c	-55	+95	40 60 80	2 2 2	500 500 500	~5	~0.01
			& 20	1	1000		
9 AUTOMOTIVE -UNDER HOOD	-55	+125	60 & 100 & 140	1 1 2	1000 300 40	~5	~0.1

(1) ΔT represents the maximum temperature swing, but does not include power dissipation effects; for power dissipation calculate ΔTe; power dissipation can make pure temperature cycling accelerated testing significantly inaccurate.
Source: Environmental Stress Screening - Solder Joint and Plated-Through-Hole Reliability Impact by Werner Engelmaier

or at different geographic locations. The cyclic dwell times, t_D as well as the number of cycles per year, depend on the source of temperature variations such as day to night or power on and off cycles.

From Table 10.2 one obtains some of the data necessary to calculate the probable numbers of cycles to fatigue failure. Additional data are design dependent. Design dependent data include substrate material, lead stiffness, product operating temperature, package size, and power off and power on joint temperatures. Equation (10.4) is used to calculate the minimum number of failure-free test cycles, N(t). In Equation (10.4) the term F(N) is the cumulative failure probability, and N is the number of individually monitored board assembled components on test.

$$N(t) = \overline{N}(t) \left\{ \left[\frac{\ln(0.5)}{\ln[1 - F(N)]} \right]^{\frac{c(u)}{c(t)}} \left[\frac{\ln(1 - 1/n)}{\ln(0.5)} \right] \right\}^{\frac{1}{\beta}} \tag{10.4}$$

N(t) is found from Equation (10.5), which results from the accelerated tests conditions shown in Table 10.3.

$$N(t) = \frac{1}{2} \left[(2N)^{c(u)} \frac{\Delta W(t)}{\Delta W(u)} \right]^{\frac{1}{c(t)}} \tag{10.5}$$

$$\Delta W \propto \frac{K_D (L_D \Delta\alpha\, \Delta T_e)}{Ah} \tag{10.6}$$

N(t) is the product of service life and the expected of annual cycles.

Shown, as an example, in Table 10.3 are the minimum number of failure-free test cycles required to meet the expected failure risk per component stated in Table 10.2. Table 10.3 assumes a board containing 32 components. In this table 8700 cycles is approximately 1 year of test time.

10.6 GENERAL DESIGN FOR RELIABILITY GUIDELINES

Based on the fatigue prediction models, several general design guidelines can be stated that a designer should follow to optimize the design for reliability of a product.

TABLE 10.3 This Table Lists the Minimum Number of Failure-free Test Cycles a Solder Joint is Expected to Survive in Order to Satisy the Reliability Expectations of the Product for that Use. This Table is Calculated for an Assembly Containing 32 Components. In this Table 8700 Cycles Is Approximately One Year of Test Time. (Source: IPC)

USE CATEGORY	ATTACHMENT TYPE	SERVICE LIFE: 1 Year			3 Years			5 Years		
		Failure-Free Test Cycles, $N(t)$, Required(5) with 32 Components on Test to Assure a Cumulative Failure Probability, x in %, at End of Service Life								
		0.0001%	0.01%	1%	0.0001%	0.01%	1%	0.0001%	0.01%	1%
1	Leadless	1600	470	140	5200	1500	440	9000	2600	760
	Leaded(6)	4500	390	33	14,700	1300	107	25,400	2200	184
2	Leadless	1400	420	127	4400	1300	400	7500	2300	680
	Leaded(6)	950	86	25	3000	270	25	5100	460	42
3	Leadless	1800	530	150	6000	1700	500	10,500	3000	870
	Leaded(6)	5200	430	36	17,200	1400	118	29,900	2600	210
4	Leadless	1800	520	154	5700	1700	500	9800	2900	850
	Leaded(6)	540	47	25	1700	150	25	3000	290	25
5	Leadless	1200	340	100	3900	1100	320	6800	2000	560
	Leaded(6)	6500	540	44	21,200	1800	146	36,900	2800	250
6	Leadless	3200	890	204	10,600	3000	690	18,600	5200	1200
	Leaded(6)	16,000	1300	100	53,900	4300	330	94,700	7500	590
7	Leadless	15,600	4700	1420	48,800	14,800	4500	82,800	25,100	7600
leo	Leaded(6)	19,000	1740	159	59,500	5400	500	101,000	9200	840
7	Leadless	1090	310	86	3700	1030	290	6400	1800	510
geo	Leaded(6)	1650	130	25	5500	440	34	9700	770	60
8	Leadless	2600	740	210	8700	2500	700	15,200	4300	1210
ΔT=20&40	Leaded(6)	4700	370	28	15,900	1250	99	27,900	2200	173
8	Leadless	5000	1430	410	16,200	4700	1350	28,100	8100	2300
ΔT=20&60	Leaded(6)	23,500	1950	160	77,100	6400	530	134,000	11,100	920
8	Leadless	7900	2300	680	25,300	7400	2200	43,500	12,700	3800
ΔT=20&80	Leaded(6)	71,200	6200	530	229,000	19,800	1700	394,000	34,200	3000
9	Leadless	13,600	4100	1200	42,600	12,900	3900	72,400	22,200	6600
	Leaded(6)	177,000	16,200	1500	553,000	50,500	4600	940,000	90,500	7800

8700 cycles ≈ 1 year test time

(5) The required number of test cycles is obtained for an acceptable cumulative failure probability, x, at the end of the service life and a number of components, n, on test from Eq.24.

(6) These numbers apply to compliant leads (diagonal lead stiffness, K_D < 100 lb/in); stiffer leads will move the required test cycles towards the leadless cycle numbers.

- Component packages should be as small as possible. Component size is one of the prime design parameters influencing reliability. This is expressed as the parameter L_D in the prediction model. Smaller component packages can be obtained by using the smallest possible terminations and using fine termination pitches.
- Design the land area and paste application method so they consistently provide adequate solder volume. The land area is the topic of Chapter 13 and paste application, Chapter 5.

TABLE 10.4 The Recommended Accelerated Test Parameters for Testing Assemblies for Solder Joint Failure. (Source: Surface Mount Council)

TEST	TEMPERATURE	FREQUENCY	WARPAGE	ACCELERATION FACTORS*	COMMENTS
FUNCTIONAL CYCLING	Realistic	< 100 cycles/day; defined dwells; $\dot{T}$ <30°C/min	Realistic	~ 10 to 20	Realistic power dissipation and heat transfer
TEMPERATURE CYCLING	0 ⇔ 100°C 25 ⇔ 100°C -40 ⇔ 0°C	≤ 24 cycles/day; defined dwells; $\dot{T}$ < 30°C/min	Somewhat higher than realistic	< 1 to ~ 10; ≤ ~ 30 with deliberate test vehicle CTE-mismatch	Very long test durations for ΔCTE < 5 ppm/°C
MECHANICAL CYCLING	Isothermal: 0°C; room temperature; 60°C; 80°C	< 100,000 cycles/day possible; ~ 1000 cycles/day realistic;	Bend radius ± 40 to 80 inches (± 1 to 2 m)	≤ 500	Excludes all CTE-mismatch effects; beware of over-stressing; somewhat higher tensile stresses; not directly useful for reliability predictions

* Acceleration factors based on mean-time-to-failure for ΔT (*test*) ≈ ΔT (*use*).

- Use leaded terminations with flexible lead shapes. Excellent choices are the gull-wing and J lead made of copper. These are expressed by the parameter K_D.
- If large leadless chip carriers must be used instead of leaded packages, then choose a substrate material that closely matches the low CTE of the ceramic package. Materials of choice for this may be aramid fiber on quartz reinforcement, low expansion metal layers, and ceramic substrates. This factor is expressed as $\Delta\alpha$.
- Design the assembly to operate at as close to normal room temperature as possible. And consider designing large critical circuits to remain powered-on. These temperature effects are expressed in the ΔT_e and the average of T_s factors.
- Implement a placement and reflow process that is capable of the least misalignment and solder void amounts. These parameters affect the solder joint area, A.

Reliable Lead Size

The solder joint fatigue model indicates that the lead flexure, K_D, package size, L_D, and the solder joint area, A, are important factors in solder joint reliability. The importance of these is shown in the results of a study conducted by McShane et. al., at Motorola (Austin, TX). In their study, they evaluated various lead materials, shapes and sizes for the PLCC (plastic leaded chip carrier), the PQFP (plastic quad flat package), and the CQFP (ceramic quad flat package).

The results of the lead material study indicates that both copper and alloy-42 materials provide reliable joints. This set of results is shown in Figure 10.14. This set of plots shows the percentage of packages displaying solder joint failures after several cycles of exposure to temperature that varied from −40 to +125°C during a one hour cycle period.

The J leaded PLCC package with leads made of alloy-42 is the stiffest configuration of the packages studied. Predictably it failed at the earliest number of temperature cycles. As shown in Figure 10.14, the PLCC package set started to experience solder joint failures at 1500 cycles. 50% of the packages failed at 2182 cycles (N50-2182 in Figure 10.12). The other packages are 68 and 132 lead PQFP packages with gull-wing lead shapes. The 132 lead PQFP package with a copper lead frame material and gull-wing lead shape is able to withstand over twice as many cycles before failure (3000 vs. 1500, and N50-5623 for the PQFP vs. N50-2182 for the PLCC). This increased reliability can be attributed to the difference in the lead joint area, A, between the gull-wing and the J-lead. The gull-wing solder area is approximately 2 to 4 times larger than the J lead solder area.

The 68 lead PQFP packages exhibit much higher reliability than either the PLCC or 132 lead PQFP. The higher reliability of the 68 lead PQFP over the 68 lead PLCC is attributed to the higher solder joint area, A, and lower lead flexure stiffness, K_D,, of the 68 lead PQFP package. The higher reliability of the 68 lead PQFP over the 132 lead PQFP is attributed to the smaller package, diagonal length (smaller sizes) L_D, of the 68 lead PQFP compared to the 132 lead PQFP. The copper and alloy-42 lead materials both provided excellent reliability results. This is displayed in Figure 10.12. The alloy-42 and copper material show solder joint failures starting at about 3500 cycles. The traditional minimum acceptable cycles before failure for high reliability applications is 1000 cycles.

As predicted by the fatigue model, larger solder joint area will reduce the probability of solder joint failure. One way to increase solder joint area is to increase the length of the foot of the lead. The results in Figure 10.13 support this and show that a long foot length begins to show failure at 9000 cycles versus 4000 for the short length. The long length is 3mm and the short length is 2mm.

References

1. Balde, J. W. 1987. The IEEE Compliant Lead Task Force. IEEE Transactions CHMT-10 (3) pp. 463-469.
2. Engelmaier, W. 1991. Designing for SMT Solder Joint Reliability. Dearborn, MI: Society of Manufacturing Engineers.
3. Engelmaier, W. 1991. Private correspondence.
4. Evans, J. and Engelmaier, W. 1990. SMT Reliability for Space Flight Applications. In *Surface Mount Technology*, Nov., 1990.
5. Mandel, C. E. N. 1989. Environmental Stress Screening. In Electronic Materials Handbook, Vol. 1. Materials. Park, OH: ASM Intrnational.

6. Marcoux, P. 1987. Implementation and Control Techniques for SMT. San Jose, CA: International Quality Technologies.
7. Marcoux, P. et. al 1991. Status of the Technology and Action Plan. Lincolnwood, IL: Surface Mount Council.
8. McShane, M. et. al. 1990. Lead Configuration and Performance for Fine Pitch SMT Reliability. Proceedings of Nepcon West '90, Feb. 1990, Anaheim, CA.

11

Design for Testability

11.0 BASICS OF ELECTRICAL TESTING PCAS

Electrical testing is the traditional way to verify the proper functioning and parametric conformance of products for the past two decades. The systems utilized, such as in-circuit or ICT and functional testers, provide a way to power on the assembly and to indicate if it is working. Depending on the tester's capabilities it may be able to offer a diagnosis of what's wrong if the assembly does not work.

Electrical testing of an assembly or an individual integrated circuit within an assembly is a straightforward exercise, provided the necessary connections can be made easily.

Making the necessary connections to an assembly for functional testing may be no more complicated than plugging the board into a socket, or soldering half a dozen wires to it (see Figure 11.1). On the other hand, if no socket connections are part of the board's design and the board is a subassembly to a larger system, then the necessary connections may be difficult to achieve.

The minimum connections to the board or the IC are power, ground, input signals, and output ports. Other necessary connections may include feedback loops, clocks, and data controls.

11.1 CONTACTING THE ASSEMBLY

The traditional method to make these connections on a board assembly is to mate the board with an array of probes. The probes are secured in the appropriate positions in a fixture called a bed-of-nails. A typical bed-of-nails fixture is shown in Figure 11.2. For through-hole and low density SMT assemblies a single side bed of nails is sufficient to provide all of the contacts. For very dense assemblies a double side contact may be needed.

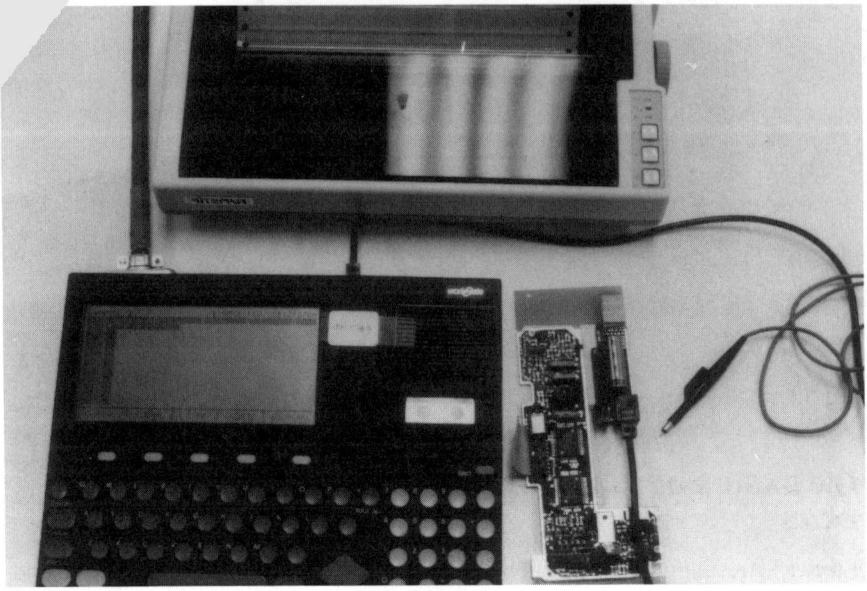

FIGURE 11.1. Photo of an assembly connected for functional testing.

The probes or nails are spring loaded spears of a strong metal such as alloy-42 that is gold plated. The metal strength is needed to prevent the probe from bending and loosing position when the board is pressed to the probe bed. The gold plating is to provide excellent electrical conductivity and corrosion resistance. Maintaining the probe strength is not a problem if the probes are large as shown on the left hand side in Figure 11.3. This large probe requires a 1.0 mm hole and 2.5 mm spacing; it is able to reliably contact a 1.0 mm test pad for at least 10,000 connections. The probe on the right needs a 0.5 mm hole and 1.3 mm spacing. It is not as strong as the other probe and is considered reliable for only 2500 connections to a 0.5 mm pad. Smaller probes are not considered practical unless they are kept very short and replaced often.

To use an extensive bed of nails approach for fine pitch assemblies is difficult and expensive. To provide room for the probes, the design must include a fanout of test pads that are spaced far enough to accommodate the necessary size probes. The smaller size probes may be used if the assembly is designed with components on one side only and the test pads on the other. This allows the probes to be shorter, thus minimizing bending and lost contacts.

If the assembly has components on both sides, or if it needs a double sided bed of nails, then the probes must be the larger diameter to provide adequate strength for the longer probe length. This larger probe needs large test pads and spacings and greatly increases the layout area. A suggested test probe pattern and test probe pad size is shown in Figure 11.4.

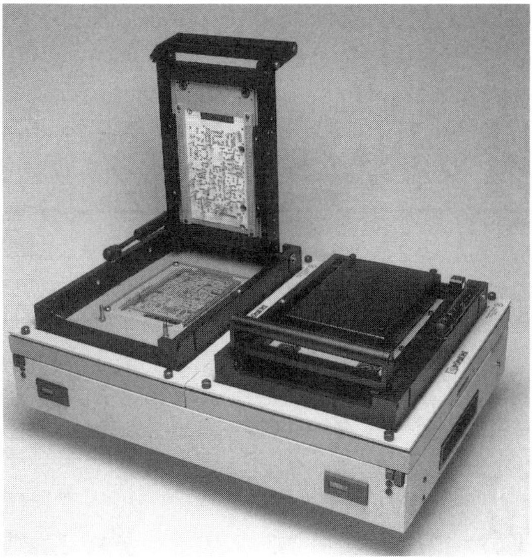

FIGURE 11.2. Photo of a bed-of-nails test fixture, as used for in-circuit testing. (Photo courtesy of In-Circuit Test/ Contact Products, Inc.)

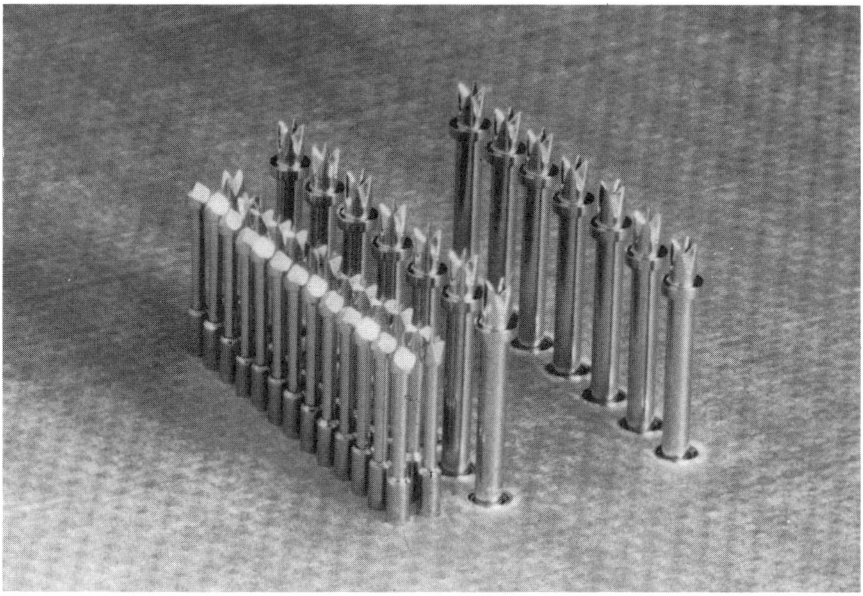

FIGURE 11.3. A photo of various test probes, known as pogo-pins because of the spring mechanisms. The probes on the right require 2.5mm spacing and the probes on the left require 1.3mm spacing. (Photo courtesy of Everett/Charles, Contact Products, Inc.)

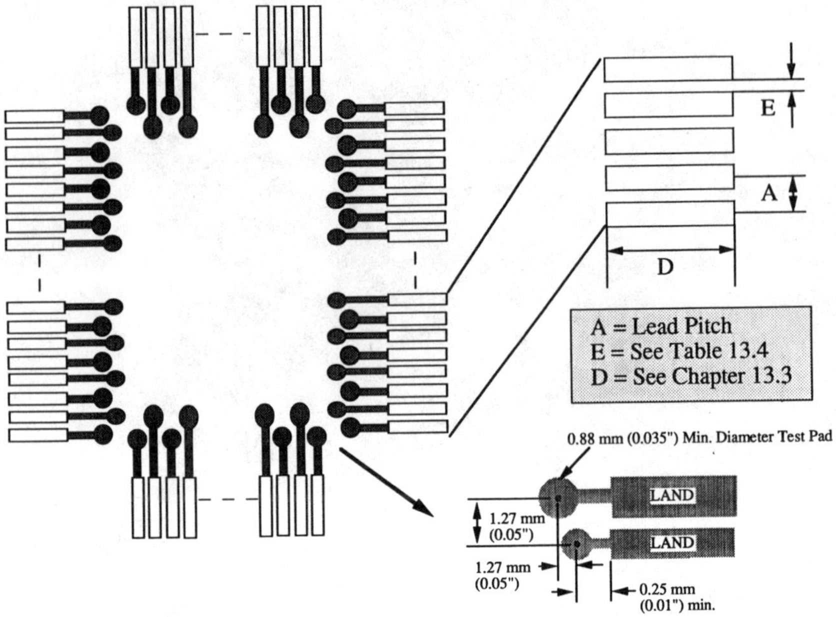

FIGURE 11.4. Drawing of the suggested test fan out pattern. (Source: Hewlett Packard)

Sometimes the pattern area can be reduced by using a plated via hole as a test pad and via, and by providing test pads for the absolute minimum number of package leads. To assure good contact to the via hole a variety of probe points is available, as shown in Figure 11.5. These improve the contact for vias that have uneven surfaces or that are not filled with solder.

11.2 DISADVANTAGES OF IN-CIRCUIT TESTING

The benefit of using a bed of nails is the ability to contact a large number of points within the assembly, to test individual components in the circuit, and to provide a diagnosis when a component does not work. The contacts allow application of power and input forcing functions such as clock pulses, input sequences, and feedback. They also allow the monitoring of the outputs of the components from the beginning of the assembly to the end. The high degree of connectivity enables a system to diagnose a failure by pinpointing the component whose output response falls outside the test system's programmed allowable limits. The ability to diagnose a problem component with this in-circuit test approach takes a few seconds or minutes, compared to several minutes or hours when functional test failures are manually probed by a technician at a bench.

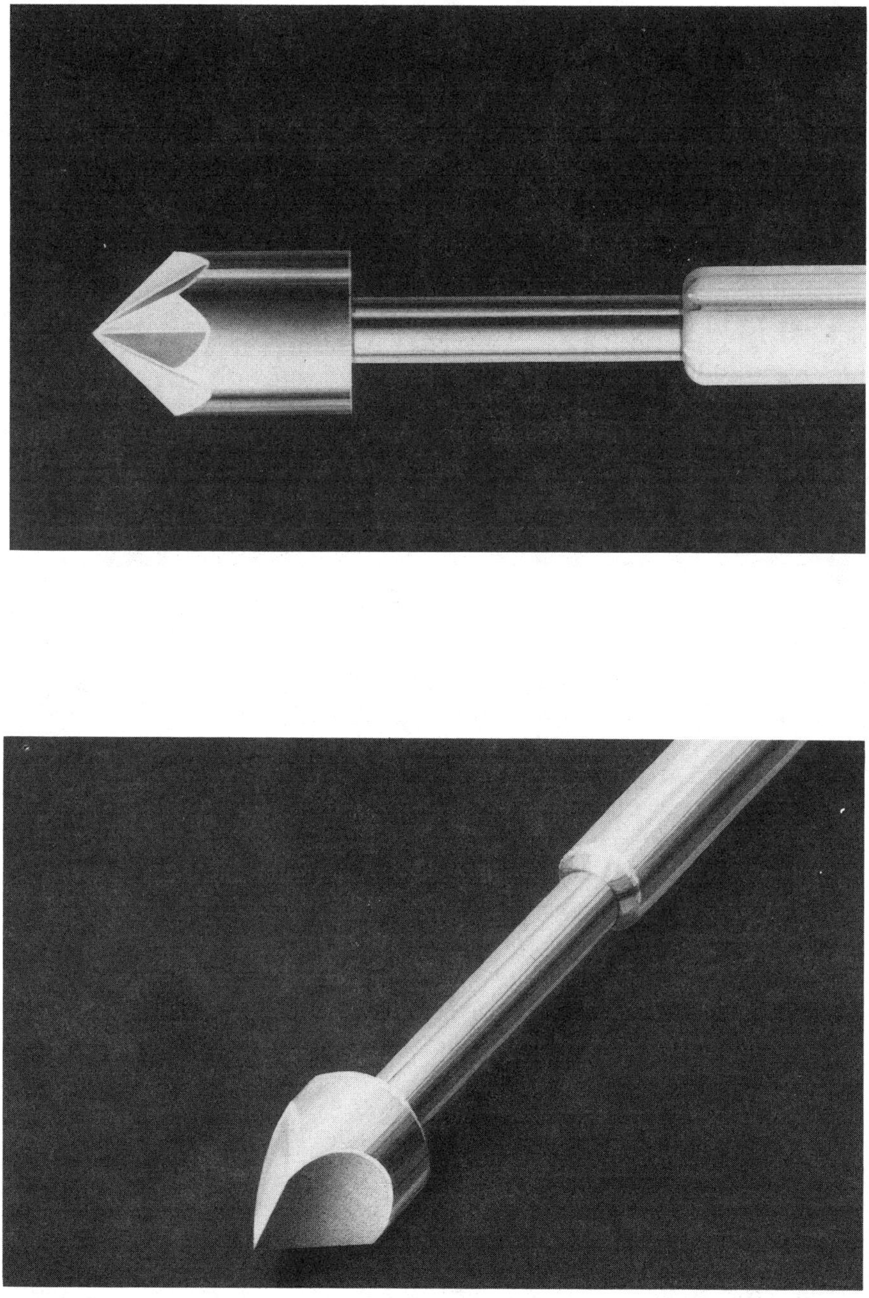

FIGURE 11.5. Photo of different pin tip shapes. (Photo courtesy of Everett/Charles Contact Products, Inc.)

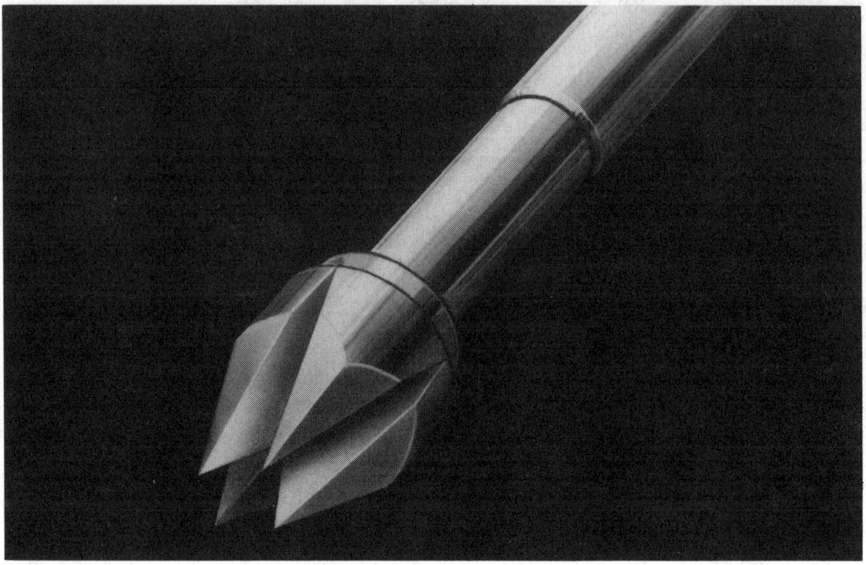

FIGURE 11.5. (*Continued*)

Unfortunately, to maintain sufficient connectivity on dense fine pitch assemblies is increasingly expensive and error prone. Presently, medium speed in-circuit testers cost between $150 to $300 per probe driver. Probe fixtures range in price from $2,000 to $30,000.

The smaller probes increase the probability of missed contacts leading to false failures. Experience has found that false failures can range from 10 to 25% of the total failures for well designed assemblies and test fixtures. The high test system cost and large number of assemblies failing incorrectly leads one to conclude that this approach is uneconomical and unreliable.

11.3 ALTERNATIVES TO IN-CIRCUIT TESTING

Several alternatives to the in-circuit test strategy exist. These range from simple functional testing to on-the-assembly testing schemes such as boundary scan testing.

The No-Test Option

Ideally, postassembly testing is needed only to catch failures that result from improper processing. If the process is very capable, then only a little waste results (in the form of failed assemblies). If this is the case, then the decision to in-circuit test would be weighed against the decision to simply scrap the few failures generated by the process. And if the money saved by not testing is applied to constantly improving the process then the amount of scrap will remain low.

Many readers will dismiss this as an impractical scenario, especially for dense fine pitch assemblies. However, by examining typical Pareto charts (defect amounts by defect type) one can see that this approach may have merit for certain products. Pareto charts for fine pitch products, such as a personal computer motherboard, a radio, an appliance controller, and a medical instrument are shown in Figure 11.6. In the cases of the computer motherboard and the medical instruments, the defect types are process based and probably justify some form of in-circuit testing to diagnose the failures. However, in the cases of the radio and the appliance controller, the failures are material based. In these cases in-circuit testing is probably less cost effective than implementing tighter supplier cooperation and monitoring. Supplier cooperation involves nothing more sophisticated than providing continuous feedback on the acceptance rates on the critical components. This feedback can be in the form of returned ICs that fail to meet the mutually agreed upon electrical parameters. The failed

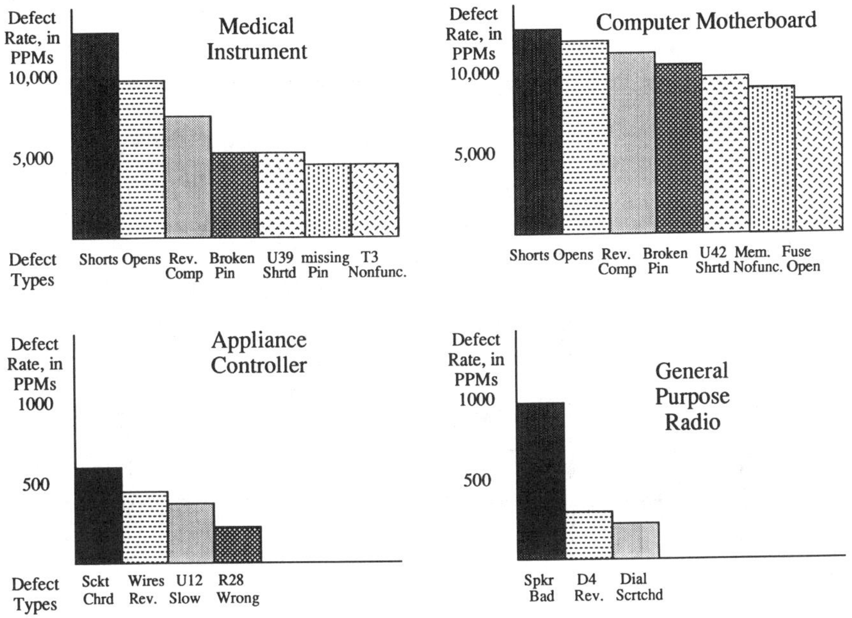

FIGURE 11.6. Pareto charts of defect rates and defect types on assemblies for four different applications.

ICs would be discovered using an IC tester that is significantly less expensive than an in-circuit tester.

Boundary Scan Testing

Another method of diagnosing individual circuit failures is to embed the test circuits into the board and the most failure prone ICs.

By including the test circuits on the assembly as illustrated in Figure 11.7 the number of probe connections is dramatically minimized. The on-board test circuits provide the opportunity to function as:

- On-board police-the-output
- Status setters

As on-board police, the test circuits monitor the preset outputs of most of the circuits or circuit blocks. When a circuit fails to supply the required output within expected limits, then the police circuit *reports* it as either good or bad to a central test monitor circuit. If all the output police report a status of good, then the assembly is considered functional. The circuit outputs are the responses to a single input or an array of inputs.

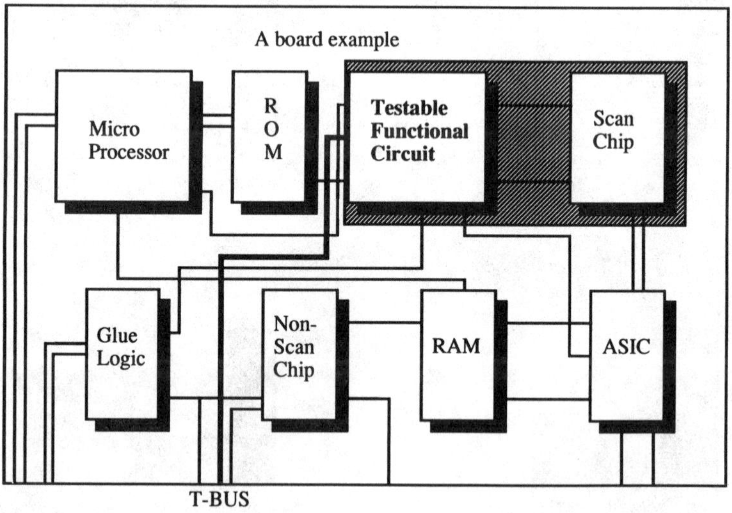

FIGURE 11.7. Test circuits may be included within the assembly. This will reduce the number of probes needed for full in-circuit test capability, but will add cost and real estate. (Source: Logical Solutions Technology, Inc.)

An early version of the police-the-output test strategy is called *signature analysis*. Signature analysis requires no additional circuitry be added to the board, but instead uses probes and a test system to monitor the various outputs. The major problem with signature analysis is that the test diagnosis of a failure is only valid up to the first bad circuit output. After that output, all the other outputs would be wrong or suspect since they were responding to the bad output. This means an assembly requires several trips to the repair bench to replace each bad output circuit. If a specific assembly is prone to have several problems, then this approach quickly becomes overly expensive because of the number of loops and handling in the test and repair areas. Also it creates errors in the material tracking system that monitors the material flow during the assembly process.

The newer boundary scan method requires additional circuitry be added to the IC and the assembly. Generally, the additional circuitry is additional gates to provide access to a test port, gates to create a test port, a scan device, and the necessary interconnecting traces. When custom ICs (ASICs, VLSI, ULSI) are used, some of the boundary scan test circuitry can be incorporated into the IC itself.

The gates for the test port are multiplexers and flip-flops that are placed on the inputs and outputs of each IC of the main circuit to be tested. The outputs of these gates are sent serially into boundary scan registers. The output status is read from the scan registers in response to signals from the test access port controller. Illustrated in Figure 11.8 is the boundary scan circuit (BSC) at each input and output, and the scan register (BSR). The input boundary scan cell is shown in Figure 11.9 and the output BSC in Figure 11.10.

The basic functioning of boundary scan testing is as follows: The input BSCs to the IC to be tested are set to a certain logic state (a logic 1, a logic 0, or a high

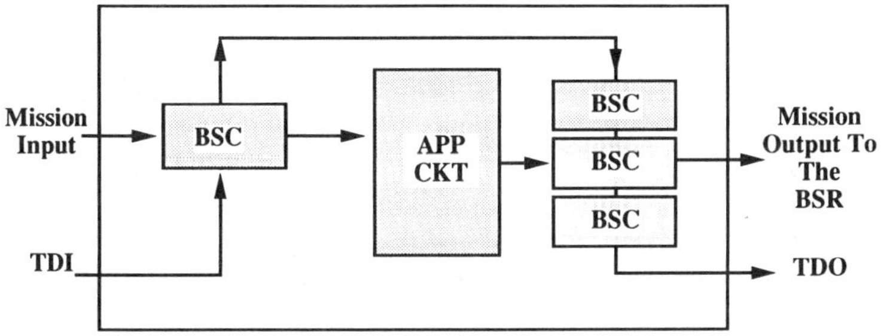

FIGURE 11.8. A boundary scan test circuit can be added to each application circuit input and output. The input BSC sets the logic state of the gate and the output BSC reads the resulting gate response which is stored in the boundary scan register, or BSR, for comparison with the expected correct response. (Source: Logical Solutions Technology, Inc.)

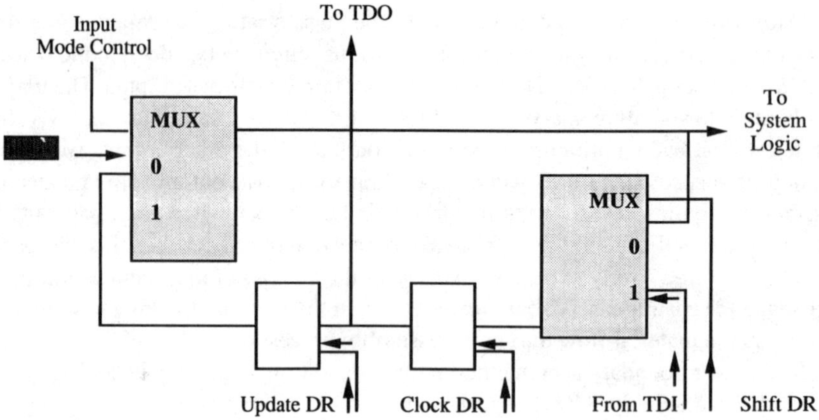

FIGURE 11.9. The input boundary scan cell. (Source: Logical Solutions Technology, Inc.)

impedance state). Additionally three test vectors can be shifted into the output BSCs, setting the outputs to one of three logic states. The output BSCs enable every output of the IC under test to be monitored simultaneously with 3 tests instead of individually. An IC with 100 output pins would require $3\times 100 = 300$ test cycles without BSC test capability.

The simultaneous output monitoring of the output BSC is done by the BSR. When the output BSCs receive a test vector, their output state is shifted into the BSR. Data shifting is controlled by the test access port controller, or TAP controller. Once the data is in the BSR, it can be compared to the expected

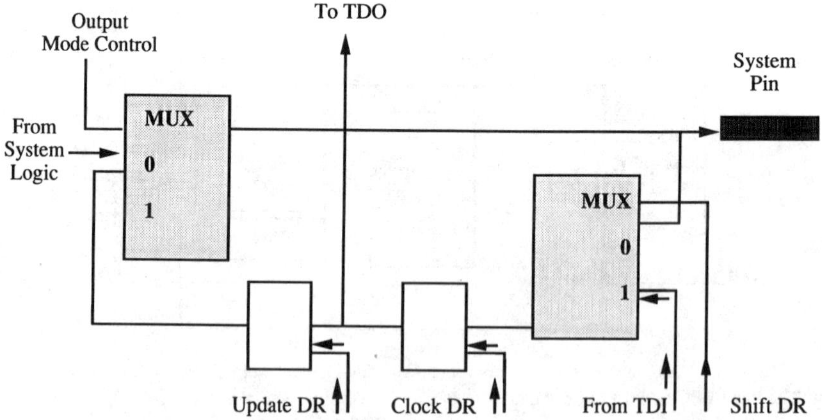

FIGURE 11.10. The output boundary scan cell. (Source: Logical Solutions Technology, Inc.)

data. Bad or nonfunctional outputs are readily identified as improper data-word comparisons.

The main benefit of the boundary scan test strategy is the ability to individually set the input states of an integrated circuit, and to read the output states. This enables each IC to be individually tested within an assembly without the use of a bank of probes in a bed-of-nails test fixture. Therefore all bad ICs will be diagnosed in a single test pass. Since the testing uses IC circuitry instead of mechanical pins it is very small.

The greatest benefit of boundary scan for size and cost considerations is the ability to include the input and output BSCs and the TAP controller circuits in the ASIC and custom integrated circuits. The cost and silicon area increase to the ASIC are very small sacrifices compared to the cost of a high speed, high pin tester and bed of fine pitch nails. Another benefit is that the test routines used by the IC fabricator and user will be the same using the same circuitry. This will give close correlation when the ICs are tested at each facility prior to final assembly, and again after assembly.

The input and output BSCs and TAP controllers are available as individual circuits from various circuit suppliers. These circuits are added to an assembly to test other circuit areas. The common approach is to cluster a group of individual components in a functional block. An input BSC is placed on the input of the cluster and an output BSC on the output of the cluster. An example of this clustering is shown in Figure 11.11.

Several boundary scan standards exist to control this testability approach. Most of these are being issued under the auspices of the IEEE Computer Society. The present standards include:

- P1149.1 Boundary Scan and Test Access Port Architecture
- P1149.2 Extended Serial Digital Testability Bus

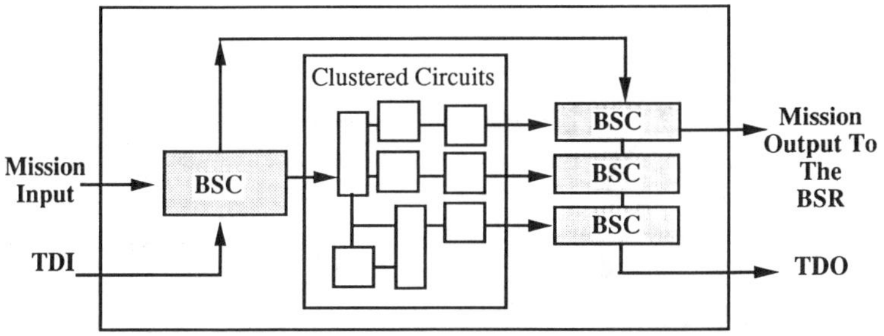

FIGURE 11.11. This diagram shows the addition of BSC circuits as a monitor of a cluster of a subcircuit. This approach reduces the number of BSC circuits that need to be added. (Source: Logical Solutions Technology, Inc.)

- P1149.3 Real Time Digital Testability Bus
- P1149.4 Real Time Analog Testability Bus
- P1149.5 VHSIC TM-Bus

The P1149.1 defines the characteristics of the input and output boundary scan cells and the test access port controller. This assures uniformity of the commercially available individual BSC and the TAP integrated circuits, and assures the uniformity of the circuits integrated directly onto the custom ICs.

The P1149.2 defines the extended serial digital testability bus. This approach is more open than the P1149.1 approach and provides additional control parts. This allows greater user control of the scan path, the scan settings, and reduced intrusion on the scan path. The P1149.2 approach supports multiple test circuits of the external boundary scan parts. The P1149.2 implementation is shown in Figures 11.12 and Figure 11.13.

Real time digital testing is controlled by the P1149.3 standard. The P1149.3 is a multiplexed approach with selective test vectoring which allows selection of the circuit to be tested. The multiplexer, when not used for testing, is transparent to the main circuit. This enables the P1149.3 circuits to be connected to circuit nodes not regularly testable. These include circuit resets, tri-state controls, and "hold" and "wait" lines. P1149.3 test circuits are designed so as not to affect the main circuit speed and performance. This allows the main circuits to be tested at normal operating speeds which is a major limitation with external test systems.

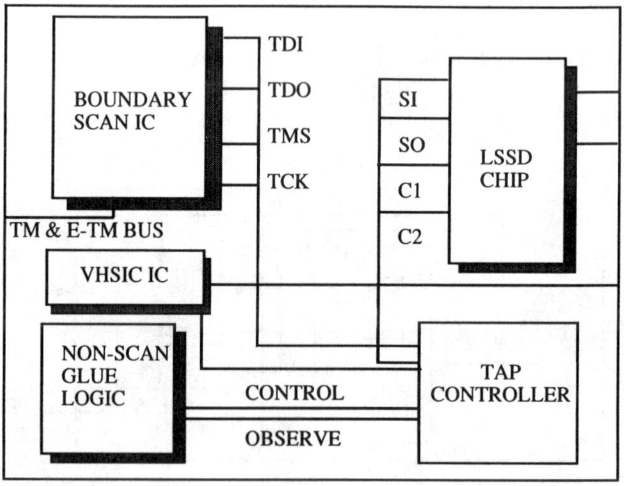

FIGURE 11.12. The implementation of boundary scan to comply with P1149.2 (Source: Logical Solutions Technology, Inc.)

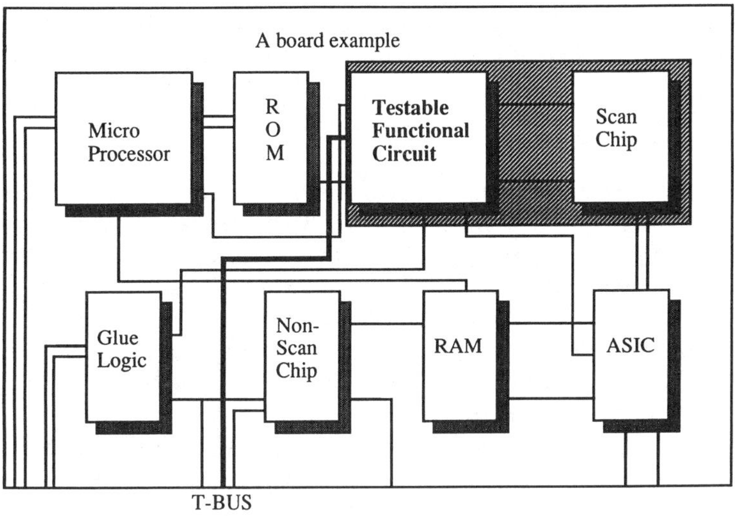

FIGURE 11.13. Boundary scan test implementation as envisioned in the P1149.1 standards. (Source: Logical Solutions Technology, Inc.)

11.4 SUCCESSFULLY USING ICT FOR FPT TESTING

If it is not possible, for valid reasons, to switch from ICT to an alternative test strategy, then there are a few rules that may help the ICT testing of fine pitch packages. The physical layout guidelines are illustrated in Chapter 13, and repeated below.

The in-circuit test guidelines for fine pitch and surface mounted assemblies are listed below. They were developed by a committee of the Surface Mount Technology Association (SMTA, Edina, MN). They have been reprinted, with permission, from the SMTA publication TP-101A.

Probing Considerations

The first guideline listed below has the most impact on the probability of spring-loaded probes reliably contacting the device under test (DUT). The criteria used in the tolerance analysis were (1) a single point (spear head) probe, (2) the worst case tolerance internal to the probes, (3) their mounting tolerances associated with DUT platen (probe platen) alignment, and (4) test pad diameter.

With larger PC boards, these tolerances become more difficult to control than with smaller boards. The tolerance buildup directly and adversely impacts test pad size. Larger diameter test pads may be required to assure reliable probe contact as the board size increases, due to artwork runout.

Probing Guidelines

Guideline #1. Provide Accurate Tooling Holes. (A) TOLERANCE FROM THE DUT DATUM TO THE TEST PAD SHOULD BE 0.05 mm (+0.002″). Tooling pin location is critical and every effort should be made to reduce this indexing tolerance. In cases where PC boards are manufactured in panels and then broken out, datum holes should be supplied both in the mother panel and in each individual board. The datum holes (at least two) in the panels are required for accurate processing and for subsequent testing and possible rework. If necessary, because of space considerations, breakaway tabs can be used to carry the datum holes for the assembly process and testing the DUT. The same datum holes must be used to fabricate, populate, and test the PC board to ensure probing accuracy.

(B) PLACE TWO TOOLING HOLES ON THE DUT AS FAR APART AS POSSIBLE WITH A TOLERANCE BETWEEN HOLES OF 0.05 mm (+0.002″). The tooling holes should be diagonally opposite for the best probing accuracy and must be part of the primary drilling operation. The tooling holes must not be indexed from the edge of the board as board profile tolerances are generally not controllable enough for an accurate datum.

The tooling holes should accept at least 3.1 mm (0.125″) diameter pins to maintain stability and alignment of the board with the fixture. If dual sided probing is required, the tooling pins may have to be long enough to go through both fixture plates.

One of the key issues with tooling pin diameter is the associated strength of the tooling pin of that diameter. One must insure that the tooling pin is not bent in the process of placing or removing the PC board from the fixture. It is more difficult to bend a larger diameter pin. This situation can be helped by using tooling pins made from a very brittle material so that the pin is broken rather than bent if improperly aligned. One is better off replacing a known broken tooling pin rather than having false failures due to poor alignment caused by a bent tooling pin.

(C) TOLERANCE OF TOOLING HOLE DIAMETERS SHOULD BE +0.08/−0.00 mm (+0.003/−0.000″). Nonplated tooling holes are preferred as they decrease the likelihood of solder buildup and make the 0.08 mm tolerance more feasible. The diameter of plated through holes is very difficult to control. Figure 11.14 shows the tooling tolerances.

Guideline #2. Probe Pads Should Be No Smaller than 0.89/1.00 mm (0.035/0.040″) Diameter. Using the tolerances in guideline #1, this is the minimum pad diameter that will guarantee probe contact. Test pad size is dependent on many factors, such as accuracy of tooling, and PC manufacturing processes, as well as the physical size of the board.

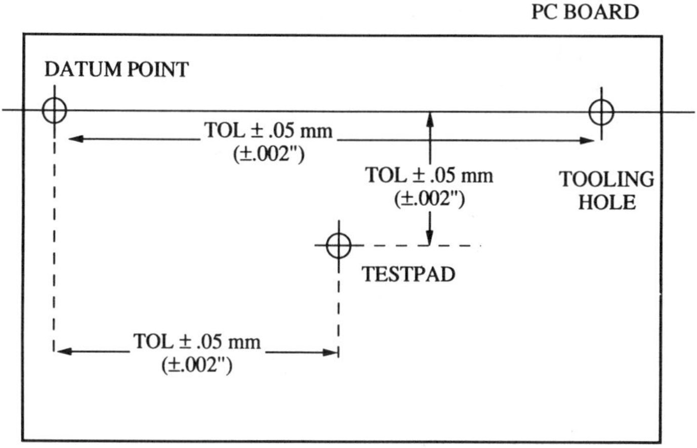

FIGURE 11.14. Tooling hole tolerances should be within these limits to minimize probe to test pad misregistration.

For small probes, the 0.9 mm (0.035″) pad size should be used, and for the larger probes, the 1.0 mm (0.040″). It is feasible to use smaller 0.61 mm (0.024″) test pads on small boards, under 30.5 cm (12″) square, but the designer should check fixture cost and tolerance specifications. Figure 11.15 shows the minimum desirable pad-to-pad tolerances.

The minimum pad geometry uses industry standard tolerance for board geometry, probe tolerance, and fixture assembly. If a different set of tolerances and/or

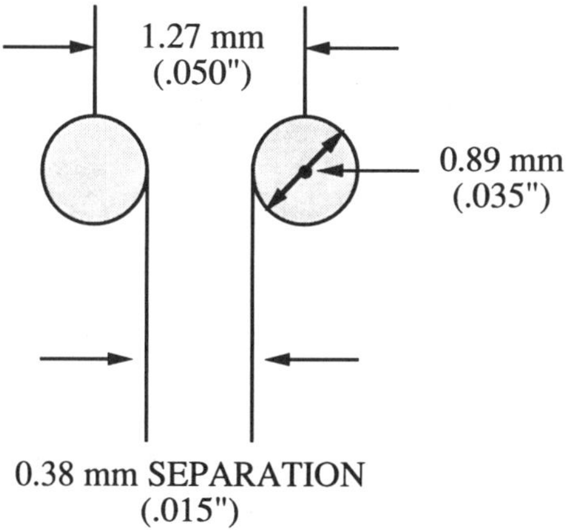

FIGURE 11.15. The minimum test pad to pad positioning.

assembly techniques are used a different pad size may be required or desirable to obtain a reliable test contact.

By performing the calculations using the data in Table 11.1 it may be possible to use a smaller pad size because tolerances are tighter, or one may find it necessary to enlarge the pad diameter because of tolerance stackup. Designers are encouraged to understand this tolerance analysis and adjust pad size as required.

Guideline #3. Component Height on Probed Side of DUT Must Not Exceed 6.4mm (0.255"). Tall components on the probe side of the board will require cutouts or relief in the platen. They can be accommodated, but the cost of special machining of the platen and loss of platen strength should be considered. Excavation of the platen also restricts test probe placement. When necessary, test pads should be placed 5.0 mm (0.2") away from tall components (greater than 5.0 mm high) to allow for milling tolerances.

Guideline #4. No Components or Test Pads Are to Be Located within 3.2 mm (0.125") of DUT Edges. This ensures that components will not interfere with the vacuum seal and seating of the probe. It also provides longer life for the gasket

TABLE 11.1 Smaller Test Pads are Possible if Certain Tolerances are Tighter. Before Using Smaller Test Pads or Reducing any Tolerance, the Process Capabilities Must be Measured and Verified. (Source: SMTA Testability Guidelines, TP-101A)

SYSTEM TOLERANCES	DISTRIBUTION	TOLERANCE	VARIANCE
skt hole to tool pin (10")	gaussian	0.002	0.000000
angle of skt hole (proj)	gaussian	0.003	0.000000
angle of skt in hole (proj)	tubular	0.0008	0.000000
bow of probe plate (10")	gaussian	0.002	0.000000
tool pin straightness	tubular	0.0008	0.000000
tool pin hole straightness	gaussian	0.00075	0.000000
tool pin in hole angle (proj)	gaussian	0.0015	0.000000
tool pin size	uniform	0.0005	0.000000
digitizing errors	gaussian	0	0
angle of plunger in barrel	tubular	0.0025	0.000002
angle of probe in socket	tubular	0	0
probe straightness	tubular	0.001	0.000000
socket install bending (.050)	gaussian	0.003	0.000000
probe target to tool hole	gaussian	0.006	0.000003
target size	gaussian	0.002	0.000000
tool hole size	gaussian	0.002	0.000000
tool pin to tool hole clearance	uniform	0.003	0.000002
sum of tolerances		+/- 0.03085	
Sq Rt of sum of the variances			0.003363 in.
pad size is 8X SRsum of var			0.026907 in.

while conveyorized assembly equipment and pick and place handling systems also generally require accessible edges.

Guideline #5. A 0.46 mm (0.018″) Annulus, Free of Components or Other Obstructions, Must Be Left around Each Test Pad. This is to prevent shorting components or test pads when a worst case tolerance situation exists and probe heads other than single point are used. It assumes that no component within 5.1 mm (0.2″) of the test pad exceeds 6.4 mm (0.255″) in height. Further, when a high component is within 5.1 mm (0.2″) of the pad, space for mounting the probe receptacle in the fixture must be provided. See Figures 11.16 and 11.17.

Guideline #6. Edge of Test Pads on Component Side Must Have a Minimum 1.0 mm (0.040″) Clearance from Components. This is to avoid striking the component which may cause probe or component damage.

Guideline #7. All Probe Areas Must Be Solder Coated or Coated with an Equivalent Conductive, Nonoxidizing Surface. Solder has proven to be an excellent material to probe. Solder oxides are soft and are easily penetrated, providing good probe contact while helping to prolong probe tip life.

Guideline #8. Test Pads Must Not Be Covered by Solder Resist or Legend Ink. If the resist or ink covers part of the test pad, the area available for contact will be reduced. It should also be noted that using a probe with a large contact head, such as a serrated or crown probe, may interfere with pad contact (see Figures 11.18 a and b).

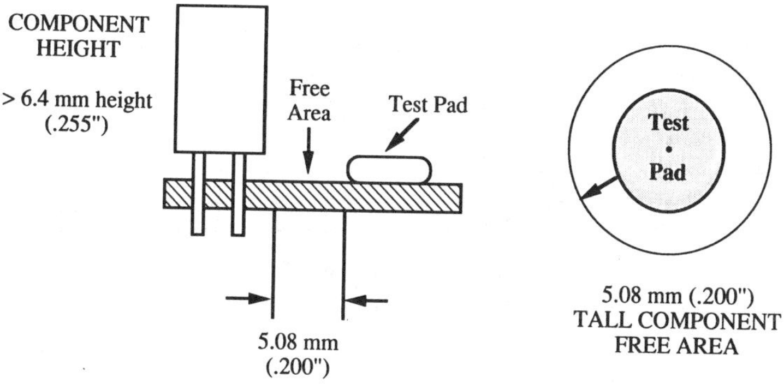

FIGURE 11.16. Diagram of the desired free area around tall components if a probe is to contact the pad from the tall component side.

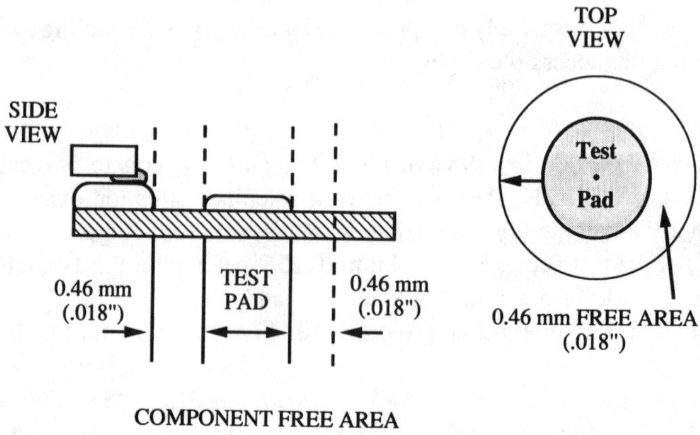

FIGURE 11.17. Diagram of the desired free area around components less than 6.4mm (0.25") in height.

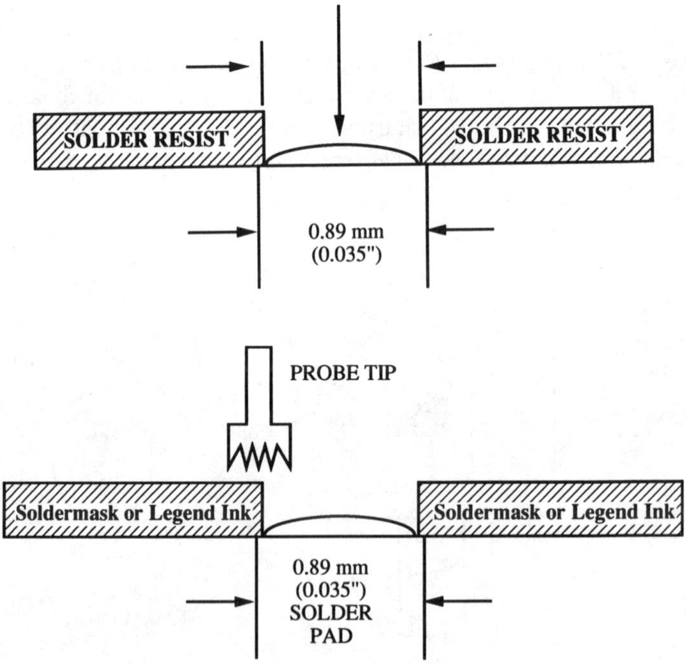

FIGURE 11.18. (a & b) Misregistered soldermask or silk screen legend ink act as insulators on test pads, especially of serrated or crown type probes are used.

Guideline #9. All Through-Holes Should Be Filled. Filling all the holes reduces fixture air leakage and prevents spear point probes from sticking in unfilled holes (vias). Unfilled holes also contribute to intermittent probe contact. When air is pulled through unfilled vias, the probes may become contaminated causing intermittent contact.

Unfilled holes may require the use of a rubber mat or other means to cover the assembly during test. This is to ensure a proper vacuum seal and fixture operation. The mat may preclude probing the board under test for fixture verification or for debugging of tests.

It must be realized that the use of a rubber mat or other materials that directly contact components on a surface mount board may cause lifted (tombstoned) components or lifted leads to come in contact with the PC board. This will mask opens that will fail after the vacuum is turned off. If it is not desirable to catch these possible defects at a later stage of test it may be necessary to change to a mechanically actuated fixture not requiring vacuum.

Where reflow soldering is used and space permits, the vias should be tented over with soldermask. Separate dedicated, solder coated test pads should then be provided.

Guideline #10. A 3.2 mm (0.125″) Annular Area Around Tooling Holes Must Be Clear of Components and Test Pads. This prevents damage to components when placing and removing the board from the test fixture. Many tooling holes are stepped for positioning the height of the DUT. The danger of shorting component leads exists even with unstepped pins.

Guideline #11. Probe Test Pads or Vias, Not Components or Component Leads. Contact pressure can cause an open or make a cold solder joint appear good. Component mounting variations may also result in missing the target, probe glancing causing fixture induced shorts, and probe damage.

Guideline #12. Avoid Probing Both Sides. Use Vias to Bring Test Points to One Side (Preferably Noncomponent Side). It is always preferable to design a PC board so that testing can be accomplished from the preferred non-component side. This allows the use of simpler, less expensive, more reliable fixtures. Troubleshooting with double-sided fixturing is very limited and more difficult.

Guideline #13. Probe Pads Should Be 2.54 mm (0.100″) Apart (Center Spacing) if Possible, in Preference to 1.27 mm (0.05″) Spacing. Wider center spacing of test pads allows the use of less costly and more reliable probes. Also drilling and wiring the fixtures will be less expensive. Smaller probes for 1.27 mm (0.050″) spacing are less reliable, fragile, and easily damaged.

If 2.5 mm (0.1″) spacing is not possible there is an intermediate solution in 1.9 mm (0.075″) probes that are more reliable and less costly than the 0.050″ probes.

Guideline #14. Investigate More Aggressive Head Style Probes. The changes in the cleaning process for SMT boards have resulted in increased contamination of the surface of the board. While these contaminant have no effect on the long term performance of the board they are difficult to penetrate for reliable test contact. There are more aggressive probe types available as well as rotating probes that make it much easier to penetrate the residues and oxides on the test contact surfaces.

Electrical Design Considerations

In addition to the probing rules stated previously, increased testability can be gained by incorporating some simple electrical guidelines.

Guideline #1. A Test Pad Must Be Provided for All Electrical Nodes. A node is defined as an electrical connection between two or more components either analog or digital. Unused pins of ICs, connectors, etc. are included. Test pads should be provided for all I/Os, power, ground, and return signals.

Each IC should also have a test pad on the power and ground nodes as close as possible to the device. For devices having more than one power or ground pin, additional test pads should be added. As a rule, a power and ground test pad should be provided with in 2.54 mm (1 inch) of each IC. Fixturing for testers in the 5 MHz and above test speeds will require power pins to be placed on each IC for good power supply coupling.

NOTE: Jumpers (zero ohm resistors), fuses, and SPST switches are treated as two lead components and are connected between two electrical nodes; therefore, each node requires a separate test pad.

Guideline #2. Do Not Rely on Edge Connectors or Circuit Traces as Test Points. Probing an edge connector plating or circuit trace may produce poor contact reliability. Gold plated fingers are also easily damaged with test probes. A better approach is to use a dedicated soldered test pad. Conductor traces can be expanded to 40 mil width if direct probing of the trace is desired.

Guideline #3. Distribute Test Points Evenly over PC Board. When the probes are not evenly distributed on the platen or concentrated in one area, the resulting high stresses may bow the DUT or platen. This may cause some probes not to make contact with the test pads and can also result in vacuum sealing problems or possible board damage. It is recommended that the probes be distributed about the board as much as possible.

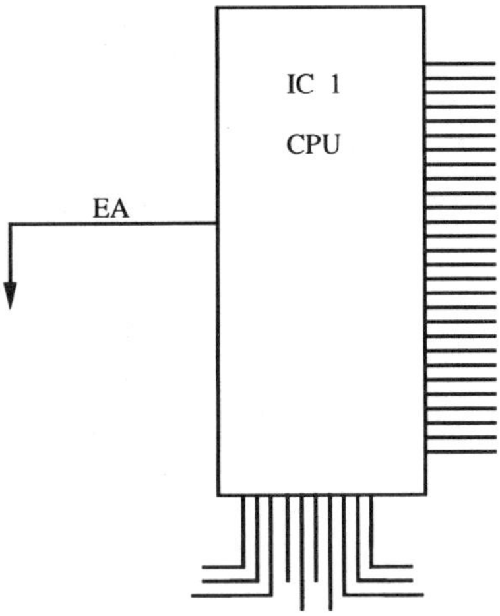

FIGURE 11.19. The external access (EA) control line is grounded to disable the microprocessor's library test routing.

Guideline #4. Do Not Wire IC Control Lines Directly to Ground or Vcc or a Common Resistor. With CPUs, there are at least three key types of pins that should be considered: external instruction access pins, initialize pins, and output enable pins.

In the example (Figure 11.19) the external access (EA) feature of the CPU has been grounded making it impossible to use the library test routine. This may force specialized testing or reduced coverage.

Similarly, control lines that are tied to a common resistor cannot be toggled independently to initialize the circuit (see Figures 11.20, 11.21, and 11.22). Use individual pulldown or pullup resistors.

Guideline #5. Clock Sources Must Be Controllable by the Tester. Disabling can be accomplished by means of jumpers or tri-state buffers.

Guideline #6. Power-On Reset Circuits Must Be Backdrivable. The RC circuits for power-on reset should be isolated from other digital components with a

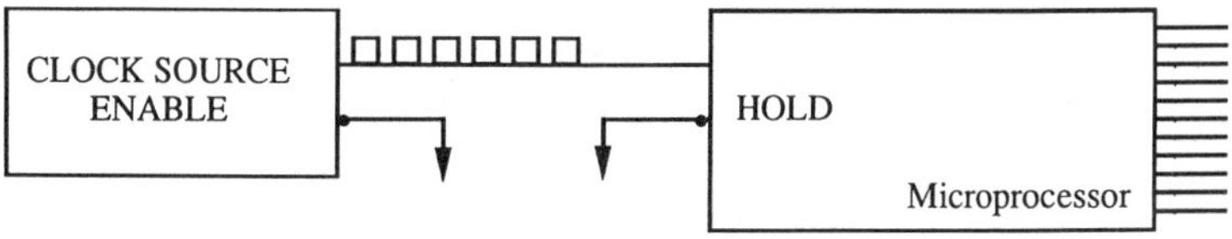

FIGURE 11.20. The clock enable and microprocessor hold lines grounded.

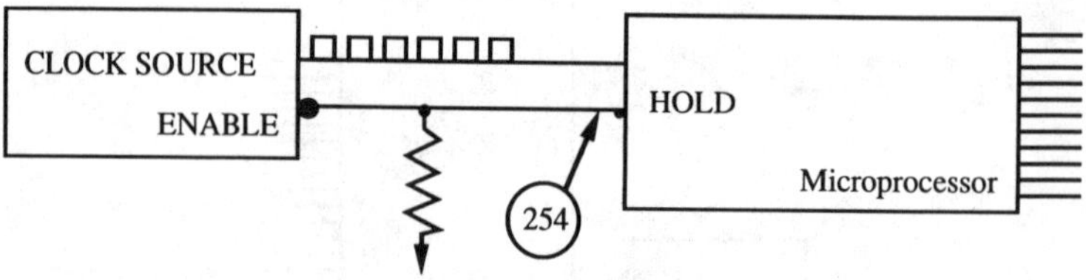

FIGURE 11.21. The use of a common pull-down resistor to control the clock and microprocessor.

digital buffer. Three simple ways to provide a backdrivable signal reset signal are shown in Figures 11.23, 11.24, and 11.25.

Guideline #7. Move Test Pads Only When Absolutely Necessary. Provided only a few test pads have to be moved when there is a design change, it may be possible to modify the test fixture rather than replace it. This can be accomplished by using a bare PC board as a drill guide. As long as the new location of the pad is at least 0.1 in. from the old location, drilling the new hole will not cause a breakthrough into the wall of the old hole.

Guideline #8. Provide a Test Pad on the Power and Ground Nodes as Close as Possible to the Device. Each IC should have a test pad on the power and ground nodes as close as possible to the device to minimize voltage drops and noise problems. For devices having more than one power and ground pin, additional pads should be provided. Designers should also ensure there is adequate decoupling for ICs to reduce noise spikes.

Guideline #9. Provide Simulation Vectors for All ASICs (Application Specific Integrated Circuits) and Other Semi-Custom or Custom Parts. Development of test vectors to exercise custom devices can take long periods of time (months to infinity); if simulation vectors are made available in a format that can be translated or understood by the test system, test development time can be significantly reduced.

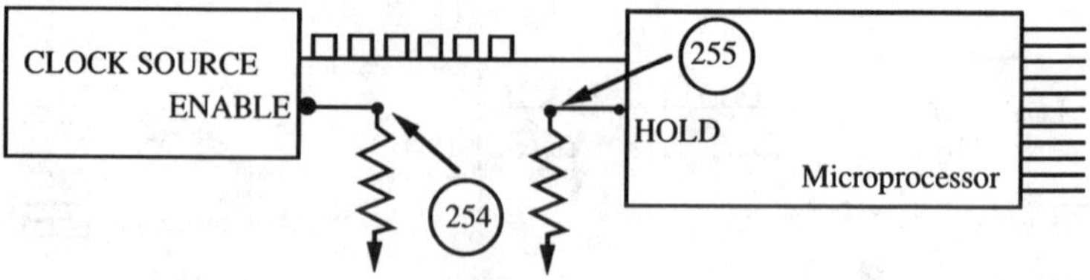

FIGURE 11.22. The use of individual control resistors.

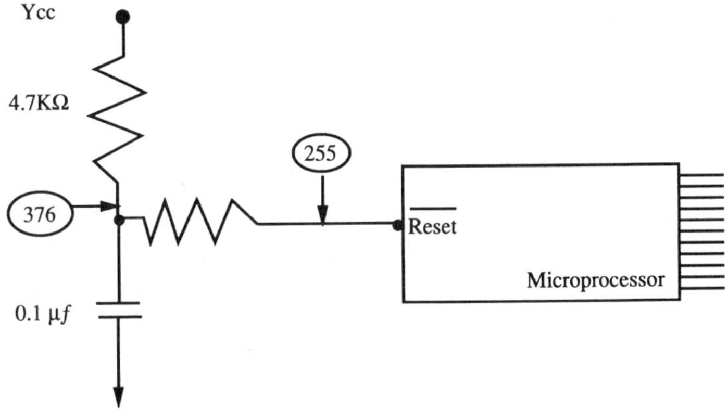

FIGURE 11.23. Using an isolating resistor on the reset line.

Guideline #10. Develop a Test Strategy for Multi-chip Modules before They Are Included in a Design to be Released to Manufacturing. The current working draft of the Guidelines for Multichip Module Technology Utilization (February, 1991) contains no testability information. This is not to imply that testing is not a concern; it is still very early in the life-cycle of this technology to have standardized on test strategies.

Guideline #11. Use IEEE 1149 (Boundary Scan) as a Testability Tool in Digital Design. See the earlier discussion of boundary scan test (in Section 11.3 of this chapter) for details.

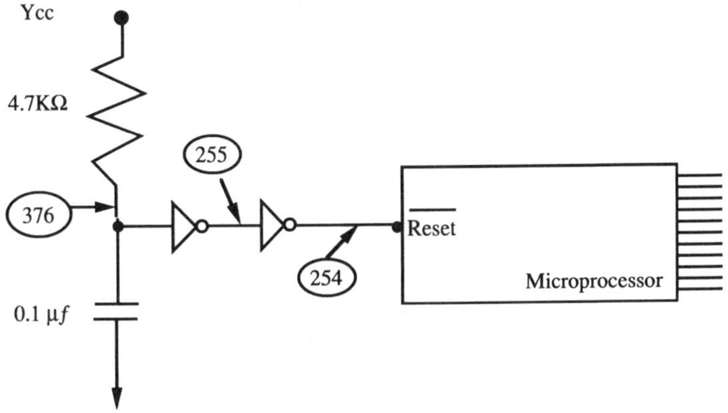

FIGURE 11.24. The use of two inverters to backdrive the reset line.

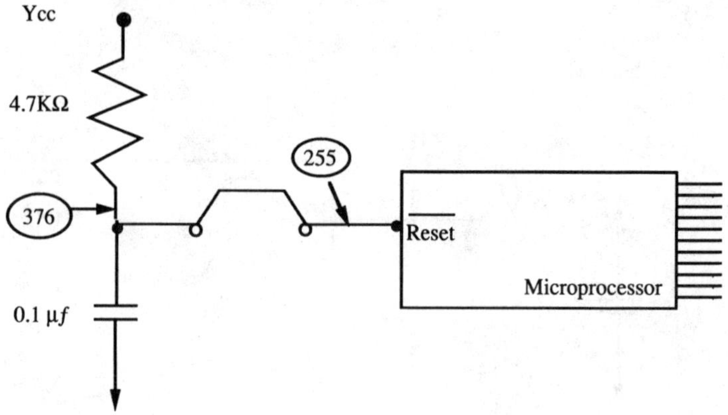

FIGURE 11.25. Use of a wire jumper to backdrive the reset line. This practice is not recommended.

References

1. Fasang, P. 1990. Boundary Scan Addresses Parametric Test Issues. In *ASIC Technology and News*, Oct. 1990.
2. McElenery, J. 1991. Testing Loaded PCBs With Care. In *Electronic Engineering Times*, Nov. 1991.
3. Turino, J. 1990. Testability Bus Standards in the 1990s. Proceedings of Nepcon West '90, Feb. 1990: Anaheim, CA.
4. Whetsel, L. and Coleman, J. 1991. Boundary-Scan Testing. In *Electronics Engineering Times*, March, 1991.
5. —1991.Testability Guidelines, TP-101A. Edina, MN: Surface Mount Technology Association.

12

Design for Manufacturability

12.0 GENERAL DFM GUIDELINES

This chapter moves from the specifics of fine pitch technology to temporarily cover a more general topic which affects all electronic assemblies regardless of the manufacturing technology used in assembly.

Design for manufacturability has received much attention in recent years. This development is a logical extension for companies which have successfully implemented quality and continuous improvement programs. Design for manufacturability is an approach, an attitude, and a toolbox to enable a new product to be put into volume production in the shortest time and at the optimum cost. The time to volume production is not the time needed to get the product to work, but rather the time for the product to be production worthy, i.e., readily producible in the needed quantities at optimum cost and quality levels.

In general, a product designed using a strict design for manufacturability (DFM), approach will cost slightly more to design, but will take less time to reach production release than a product designed using conventional approaches. The chief rewards of this delay are reduced production cost and higher profit. This is illustrated in Figure 12.1. An additional reward is avoiding the high cost (in dollars and pride) of field service. The comparative cost of proper initial assembly versus field repair is shown in Figure 12.2.

In an internal study at a large instrument manufacturer in the late 1980s, it was found that 70 to 80% of the product cost in manufacturing was established and set during the design phase of product development. Even more dramatically, the designer's choice of components, suppliers, PC board layout, and assembly technology affects customer acceptance, reliability, and field service cost. Any extra cost in product design experienced by implementing design for manufacturability can be justified. There is a widely quoted analysis of lost profit due to delays of the product to market. McKinsey and Company studied a product having a 5 year life

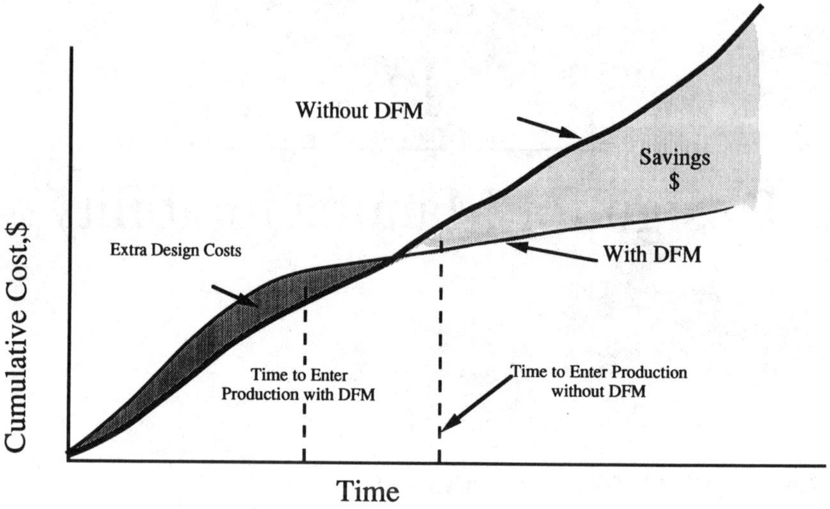

FIGURE 12.1. Design for manufacturability increases the design cost of a product, but reduces the production cost, generating greater profitability.

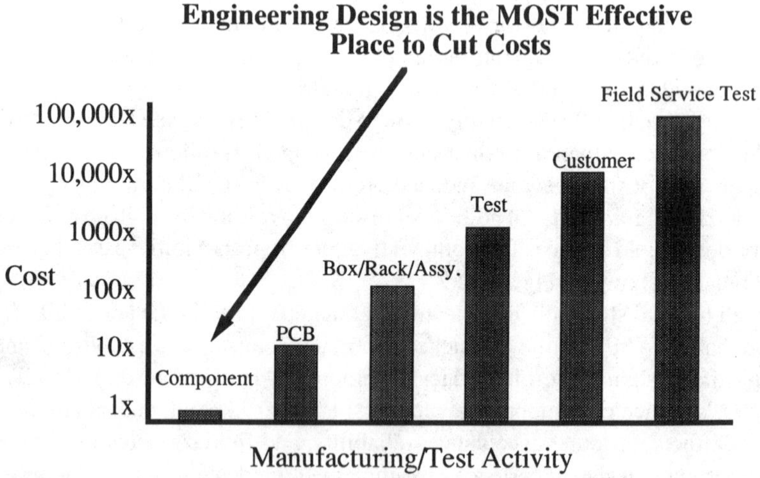

FIGURE 12.2. Problems quickly become very costly if not caught in the design phase. (Source: 1989 Proceedings of Design for Manufacturability and Concurrent Engineering.)

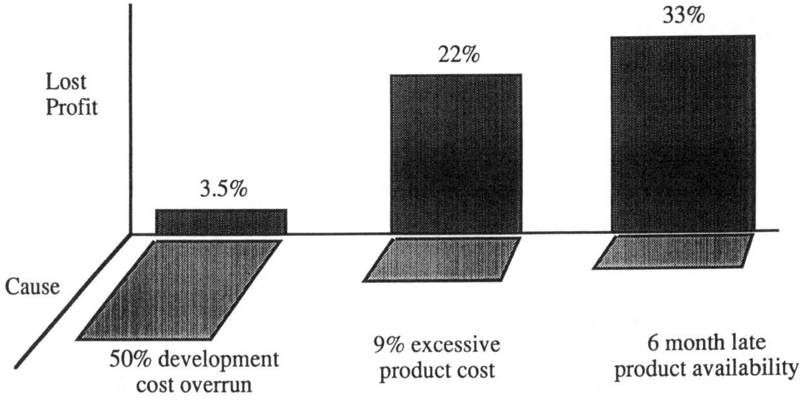

FIGURE 12.3. The effect on a product's profits resulting from a) if the product is 6 months late into production, b) if the product cost 9% more than the optimum and c) if 50% more were spent in design than budgeted. (Source: McKinsey and Company)

in a market growing at 20% per year. They assumed that the product's price would drop by 12% per year. The lost profit results are shown in Figure 12.3. The study found that 33% of the profit potential to be earned from the product was lost if it was 6 months late in shipping to (and staying with) the customer. A loss of 22% of the profit potential resulted if the product was released on time, but cost 9% more in production than the optimum cost.

If, in the design phase, greater cost is applied to employ those resources needed to assure design for manufacturability, and if these resources result in a 50% development cost overrun, but the product is released to production on time and with optimum production cost achieved at the initial release, then the loss is only 3.5% of the profit potential.

The greatest design obstacle to surface mount and fine pitch technologies is finding capable IC package suppliers. Increasingly, IC suppliers are using outside contractors to package the integrated circuits. This complicates and delays the responsiveness of the integrated circuit supplier to fixing the problems encountered with surface mount and fine pitch packages. These problems are typically in the areas of lead planarity, solderability, and ease of placement.

12.1 THE RULES OF DFM

The basic precepts of a DFM approach to printed wiring boards include:

1. Use standard or approved parts and materials.
2. Design for ease of orientation.
3. Determine the capabilities of each process step.

4. Eliminate multiple solder and cleaning steps.
5. Eliminate adjustments.
6. Minimize the number of different part values and sizes.
7. Publish up-to-date design guidelines.
8. Design for variability.
9. Identify defect rates by cause.

Each of these precepts will now be discussed.

Use of Standard Parts

Using standard parts reduces the cost, delay and risk of adding a new and possibly unsafe part to the parts catalog system. Most medium and large companies have slow, intensive systems that a part must move through before it can be added to the standard part library. If the system is circumvented, then problems occur. Simple oversights, such as using a part that melts in the reflow oven or dissolves in the cleaning agents, are hard to excuse when they could have been caught during the correct part certification process.

Some companies allow parts to be added to a parts library on a conditional basis. This adds flexibility to the parts library while still providing checks and balances to prevent unwanted or unsafe parts from becoming permanent members of the library. Often minor modifications to a standard part of one technology can enable it to be utilized in another technology. For example, the leads of a through-hole component can be trimmed and formed so it can be surface mounted, thus eliminating the need for an additional soldering step.

Design for Ease of Orientation

Design for ease of orientation applies to both automated and manual assembly. The less a part has to be twisted or turned to get it into place the easier. This also applies to the assembly itself. Uniformly oriented parts make inspection easier and reduce assembly handling time and effort.

Specifying easy to read markings on components and matching indicators that are stenciled on the board are helpful. For example, the notch indicating the top of an integrated circuit in a dual-in-line package (DIP) or the chamfer on the side of a small outline (SO) package are not as easy to detect as a dot placed adjacent to pin 1 on the package and a stenciled label on the board.

Polarized parts such as ICs, diodes, and electrolytic capacitors should always be oriented in the same direction.

The aforementioned techniques may not be of obvious benefit when human inspection is utilized, but with the growing use of automated inspection, they save programming and inspection throughput time.

Parts orientation is particularly important in soldering. During wave soldering, for example, the body of the package which is glued on the wave solder side of the assembly will briefly disrupt the shape of the wave. Any terminations behind and too close to the wave disruption will have trouble soldering properly. Generally, orienting the terminations so they enter the wave uniformly and are spaced sufficiently away from other packages eliminates the problem.

Determine Process Capabilities

Process capability is defined as the ability of a process to produce, within predictable or acceptable specification limits, without generating waste. The index of inherent process capability is Cp. Cp can also be defined as the allowable process limits divided by the actual measured limits.

The equation for calculating the inherent process capability, Cp, is shown in Figure 12.4. In this equation the difference between the upper and lower specification limits is divided by the process width, which is expressed as 6 times the

The Index Of Process Capability Is Calculated As:

1. The "Inherent" Capability, Cp

$$Cp = \frac{\text{Specification Width}}{\text{Process Width}} = \frac{\text{Upper spec limit - Lower spec limit}}{6\sigma \text{ (Actual Spread)}}$$

2. The "Non-Centered" Capability, Cpk

$$Cpk = \text{Minimum of } \frac{\text{Average - Lower Spec Limit}}{3\sigma \text{ (1/2 Actual Spread)}} \text{ or } \frac{\text{Upper Spec Limit - Average}}{3\sigma \text{ (1/2 Actual Spread)}}$$

3. The "Absolute" Capability, k

$$K = \frac{2 \text{ m - average}}{\text{Upper Spec Limit - Lower Spec Limit}} \qquad M = \frac{USL + LSL}{2}$$

4. Relationship of the Various Capabilities

$$Cp = \frac{Cpk}{(1 - k)}$$

FIGURE 12.4. The index of process capabilities for various conditions. The relationship of Cp to Cpk and absolute capabilities, k.

standard deviation, sigma. Other process capability indices exist to account for noncentered distributions and to calculate the absolute process capability. These indices are known as Cpk and k, respectively.

The allowable index of process capability value depends on the process and on subsequent processing. Rough values of acceptability have been advanced from many sources.

Authors Ekvall and Juran recommend the following classification Cp and Cpk values:

- Less than 1 is "not adequate"
- Between 1 and 1.33 as "adequate" but the process requires close control to prevent approaching inadequacy
- Greater than 1.33 is "more than adequate"

Montgomery recommends minimum Cp and Cpk values depending on the maturity of the process:

- Cp = 1.33 minimum for established processes
- Cp = 1.50 for new processes or when the established process is "critical"
- Cp = 1.67 for new processes whose variation will be "critical"

To illustrate Cp, assume an established process that is always well within the allowed limits. This means the computed Cp value is greater than 1.33. For an index value of greater than 1.33 it is said that the process will always produce, within the specification limits, without any waste.

The mathematics of the Cp calculation shows that a Cp greater than 1.33 will probably never allow a defect to occur, hence the process is capable of creating no waste. If the Cp is between 1.0 to 1.33 then there will probably be some minor waste since the process is barely capable. The slightest shift in the process, resulting from events such as lower quality materials or machine wear, will drop the Cp and create greater waste. Representations of the consequences of different Cp values are shown in Figure 12.5.

The expected limits may be the machine or factory limits that are specified in the design guidelines. In essence, as long as the parameter (trace width, placement area, stencil opening) stays within the true capabilities of the process, there should be no problem, and no yield loss (waste). With surface mount and fine pitch this is more complex than through-hole assembly technology because some mutually dependent capabilities must be accounted for, e.g., the relationships between stencil opening, solder paste rheology, and solder runout.

The process capability index is a very valuable tool as it effectively and quantitatively communicates the various process boundaries to the designer. The index, Cp or Cpk, is easily calculated if a database of the critical process variables

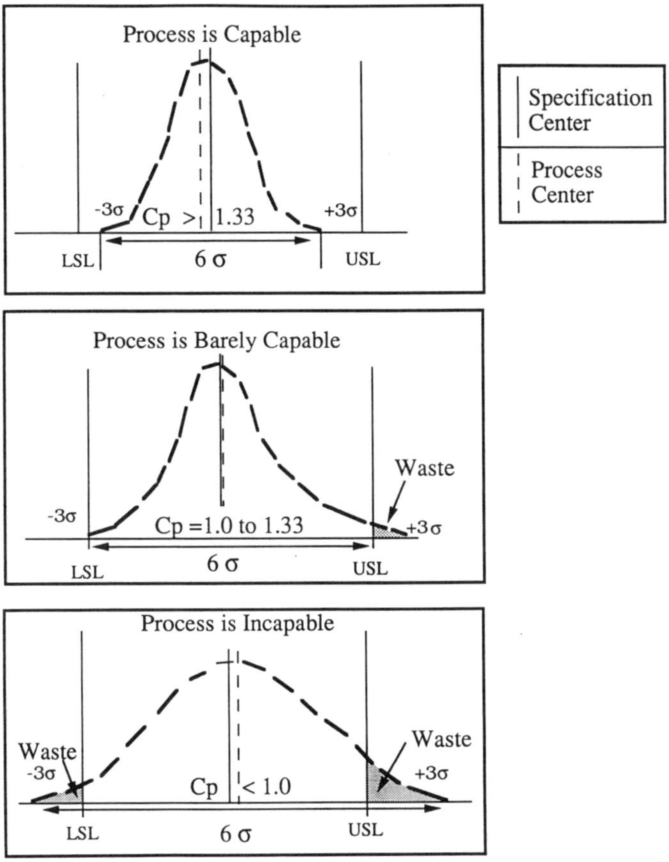

FIGURE 12.5. Process shifts and marginal process capability result in waste, rework, and quality problems.

and a realistic specification limit are available. The critical process variables are determined by testing each element of the process for statistical significance. Once these are identified, then they should be measured regularly and the measurements logged into the database.

Eliminate Multiple Solder and Cleaning Steps

This means to minimize the number of different technologies used to assemble components on the board. The fewer chances for error and yield loss, the better.

Take, for example, a simple-single sided surface mounted assembly. The process steps are to stencil paste, place components, reflow the solder, and clean the assembly. The addition of one through-hole part, such as a connector, causes the assembly to require wave soldering or manual soldering. The extra process adds cost, risk, and delay. This does not mean avoiding a through-hole part which is more reliable or sufficiently less costly than an SMT equivalent. But the tradeoffs to extra process steps should be examined before the design is completed and the extra process steps are required.

Another example of avoiding multiple steps is the processing of a two sided SMT assembly. Putting components on both sides of the printed wire board is one of the functional benefits of SMT. However, it may cause manufacturing and reliability problems if the assembly and the parts assembled on the first side of the board go through the reflow oven a second time.

One option is to use an adhesive with the solder paste to hold the components on the first side of the board. The adhesive holds these components in place while the second side parts are placed on the board and the assembly is transported through the reflow solder system. When using this technique, it is best to design the board so that all the large, heavy components are on the second assembly side. The adhesive offers a significant advantage for assemblies with temperature sensitive parts by avoiding the extra thermal energy of a second pass through the oven. However, on less sensitive assemblies, the adhesive may contribute more defects than added value. Again the tradeoffs need to be examined.

Eliminate Adjustments

Eliminating mechanical adjustments generally means eliminating potentiometer and manual switches from the assembly. To adjust these requires human labor, which adds cost and delay. These parts can be eliminated by specifying tighter tolerances on other parts or by including electronic adjusters. Examples of electronic adjusters include phase lock loops, feedback loops, voltage regulators, nonvolatile memories, and autoconfiguration registers.

The elimination of adjustments is not limited to the electrical adjustments of potentiometers and switches. Adjustments to component misalignment after placement is an example of a very costly process that may add no more benefit than to make the product look pretty. If certain parts have a tendency to misalign during soldering (tubular resistors are a good example), the footprint designs should be designed to allow for a reliable solder joint even though the part is misaligned.

Adjustments to the lead alignment is very costly. Desoldering the leads can often result in lifted pads because of the different desoldering rates of the pads. The different heating rates of the pads are due to the different amounts of copper (either traces, via holes, or ground plane) attached to the pad. Replacing lifted pads is not impossible, but costly.

Minimize the Number of Different Parts

Minimizing the number of different part sizes and values brings significant savings in the cost of component placement and product reliability. Few placement machines have methods for verifying that the operator (always a human!) has loaded the correct value. With small SMT unmarked chip capacitors and resistors in taped reels, the probability of a misinserted reel feeder goes up greatly with the increasing number of different feeders. By minimizing the number of different values, one can minimize this problem, and potentially the number of different placement machines on the factory floor.

Placement machines have a finite number of feeders for presenting the component to the placement quill. The placement quill is a vacuum pipette that holds the component during pickup, transport to and placement on the board. If the maximum number of component feeders is exceeded then the dilemma is created of whether to buy a new placement machine (at a cost of several thousand dollars), manually place the extra components off-line, or persuade the designer to reduce the number of different parts.

Publish Up-To-Date Design Guidelines

Written design guidelines along with a standard part library form the starting point of any DFM system. In some companies this may be the entire DFM system. Written design guidelines are marginally effective if they are based on somebody else's processes, are written with difficult terms and expressions, use poor graphic examples, and are not current. Several companies are very diligent in the preparation and maintenance of their design guidelines. Most are very remiss.

Design guidelines describe the reliability, quality, and manufacturing requirements of the company and general customers. The guidelines, at a minimum, list the allowable printed wire board parameters (such as coatings, platings, and materials), the minimum part spacings, the solder joint features (pad and land). The guidelines should also define the critical process steps and how to design well within the capabilities of these processes. These include accurate phototool generation for the board and the stencil fabricator, the placement coordinates for the placement machine, and the hole locations for any test and tooling fixtures. The best design guideline maintenance systems have copy and revision controls.

To help keep the guidelines current and relevant, several companies are experimenting with other ways to provide the designer the guidelines. One increasingly popular way is to use the power of low cost mini and personal computers. This enables some companies to publish and maintain the guidelines on a network of computers. The guidelines document can have embedded tables of tolerance and feature limits that can be directly accessed by the design engineer's computer aided engineering (CAE) and computer aided design (CAD) software. With this feature the guidelines document is both a reference to the engineer and a controller of the design tools. Advanced features of the computerized design guidelines include the ability of a software program to calculate the estimated product cost and yield using the allowable design guidelines, parts and materials. This type of program can be expanded to estimate the cost and yield of any "non-allowed" parts, materials and design features that may be desired by the designer.

Design for Variability

Design for variability does not mean to design for every conceivable variation or event that can occur to the product. This requirement means to allow room for the significant, uncontrollable, and predictable variations. The predictable variables and the magnitude of variation are provided by measuring the process capability, Cp. In many cases, the process can be controlled tightly or is inherently capable, so there is no need to design for variability. Other processes are less controllable or would be unreasonably expensive to control.

A good example of an uncontrollable variability is the lead width variation on fine pitch integrated circuit packages (see Chapter 6). The suppliers set an industry limit on the lead width tolerance. It is unreasonable, in most cases, to try to make the tolerances tighter. The simplest solution is to design pads so they are wide enough to reliably hold the lead regardless of its width.

Another example of uncontrolled variability is when the lead has variability in its planarity and alignment. The design for this variability is not as easy or obvious as in the preceding example. In this case, manufacturing experiments would show that adding more solder paste will compensate for the planarity variations. However, more paste increases the possibility of shorts. To minimize the shorts, the pads can be designed to be longer than normal so that the extra pad length wicks away the excess solder and breaks the solder short. Another example of designing for variability is the printed wire board's tooling holes. Adding localized placement location (fiducial) marks adjacent to the fine pitch packages enables the use of a vision system to guide the placement quill. With the help of the fiducial marks (shown in Figure 12.6), the placement accuracy will be assured to be within ± 0.03 mm (± 0.001 inch). Without the fiducial marks, part placement may vary as much as the tooling hole variation. PCB tooling holes may vary by ± 0.25 mm (± 0.010 inch) or more, which creates unacceptable part registration.

Identify Defect Rates by Cause

Identifying the causes of defects goes beyond the task of researching a problem and engineering a solution. The task should involve the determination of cause and effect. The conventional cause and effect approach segments the effects of into six

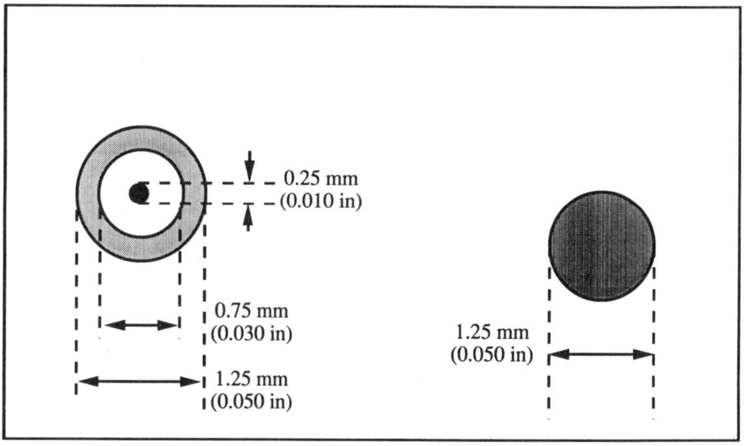

FIGURE 12.6. Fiducial marks etched in the copper foil significantly improve placement accuracy when the placement machine is equipped with a vision system.

different defect causes: people, methods, machinery, materials, environment, and management (see Figure 12.7).

This conventional diagram unfortunately ignores two important defect causes, one of which is the topic of this text. The first additional cause is design, and the second is interactions. The modified or DFM cause and effect diagram is shown in Figure 12.8. The causes due to design are to highlight effects that are due to the inherent limitations, mistakes, or oversights in the design guidelines, equipment, or other design controllable practices.

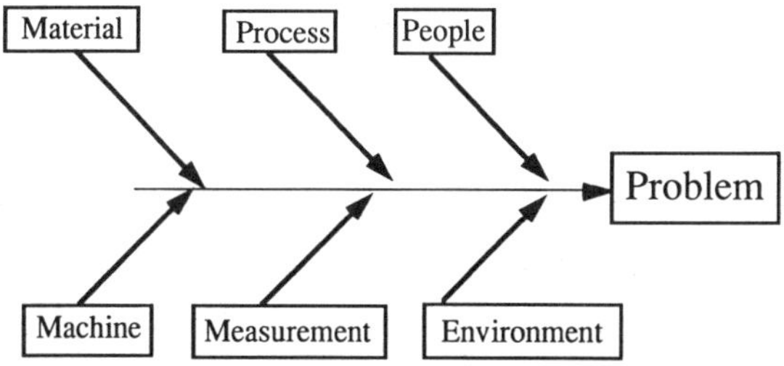

1. State the problem.
2. Draw a fishbone diagram.
3. Brainstorm causes and assign to proper fishbone category.
4. Determine most likely cause, by consensus agreement, gather data and analyze to verify.

FIGURE 12.7. The conventional Cause and Effect diagram, also known as the "fishbone" diagram.

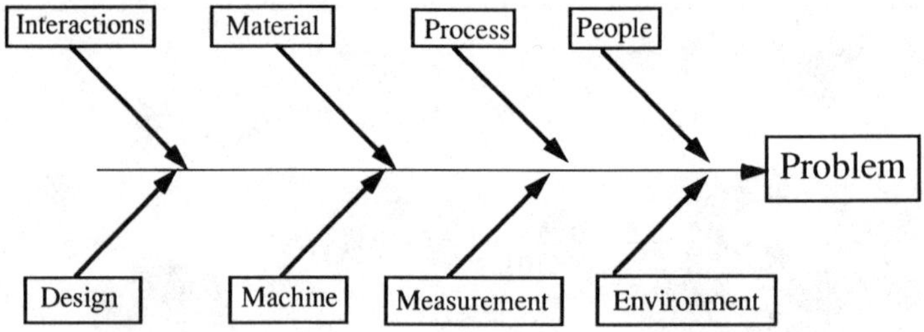

1. State the problem.
2. Draw a fishbone diagram.
3. Brainstorm causes and assign to proper fishbone category.
4. Determine most likely cause, by consensus agreement, gather data and analyze to verify.

FIGURE 12.8. The modified Cause and Effect diagram to isolate design related and interaction causes. These causes are usually buried in the conventional diagram use.

The interactions category of causes will, when the defect analysis exercise is begun, be a bucket for miscalculation effects that are too complex to attribute to any of the other obvious causes. Over time, the effects in this bucket will be the most difficult to analyze and control.

12.2 IMPLEMENTING DFM

The successful implementation of a design for manufacturability system requires climbing over several hurdles and developing a thick skin. DFM is much talked about, but, a uniform definition does not exist. To some, DFM is simply a book of design guidelines and a parts library. To others, DFM is a computerized model that allows the designer to enter data such as material lists, package spacings, and test fixture information, from which, the DFM model predicts the product cost and yield for the matrix of options available to the designer.

Whatever the sophistication level of the model, the primary goal is to provide to the designer guidelines and boundaries as quickly as possible and as early as possible in the design cycle so that a new product can be produced optimally and quickly the first time.

The author's DFM implementation steps are called "PPMS for DFM: people participation, predictive tools, management commitment, and suitable parts and materials (Figure 12.9).

People Participation

Enlisting the support and commitment of the people in the company to the goal of implementing DFM is paramount. Regardless of the degrees of automation of the

* People Participation

* Predictive Tools

* Management Commitment

* Suitable Parts and Processes

FIGURE 12.9. The PPMs of DFM.

design for manufacturability process, there will always be the need for human involvement. DFM involves the evaluation of several tradeoffs, can only be accurately judged if sufficient and valid information is available. The information needed includes cost, ease of use, and availability. For example, it is of less apparent value to design in a part that costs 50% of a competing part if it is not as readily available as the more expensive part. It is also unwise to use a part that requires an assembly process that is very variable and difficult to control when compared to an apparently more costly part. People are needed to generate the information of the tangible and intangible value, advantages, and disadvantages of the various tradeoffs.

As viewed by most companies, design for manufacturability is new and in experimental development. The benefits of DFM are easily accepted by everyone in the company. The realization of these benefits is dependent on the willingness of the people, who are required to collect and analyze data, to decide on the tradeoffs, to design the product, and to accept the discipline of the DFM system.

The writing or updating of the design guidelines manual is the first task. The writing requires the participation of a team of engineers, who are called upon to impart their years of experience in this document. The design guideline manual is the benchmark of manufacturing, reliability, and quality boundaries. Additional expertise in authoring this manual should come from management, test, purchasing, and external expert groups. The line workers should also be asked to contribute with inspection, rework, and repair problems that can be resolved by following certain design guidelines. After the other groups have authored their DFM requirements it is the group's responsibility to blend these requirements and to resolve any conflicts and overlaps. Next, the manual should be presented to a review team of designers. Their job is to review it and to suggest wording changes that make it easily understandable by the other designers. Also, this team can challenge any manufacturing boundaries that are set too tightly or unrealistically in the designers' eyes. Once consensus is reached, the manual is issued to the designers as the benchmark guideline.

A design review at several points in the development cycle is the next people participation area. To an extent, these are judgment sessions where the manufacturing, test, and quality engineers are invited to see if the designer is faithfully adhering to the design manual limits that he has been asked to obey. However, if the reviewers enter too many design reviews with this attitude, the design engineer may assume the role of a criminal defendant and little positive rapport is developed. Also, as a consequence, fewer follow-on design reviews will be conducted.

The reviewers should judge adherence to the design manual, but they should also provide information. Information of process changes, new equipment capabilities, and new materials should be a major agenda item of these reviews so the new product can take advantage of these improvements.

Two simple techniques that help stimulate the reviewers and the designer are the gift check and the new idea reward. The gift check is a game to reward the reviewers for every mistake they find. A gift is specified by the designer for each mistake. Gifts to be paid by the designer, can range from a piece of gum to a brand new car depending on the designer's confidence in his design and the design's complexity.

The new idea reward is a trophy or cash award (say $5.00) that can be awarded at the discretion of the designer to the review team member who contributes what the designer feels is the best new idea or improvement suggested at the design review. The trophy or $5.00 does not stay with the award winner, it is returned at the next review and awarded for the best idea at that meeting.

These two tools have worked well in the author's experience, as they reduce the tension of a very tension causing task. They also stimulate the creativity of the participants. Even a simple piece of gum motivates some reviewers to challenge the smallest design manual infraction.

Predictive Tools

The design manual needs to be a realistic, dynamic, and living document. If the people who produced it allow it to go unchanged for long periods (this may be as short as one month depending on the changes in manufacturing and new product developments) then the manual becomes obsolete. This gives a designer reason not to follow the design guideline manual criteria.

The manual needs to be current. Predictive tools must be installed in the critical areas of the factory to achieve this. The job of these tools is to predict or simply measure changes in the process, the materials, customer requirements or customer satisfaction.

The simplest example of a predictive tool is the tally sheet. Tally sheets are commonly used to record the visual defects found on an assembly after the soldering and cleaning steps. Several tally sheets from closely related manufacturing periods or from similar products can give an accurate picture of the problems occurring. These problems should be ranked in order of most frequent occurrence. This will require the tabulation of the sheets and the calculation of a Pareto chart. A Pareto chart is simply a ranking of the problems by descending order of the number of occurrences. The tally sheet creators must use a common language that is specific to a problem type in order for the tally sheet to be accurate. Simply listing on the tally sheet that an assembly is "broken," "it doesn't work," or "U39 is bad" are invalid entries since they are not specific. An example of a properly prepared tally sheet is shown in Figure 12.10.

Creating a tally sheet by itself does not make it a predictive tool. In the case of an inspection tally sheet and accompanying Pareto chart, someone must now analyze the causes of the top few problems that cause the majority of the problems. From the analysis of Pareto, this means to analyze "those 20% of the problems that cause 80% of the defects". Pareto ranking is also known by many as the 80-20 rule. A Pareto chart of common electronic assembly defects after soldering is shown in Figure 12.11.

The analysis of the cause of the problems determine whether the problems are caused by shifts in the process, or in the design practices. Shifts in the process are the responsibility of the manufacturing engineers to repair. For example, if a problem of part misalignment is due to belt wear on the placement machine heads, those belts must be replaced. It would be unacceptable to impose tighter tolerances in the design guidelines manual to compensate for this wear. If, however, the

Defect Type	Product 1	Product 2	Product 3	Product 4
Solder Opens	20	10	15	6
Solder Shorts	13	18	28	2
Misalignment	31	19	1	3
Tombstones	5	14	5	0
Missing Comps.	13	3	4	0
Solder Balls	14	28	2	0
Reversed Comps.	11	2	13	1
Damaged Board	1	0	0	0
Flux Residue	0	0	0	0
Damaged Comps.	4	5	9	1

FIGURE 12.10. An example of a simple tally sheet.

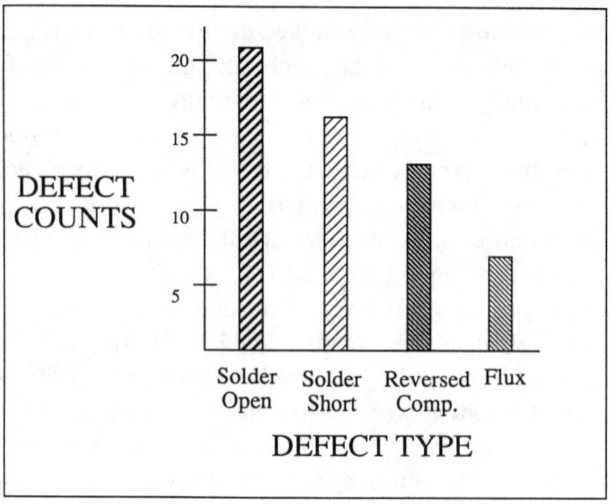

FIGURE 12.11. A Pareto chart is a powerful tool since it provides focus and "marching orders" for the problem solvers.

problem is due to something that was overlooked in the initial design guidelines manual or the use of new components that react differently to the process, then the design manual must be changed.

Another valuable predictive tool is the capabilities experiment. Several techniques have been developed for the tool known as "design of experiments." These techniques can measure the causes of a problem by varying several of the variables individually while holding others constant for a relatively small number of sample sizes and tests. The basic technique is to vary the input variables one at a time and to measure the output or process response. The variables that cause the highest output responses are the best candidates to impose tighter design controls, or to avoid, or to develop a more capable process.

Capabilities experiments can and should be conducted for every major step of the process. These experiments serve two purposes. The first is to develop a list of those variables that the process responds to by creating test problems and measuring resultant waste. Second, the experiments predict how much waste is created due to a known cause. This second purpose recognizes the fact that the process and materials will never be perfect, but their contribution to waste can be minimized by restricting certain input variables and reducing the magnitude of the variation in others.

Both the tally sheet/Pareto analysis tool and the capabilities experiment tool are simple to implement. They require only people participation and dedication. Unfortunately, that participation on an ongoing basis is difficult to maintain. Manually producing tally sheets and Pareto charts is a tedious process, even for small quantities of defects. Capabilities experiments are usually performed by manufacturing engineers whose priorities always seem to change with other company imposed demands. Fortunately, the power of inexpensive computers allows most of data organizing and analysis to be automated.

Management Commitment

No design for manufacturability system will survive if the company's management does not support it. Like a child, a DFM system needs supervision, nurturing, and growth opportunities. Simple verbal approvals are not enough.

Over the past decade, managers have found that improving product quality requires their constant vigilance. DFM requires the same, but the rewards are higher (see Figure 12.12). DFM provides reduced development time, lower product cost, and, because of easy transition into production, less emotional cost. The most successful DFM systems in use today are in companies where the managers first sat down to learn and understand the DFM process and then assigned task teams with the proper responsibility and authority to define a system. They actively supervised its implementation, and they contributed with questions and new ideas. Most importantly, when the team requested money for implementing better tools and justifiable automated tools, the management approved their requests.

The managers of these companies made sure that everyone in the company was

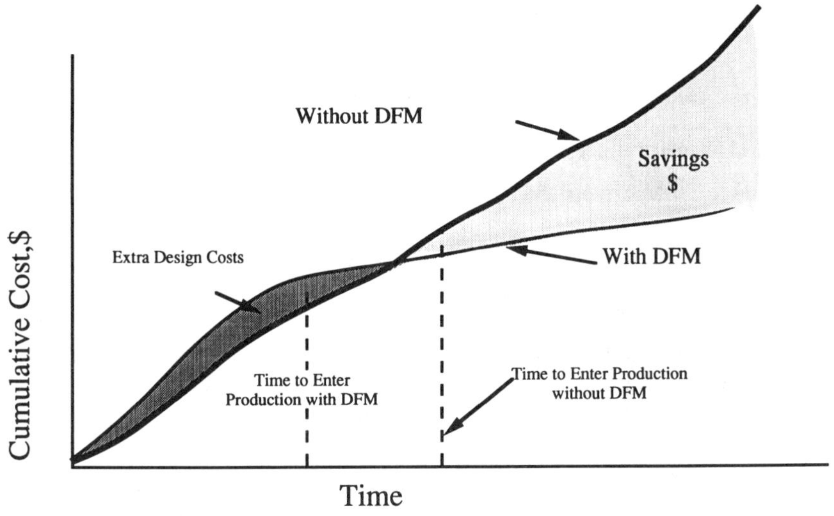

FIGURE 12.12. Consistent improvement in the use of DFM results in significant profit improvements.

aware of the DFM program, its benefits, the company's expectations, and its progress.

Suitable Parts and Materials

Suitable parts and materials are determined by their quality, reliability, ease of use availability, tolerances, and, lastly, cost.

The acceptability of a new part or material cannot be made solely by the designer. The assembly processes are too complex and customer requirements too demanding to risk even the slightest oversight in a part feature. Testing of the above determinants and comparison with other tradeoffs are a must before acceptance of the part or material into the parts library.

The unacceptability of a new part may be for a reason of economics beyond the designers knowledge. A good example is the number of different values of resistors and capacitors used in all of the company's products. The piece price of different resistor values are the same. But if too many different values are allowed, the company will need to buy several placement machines so that enough part feeders are available without the need for retooling.

12.3 DFM SUMMARY

A design for manufacturability system provides the electronic producer the opportunity to achieve cost and development benefits and timeliness beyond the benefits of previous quality and producibility improvement programs. A DFM system addresses a root cause of defects usually ignored by these other improvement programs.

The steps to implementing a DFM system may vary from company to company but the benefits should remain the same. These benefits are higher quality, higher reliability, faster time to production, and lower cost.

A suitable DFM system may be a written design guidelines manual for those companies able to maintain it properly. This has not proven to be the successful DFM system for most companies.

Utilizing the power of inexpensive computers and software, companies now have tools that are able to measure and predict the quality and cause of defects and waste. This information can be used to implement solutions such as more capable processes and machines. The information also enables the evaluation of tradeoffs and evolution towards using only suitable parts and materials.

As the tools of DFM become more advanced trending towards the "Printed Wiring Assembly Dream Factory" the system is still critically dependent on the people participation. Their acceptance and adherence to the disciplines required by their company's DFM system is the key to success regardless of how sophisticated the system is compared to others.

Reference

1. Bancroft, C.E. 1989. Design for Assembly and Manufacture. Electronic Materials Handbook, Vol. 1. Materials, Park, OH: ASM International.
2. Marcoux, P. 1991. Design for Electronic Printed Wire Assemblies. Manufacturing Engineers Handbook, Vol. 6. Dearborn, MI: Society of Manufacturing Engineers.
3. Marcoux, P. 1992. PCA Design Guidelines Manual. Sunnyvale, CA: PPM Associates.
4. Parden, R. 1990. Private correspondence.
5. Spitz, L. 1990. Design for Manufacturability. In *Electronic Packaging and Production*, Nov., 1990.
6. Stout, G. 1990. Private correspondence.
7. Wojslaw, C. 1990. Private correspondence.
8. —1991. Time-To-Market. *In Evaluation Engineering,* May, 1991.

13

Specific Design Guidelines for FPT Packages

13.0 THE ABILITIES EXPECTED FROM DESIGNERS

This section contains design guidelines to satisfy the various "abilities" that a product must meet. The guidelines are derived from the author's and other's experiences of the past several years involvement with fine pitch and surface mount technology. The abilities that are incorporated in this section are:

- Designability
- Manufacturability
- Repairability
- Reliability
- Testability
- Marketability

This chapter is organized in the following manner:

1. Package selection
2. Package placement on the board
3. Land pattern sizes
4. Trace routing and protection
5. Thermal management
6. Fabrication of the board
7. Fabrication of the stencil

13.1 PACKAGE SELECTION

The integrity of the integrated circuit package is a fundamental determinant of the product's reliability, cost, and manufacturability. This section summarizes the key issues to consider when a user has more than one package option. If there is only one package option, then these issues will provide a wish list for the user to present to the supplier. The key issues are summarized in Table 13.1 and expanded upon in the text.

Package Designability

The key issues with the package for designability are package area and ease of routing. The area of the TAB, TSOP, and guard-ring packages favor this issue since they are smaller than the QFP style packages. The printed board area versus lead count and package type is shown in Figure 13.1. The ease of routing traces from the integrated circuit package to other components is determined by how many circuits need to be connected, their location, the pin-out arrangement on the IC, and how many leads are bussed. In general, the packages with leads on two sides are easier to route than the quad sided leads. And the dual-sided packages occupy less area in lower pin-counts, as shown in Figure 13.2. The dual-side packages

TABLE 13.1 Summary of Key Integrated Circuit Package Issues.

ABILITY ISSUE	DESIRED FEATURES
Designability	The smallest package occupying the least board area. Easy board trace routing to the package leads. The lowest cost, least hassle, "no-complaints" package.
Manufacturability	Many consistent, reliable package sources. Lead planarity is within 0.6mm (0.002"). Package dimensions are toleranced to fit tightly within the allowed variation of the rest of the process so no yield loss occurs due to package size variation. The package doesn't absorb moisture and experiences no adverse effects regardless of the solder reflow method used. The leads are readily solderable after exposure to time and normal handling environments.
Testability	The package provides sufficient leads to allow boundary scan circuits to be placed and accessed within the IC. The package is easily tested with the AOI equipment used in production, this includes placement and "Pin 1" locator marks.
Repairability	The leads are easily heated for package removal and repositioning. If the leads are damaged in handling, they should be easily repaired.
Reliability	The package area is small. The length of the lead foot is long enough to guarante minimum use conditions. The lead pitch is the finest that manufacturing can assemble. The package is easily cooled to keep the IC and the solder joints close to ambient temperature.
Marketability	The package adds minimal cost, doesn't degrade the product performance and enhances the customers value perception.

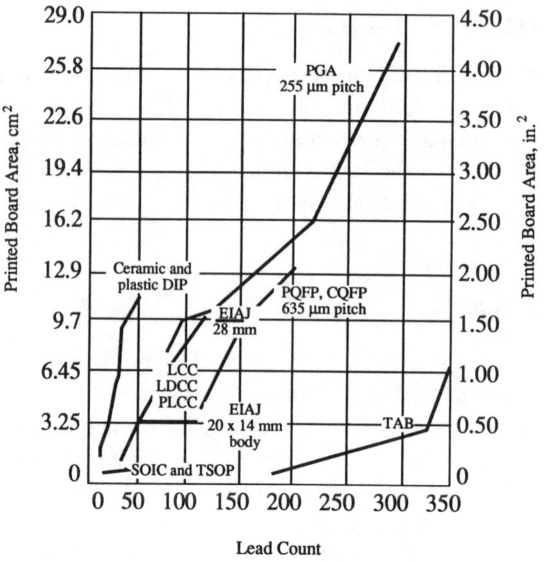

FIGURE 13.1. The approximate printed circuit board area used by package type and lead count. (Source: National Semiconductor)

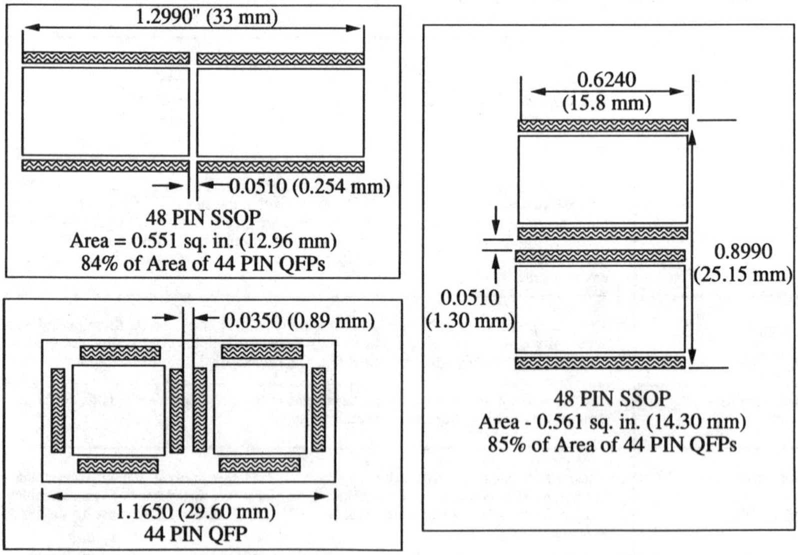

FIGURE 13.2. Dual-side leaded packages occupy less area for low lead requirements than quad-side leaded packages. (Source: Texas Instruments)

include the TSOP and VSOIC. The TSOP is already the favored fine pitch package for memory devices since the dual side lead configuration allows the packages to be placed side by side and traces routed to the inputs on one side and the outputs on the other. This is illustrated in Figure 13.3. Unfortunately, the dual side lead configuration limits the pin-out of the package. Therefore, the quad side lead configuration is needed.

Traces that are too difficult to route to leads on the quad sided packages can be connected using vias on the board to route the traces under the package and intersecting traces. The vias connect the traces to the various subterranean signal layers within the board's multilayers. In some cases the vias may extend partially through the board. If the via proceeds from the board's surface to a sublayer, it is called a semi-blind or blind via. If it proceeds from one internal layer to another internal layer, it's called a buried via. Blind and buried vias are represented in Figure 13.4.

The quest to find an integrated circuit package that offers the lowest cost, least hassle, and no complaints quickly points to the heart of a designer's life long role. Design is a task of resolving conflicts. Fine pitch packages resolve some previous conflicts, but unfortunately, the current selection of fine pitch integrated circuits does not provide an easy no-complaints solution. Table 13.2 lists the author's opinion of the package choices to meet the various design features discussed.

FIGURE 13.3. Illustration of a dense memory card showing the advantage of dual-side leaded packages.

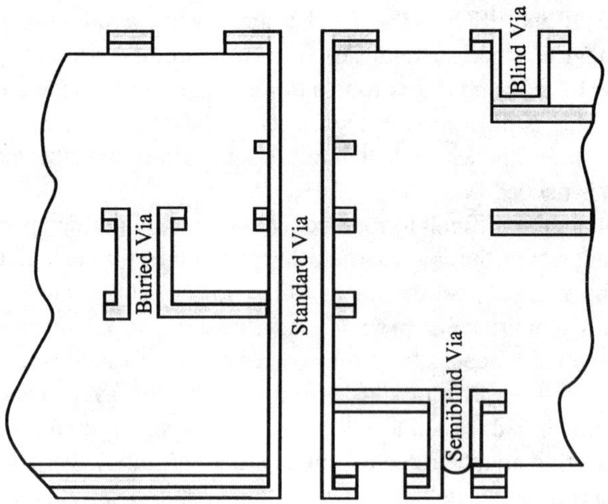

FIGURE 13.4. Blind, semi-blind and buried vias are techniques for reducing board area and layer counts. Their use needs to be weighed against cost expectations and the fabricators process capabilities. (Source: J. Fjelstad, Elf Technologies, Inc.)

TABLE 13.2 The Preferred Package Choices to Meet Various Design Features.

Feature	Package Choice
Smallest Package Area	TAB, TSOP, Guard-Ring Package
Dual-Side Leads For Easy Routing	TSOP, VSOIC
Quad-Side Leads For Maximum Lead Count	QFP, PQFP, VSOP, Guard-Ring, TAB
Lowest Cost, Least Hassle, "No-Complaints" Package	To Be Determined

Package Manufacturability

A package suitable for manufacturability needs to meet several demands. Foremost, the leads shall be solderable. This means, for one thing, that the lead base metal of copper or alloy-42 must be coated with a solderable metal that maintains its solderability after conditioning and storage. The solderable metals of choice are tin, tin-lead, and gold. Gold is best for long-term storage but is the least acceptable in most applications because of its higher cost and need for pretinning before board attachment. There are differing opinions among the industry's soldering experts as to the precise definition of solderability. The most functional definition and standard is that the leads shall be easily wettable with solder using the flux and reflow heat source of the user's choice. Wettable means that the solder forms a metallurgical bond to the majority (greater than 95% is a reasonable expectation) of the lead area to be contacted.

Another issue of solderability is that the leads be within the planarity tolerance so that they make contact with the solder. Unfortunately, solder suppliers have not found a solder with the ability to jump up and pull nonplanar leads into contact with solder. For surface mount pitch applications (lead pitch greater than 0.65 mm) the assembler can deposit a higher pillar of solder paste than is typical for fine pitch applications. Surface mount pitch (> 0.65mm) allows the package lead planarity to vary as much as 0.1 mm (0.004″). Fine pitch lead planarity is constrained to a maximum of 0.05 mm (0.002″).

The fine pitch packages most robust to meet this tough planarity requirement are those with the leads constrained during normal pre-assembly handling and shipping. The constrained lead packages are the TAB and guard-ring types. These leads remain constrained and protected until excise, form, and placement.

The VSOIC, PQFP, and VSOP packages have more robust leads than the others. And, they are shipped in tubes so there is minimal handling damage. Shipping damage to the leads of the QFP style packages was a major problem until new shipping (or matrix) tray standards were adopted by the EIAJ and co-adopted by EIA-JEDEC in 1990. The newer trays provide excellent support for the parts during handling. But they do not solve the potential for lead damage to the package during testing, burn-in, or tinning.

Of the two solderability issues, lead coating and lead planarity, the lead planarity problem accounts for 80 to 90% of the current assembly solder problems. Electrical problems of nonfunctionality or out-of-specification parts generally far exceed the solderability problems especially on custom ASIC devices. Electrical problems on custom ASICs can have defect rates of 1,000 to 100,000 parts per million (ppm) versus solderability problems of 5 to 500 ppm.

Dimensional tolerances on the package and lead sizes can readily add up to a level that the product is difficult to assemble consistently in volume (see Chapter 6.1, Table 6.4 which indicates that the current lead variability is ± 0.08 mm). The placement machine and printed circuit board fabrication process have an additional finite amount of variability. When a package supplier proposes package tolerances and lead sizes to a user, the placement and board variations must be considered. So a variability budget can be developed.

The development of a variability budget follows the line of any budget discussion. If the budget spending must be spread around to several spenders and if some of those are big spenders, then some of the other spenders will be shortchanged or will have to do without.

Component tolerances are set by the package supplier. If the supplier wants to establish these as a standard with either EIAJ (Japan) or EIA-Jedec (U.S.), the supplier only needs to find one more supplier and a user to cosponsor those dimensions and package outline for registration. The registration is sent to the member representatives of the two package standards organizations. The member representatives are usually employees of member companies of EIAJ or EIA. Generally the representatives are technical marketing personnel of integrated circuit suppliers who have limited knowledge of the user's requirements. After a majority of the member representatives vote to approve the proposed outline it is registered and becomes, for all practical purposes, a standard (even though EIA stresses that it is only a registered outline). Sadly, the package tolerances are set and approved by a very small body of people who have only a passing knowledge of user needs. This situation has improved with the creation of the Surface Mount Council in 1987. The Surface Mount Council brings members of the EIA-JEDEC together with members of the IPC (Interconnects-Packaging-Connections Association, Lincolnwood, IL) and representatives of the five U.S. Government organizations (Army, Navy, Air Force, Department of Defense and NASA).

The SMC has enabled EIA-JEDEC to get a technical review from an august user group before finalizing any proposed package registrations. Along with the tolerance review, the SMC will issue suggested user guidelines, such as land pattern dimensions, through the industry standard design guideline document, the IPC-SM-782. A similar international surface mount council was started in 1991. This council includes representatives of Europe, Japan and the United States. This council is chartered with the review of package tolerances from suppliers in one country by users in another, so reasonable dimensions can be established.

A summary of the EIA package outline numbers is listed in Appendix B, representative package dimensions of some of the fine pitch packages are shown in Appendix C.

All plastic packaging compounds are capable of absorbing and storing moisture. The moisture may cause package bodies to crack during solder reflow. Therefore, it is recommended that fine pitch packages be prebaked by the supplier and delivered in whatever packaging bags the supplier uses to prevent reabsorption during shipping and storage.

Package Testability

As discussed in Chapter 11, electrical test of fine pitch packages assembled to a board is a difficult task if test routines and diagnostics beyond a functional "go-no-go" test level is desired. It is suggested that on-chip and on-board boundary scan circuitry be incorporated to provide the test capability. Adding boundary scan circuitry to the chip during the design of a custom circuit adds a minimal amount to the silicon area and cost. It also increases the number of package leads needed to interconnect the circuits. At the component level, the extra area and cost may seem difficult to justify. However, when the cost of in-circuit testing and the number of electrical rejects after assembly is factored into the total product cost, one will find on-chip and on-board boundary scan test circuits easy to justify.

If board level in-circuit testing must be used, then a test probe pattern as shown in Figure 13.5 should be used. The test points are circular pads located away from the land area. The pads can also function as vias to connect to other layers of the board. This indirect approach avoids the possibility of a probe point pressing a lifted, unsoldered lead to the land, if the test probe is designed to contact the package lead directly.

The minimum test pad size recommended in Figure 13.5 is 0.8 mm (0.031″). This is the smallest pad that can be probed without a significant probability of probe misses. Smaller pads suffer from high miss probabilities (see Figure 13.6).

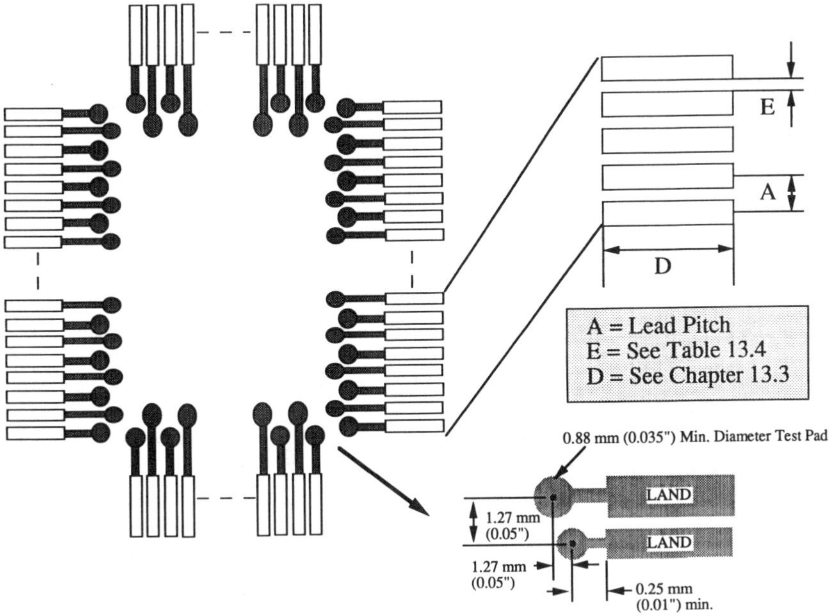

FIGURE 13.5. The suggested footprint of land and test pad and via locations.

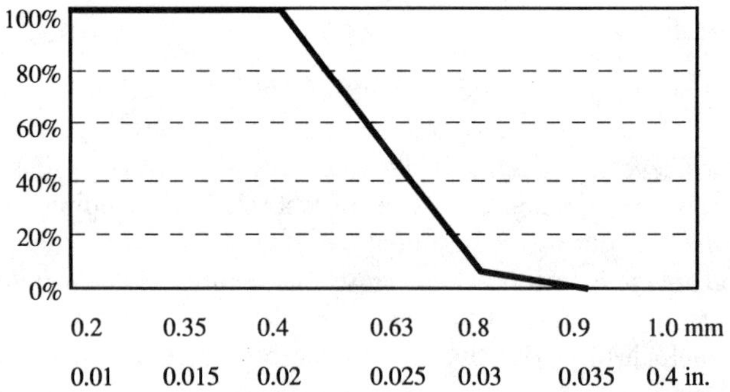

FIGURE 13.6. Smaller test pads and vias are possible, but are not practical unless tolerances and process capabilities assure no mis-probes. (Source: IPC-S-782A)

Inspectability is another important aspect of testing. A significant number of assemblies require rework due to incorrectly oriented packages. Regardless of the source of the problem, it can be minimized if there is an easy and consistent mark on the package that indicates the proper orientation. Some designers are of the opinion that the marking printed on the package is sufficient. However, the marking is not always printed on the package in the same direction.

The best mark is one that is molded into the package. There are no industry standards for molded package orientation indicators. As can be observed from the package indicators shown in Figure 13.7, there are several variations. By simply asking the assemblers for a recommendation of the preferred markings, a designer can do much to help direct suppliers into adopting a standardized set of markings.

As automated optical inspection (AOI) becomes common usage, the choice of molded markings will need to fit the AOI systems capabilities. Generally top side markings will not be visible to these systems, so the use of slanted corners, enlarged and marked end leads, and chamfers are preferred.

The approaches of a slanted corner and enlarged end pin marking will be observable by the vision system on the placement machine. The chamfer mold will not be observable but is valuable for those integrated circuit packages shippable in tubes. The supplier uses a chamfered tube so the package can be oriented into the tube in only one direction.

Package Repairability

The removal and repositioning of fine pitch packages that are electrically bad or misassembled requires the skills of a special operator and expensive rework equipment. The individual's skills must be a blend of an engraver and a microsurgeon. People with these skills are usually not inexpensive for any long period of time. Therefore, it is best that a designer choose a circuit, a supplier, and a package

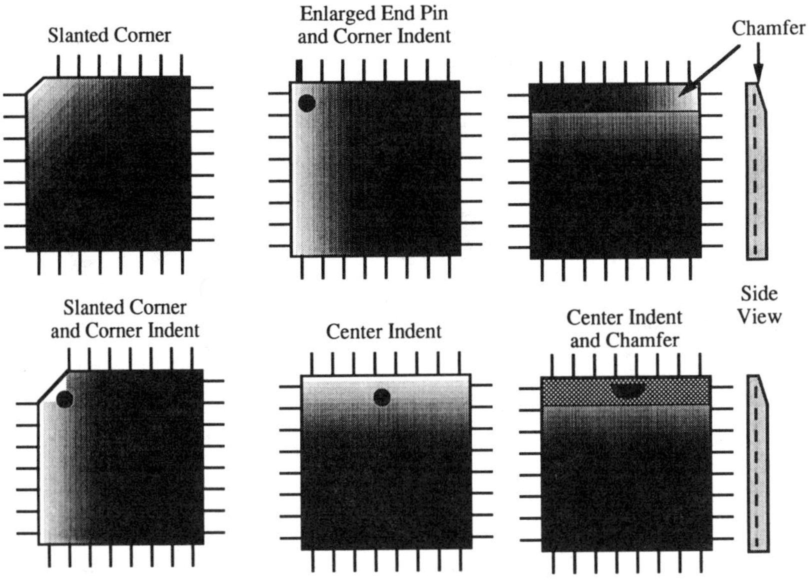

FIGURE 13.7. Fine pitch IC packages lack pin 1 marking standards. As a result, several variations exist.

that, according to the consensus of everyone involved in assembling it to the board, has the lowest probability of needing repair.

Package Reliability

The best package choice to optimize reliability is one that has the smallest package area. However, the larger packages are also very reliable if the leads are long. The length of lead should provide for adequate bending distance so the lead has high flexure, and a long enough foot so there is plenty of solder area. In general the length of the bend is determined by the package height, the standoff height, and the bend radii. These are illustrated in Figure 13.8a. Alternative lead bends are shown in Figure 13.8b.

The preferred foot length, L_{foot} in Figure 13.8a, is twice the lead width for most applications. However, some severe applications such as under the hood automotive applications will benefit from longer leads. The best way to determine the safe length is to calculate the required area using Equation (10.3), the fatigue prediction model of a leaded package, solved for the solder joint area, A, instead of the solder joint failure probability. To solve for A, the designer must determine the key parameters such as package size, the temperatures, and the lead flexure. The desired failure probability must also be assumed. This calculation also provides the basis for the land pattern area as will be described in Section 13.3.

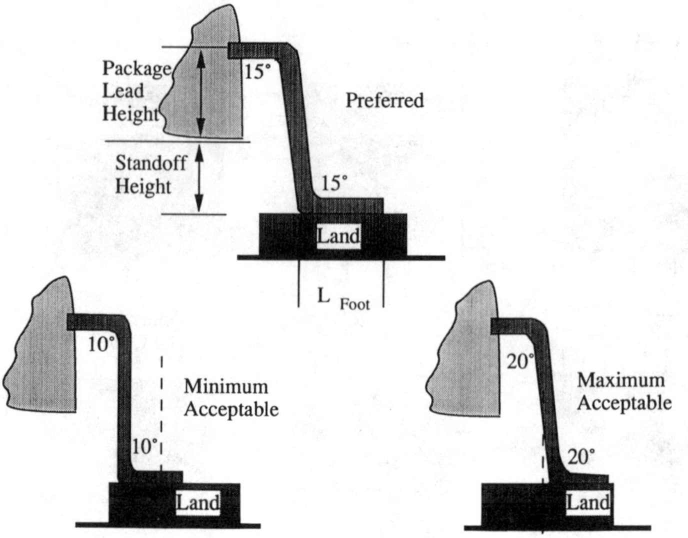

FIGURE 13.8a. The preferred gull lead bend and tolerance to maintain optimum flexure. (Source: IBM)

Cooling the package and solder joints is a valuable way to improve reliability. As described in Chapter 10, the package and solder joint temperatures contribute greatly to solder fatigue. This is much more the case with fine pitch packages since they generally house larger power consuming integrated circuits than smaller through-hole packages. Shown in Figure 13.9 are the average thermal resistance

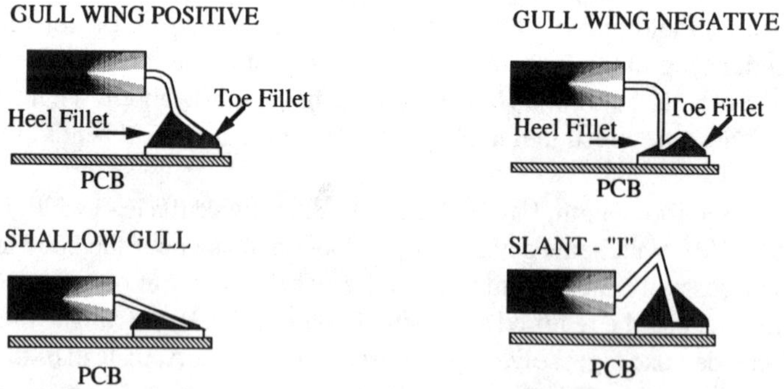

FIGURE 13.8b. A diagram of alternative "gull-lead" and "I-lead" shapes for fine pitch packages. (Source: National Semiconductor)

Average Thermal Resistances of Various Packages						
Package Type	Average Thermal Resistance, Theta Junction to Ambient, in °C/W					
		Air Flow, Linear Feet per Minute				
	Free Air	250LFM	500LFM	750LFM	1000LFM	DIE SIZE
						Sq. Mils
28 PLCC	80.7	60	46.2	43	40	136 X 136
120 QFP	62.5	57.8	43	41	38.6	232 X 232
160 QFP	55.3	41.2	35.1	33	30.8	294 X 294
24 DIP	67	62	47	44.5	42	90 X 90

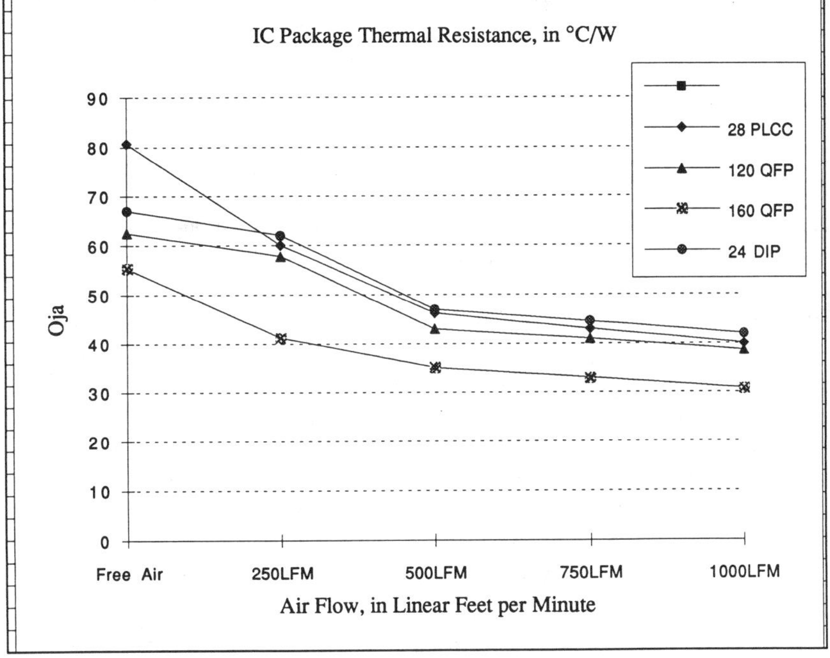

FIGURE 13.9. The average thermal resistance values of different IC packages and lead materials. (Source: Motorola)

values for a PLCC, DIP, and QFP package examples. The QFP packages display a slightly lower thermal resistance than the other packages. Part of the lower resistance may be explained by the additional heat flow through the dense copper leads in the QFP package.

Thermal resistance depends on the thickness and the thermal conductivity of the materials conducting the heat. The fine pitch packages being thinner and of the same plastic material as the thicker DIP packages should conduct heat readily to the package surface. Which means metal heat sinks glued to the top of a fine pitch package are good heat removers. However, the metal leads (especially copper) are very close to the chip and because of the high lead count provide a path of lesser thermal resistance.

Since some of the heat flows quickly through the leads and is collected at the small solder joints, the fine pitch solder joints will be hotter than for the same circuit operating in a DIP or pin grid array (PGA) package. The DIP and PGA having larger through-hole solder joints will spread any lead heat over a larger area. Therefore, a means of removing the heat from the fine pitch solder joints is needed for the more severe applications where metal heat sinks glued to the top of the package are not adequate. In these situations it is necessary to use the board as a heat sink, which is the topic of Section 13.5.

13.2 PACKAGE PLACEMENT ON THE BOARD

After all of the components, their packages, and suppliers are chosen and approved for the product, the board layout can begin. The initial phase of board design is package placement and orientation. When placing the packages on the CAD screen, it may be adequate to place the footprint pattern for a given package directly beside the footprint pattern of another package. This provides adequate package to package spacing if the footprint area includes adequate room for vias, test pads, and the various traces that will route through the package area. The footprint area will vary depending on the trace width and space, the number of board layers and the number of vias needed for interconnection and incircuit testing (if used). It is difficult to estimate the added area these require. Experience has shown that the footprint area can be 120 to 200% of package area. Despite the above variables, the CAD operator's skill and patience contributes another variable.

To allow for the highest packaging density most CAD footprints contain the absolute minimum amount of interconnecting area. This would include the package and lead area, the lands, and the vision/placement fiducials (see Figure 13.10). The CAD operator adds the necessary via pads and connecting traces during the interconnection phase of the board design.

The gull-wing leaded packages do not require additional space between the packages for human visual solder joint inspection. However, the J lead does and it is recommended, as shown in Figure 13.11, that a minimum spacing from one package edge to the next package edge of 3 mm be provided for human inspection. This is not required for many automated optical inspection systems.

The conventional norm is to place the integrated circuit packages in position based on the flow of the schematic. It is also common practice to place the circuits

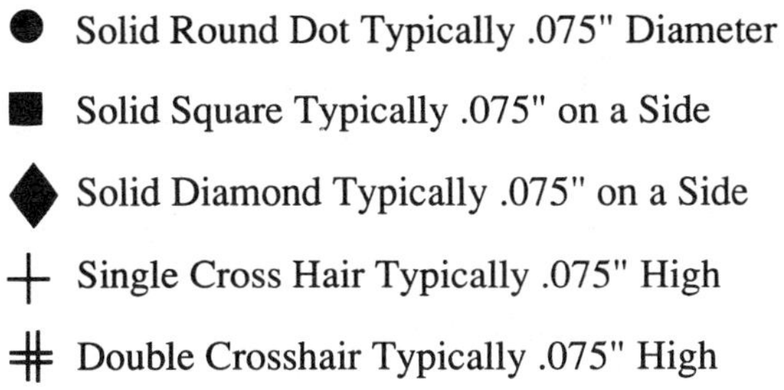

FIGURE 13.10. The various fudicial marks required by existing fine pitch capable placement systems.

by common operating frequencies. These conventions are illustrated in Figures 13.12 and 13.13.

Another convention is that the packages be oriented so that pin 1 is always in the same direction. This is usually toward the *top of the board,* the top being however the operator has it positioned on the CAD monitor. This convention is not required if the packages are inspected for orientation during placement and postsoldering by the vision systems. The placement machine and inspection systems are easily programmed to place and inspect the components on non-90° angles. Some systems are capable of placing parts on angles of 1° increments at accuracies of ± 0.125°.

It is generally not recommended that IC packages be placed on non-90° angles due to the complexity of processing. However, some components such as resistors that need to be thermally matched may benefit by non-90° placement.

The inexpensive ceramic chip resistors used on surface mount assemblies may have a high thermal coefficient of resistance, or TCR. The TCR for these may be as high as 500 ppm/°C. If the resistors are aligned in the conventional

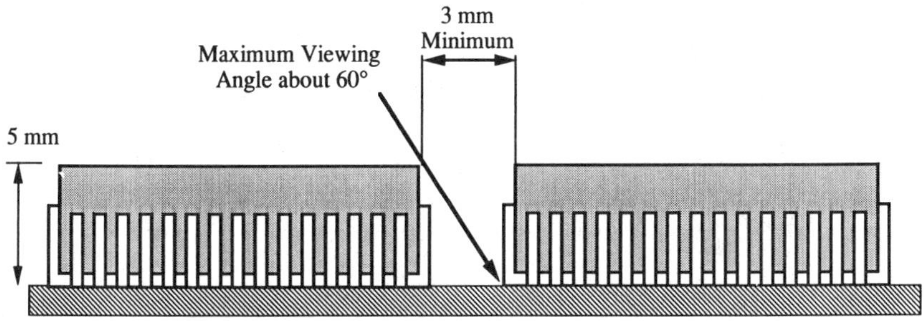

FIGURE 13.11. The "J" leaded PCC package needs extra space between packages to allow operator inspection and rework.

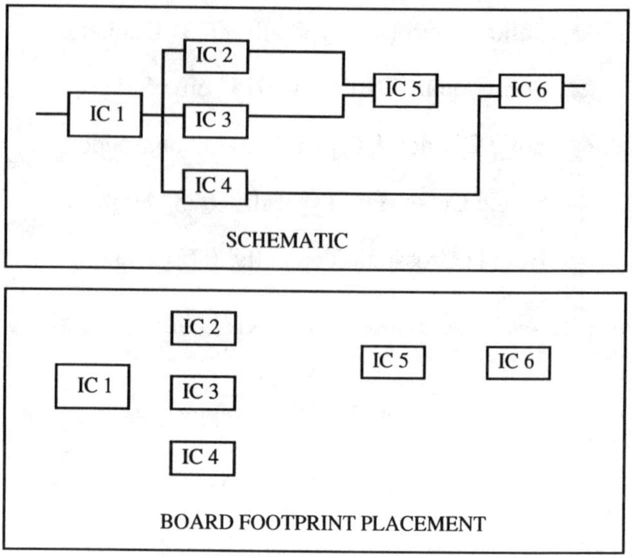

FIGURE 13.12. Components should be placed on the board based on the flow of the schematic.

90° angle, as shown in Figure 13.14, and experience the thermal gradients shown, then R1 and R7 may differ in value by 0.5% (10°C × 500 ppm/°C) due to the temperature difference. However, if R1 through R7 are to aligned along the same isothermal gradients, using non-90° placement, then they will track in value as the temperature changes. The non-90° placement to achieve thermal matching is shown in Figure 13.15. Several CAD software packages offer thermal gradient routines that will display the predicted temperature contours on the board.

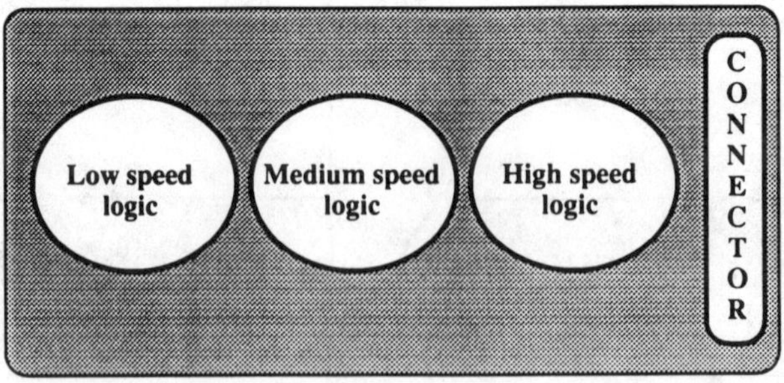

FIGURE 13.13. Circuits operating at similar frequencies should be clustered together in the board layout.

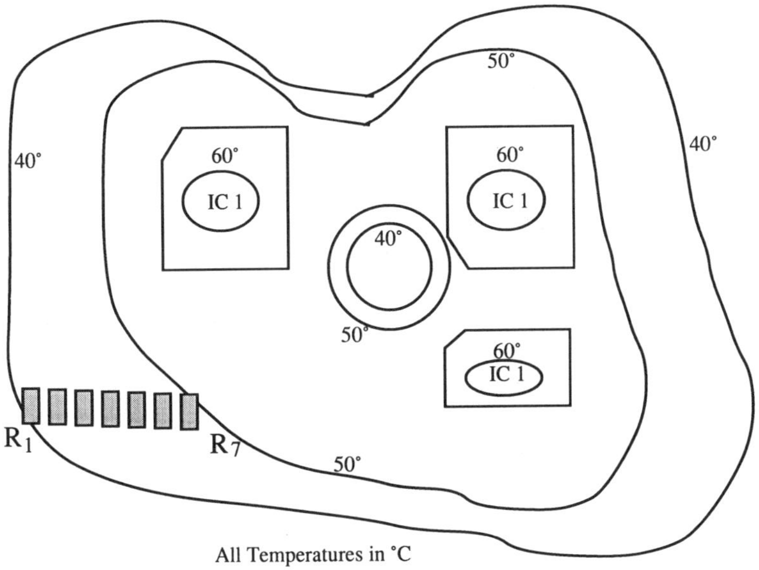

FIGURE 13.14. Resistors, in critical applications, won't match or track if the thermal differences between them is great.

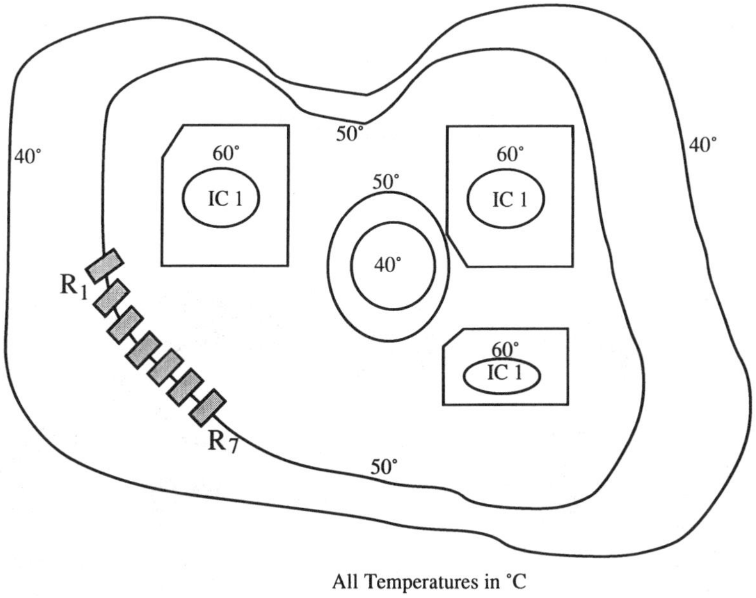

FIGURE 13.15. Resistor tracking and matching can be improved by aligning the resistors to the iso-thermal plane.

13.3 LAND PATTERN SIZE

The land pattern is the board solder joint area for each surface mounted lead and termination. The term *land* was chosen to distinguish the surface mount joint from the through-hole joint. The through-hole board feature is called a *pad with a hole*. Any holes too close to the surface mount joint rob the joint of its precious solder paste during reflow. So the land nomenclature developed to remind designers to keep holes suitably away from the area where the package lead *lands* on the board and is soldered. The ideal land size has sufficient area to assure the optimum mechanical support and electrical conduction. Additionally the land can be used to help in assembly by wicking excess solder away from bridged leads.

The land should be as wide as possible. The main limitations on the land width is paste printing and the copper etching of the board fabrication process. For fine pitches down to 0.3 mm the board fabrication process is the limiting factor (see Table 13.3). Therefore, the maximum land width is found by subtracting minimum trace space distance from the package lead pitch. For example, if the lead pitch is 0.5 mm center to center and the minimum trace distance guaranteed by the fabricator is 0.15 mm (0.006″) then the maximum land width is 0.5 mm − 0.15 mm = 0.35 mm.

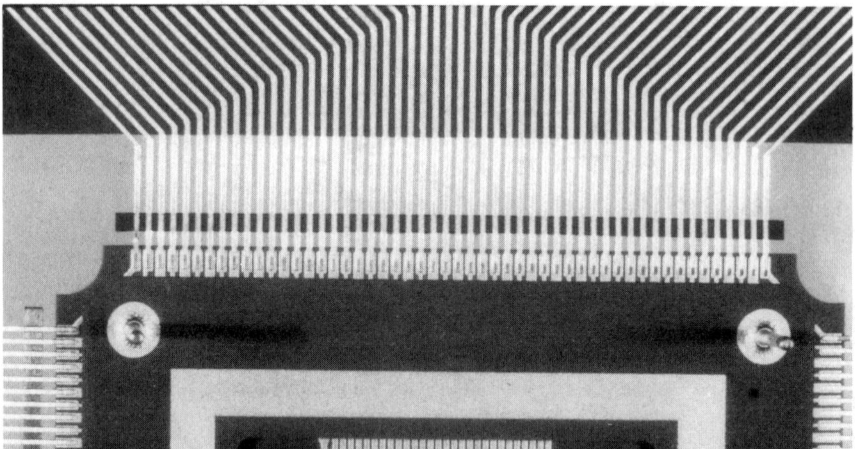

FIGURE 13.16. Photo of a fine pitch package properly soldered onto a printed circuit land area. (Photo courtesy of International Micro Industries)

TABLE 13.3. When the Lead Pitches Become Very Fine, the Land Width May Possibly be Narrower than the Packages Lead Width.

Pitch	Lead width	Trace space	Maximum Land width	Land to lead overlap
TSmin = 0.15mm (0.006")				
0.65	0.3	0.15	0.50	0.20
0.5	0.25	0.15	0.35	0.10
0.4	0.18	0.15	0.25	0.07
0.3	0.13	0.15	0.15	0.02
0.25	0.13	0.15	0.10	-0.03
0.2	0.13	0.15	0.05	-0.08
0.15	0.1	0.15	0.00	-0.10
TSmin=0.18mm (0.008")				
0.65	0.3	0.18	0.47	0.17
0.5	0.25	0.18	0.32	0.07
0.4	0.18	0.18	0.22	0.04
0.3	0.13	0.18	0.12	-0.01
0.25	0.13	0.18	0.07	-0.06
0.2	0.13	0.18	0.02	-0.11
0.15	0.1	0.18	-0.03	-0.13

The minimum land width should be equal to the maximum width of the foot of the lead, provided the land to land spacing is not less than the allowable board fabrication trace space distance. For example, in Figure 13.17, the nominal and maximum lead foot width is shown for a 0.3 mm pitch package. If the minimum trace distance allowed by the chosen fabricator is 0.15 mm (0.006″), then the land area can equal the foot area. However, if the minimum trace distance is 0.18 mm (0.008″) then the land width will be less than the foot width by 0.01 mm (0.0004″).

The smaller land width is allowable if:

1. The variation in placement and lead location does not cause the lead edge to the adjacent land to be less than the electrically safe distance
2. The wider leads do not cause solder bridges during reflow
3. The small distance does not allow dendrite growth during use.

The placement and lead variations are functions of the capability of the assembler's placement machine and the package supplier. The solder bridging is a function of the package lead pitch, the size of the paste particles, the stencil, and the length of the land. The size of the paste particles and the stencil are covered in Chapter 5.

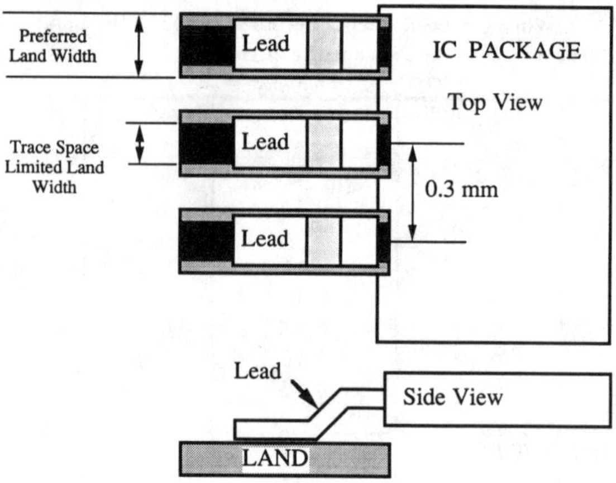

FIGURE 13.17. Land widths can vary from wider to narrower than the package lead as a result of the board fabricators capabilities and lead width tolerances.

If the land is longer than the foot of the lead, then it may allow some of the solder to flow away from the lead width. This can draw away solder paste that has bridged the gap between two leads. If this solder remains, it will short the leads. How much solder will be drawn away is a function of the land length, the stencil opening length and the solder's flow characteristics.

In practice, a land 1.6 to 2.5 times longer than the foot provides excellent solder debridging for lead pitches down to 0.3 mm, provided that the stencil opening length is equal to 50 − 60% of the foot length and that the paste be Type III and L1 activity level flux. This configuration is shown in Figure 13.18.

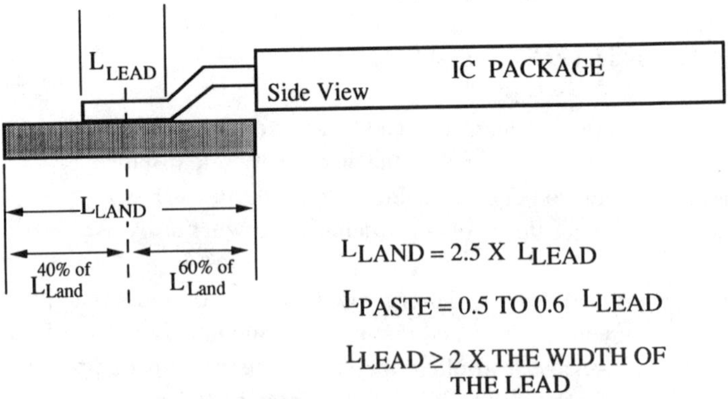

FIGURE 13.18. The recommended land length for fine pitch solder joints.

The land should be placed such that 60% of the land length is behind the center line of the lead foot. This allows solder to wick up the heel of the lead and increases the effective solder joint area, thereby increasing the solder joint resistance to fatigue failure. Solder wetting on the toe and sides of the lead are good indicators of solder wetting, but the significant mechanical connection is the area under the foot and heel of the lead.

To summarize, the area of the fine pitch land (see Figure 13.19) is calculated using the following equations:

Land Length, L = 2.5 × Lead foot length
Land Width, W = Maximum lead width or the lead center to center pitch minus the minimum allowable trace space distance (see Table 13.4).

For lead pitches below 0.3 mm the solder paste and trace space combine to form major limitations to assembly. To eliminate the solder paste limitation many assemblers have a solderable metal (such as tin-lead solder) fused onto the 0.3 mm and less pitch areas prior to assembly. Often the fabricator is willing to provide this service if specified in the fabrication instructions. To apply the metal the fabricator stencils a very fine particle solder paste and reflows it onto the selected areas.

Solder mask between the land areas is a customary feature to protect any traces routed between the lands, and to help discourage the formation of solder bridges. Soldermask between the lands serves an equally important purpose when applying solder paste. The solder mask supports the stencil and seals the area between the lands to prevent solder bridging, thus functioning much like a gasket in machine and engine parts. This is illustrated in Figure 13.20.

Some soldermasks, such as wet and dry film masks are either too variable or too thick to be dependable. For these masks, the areas between the pads should be left free of the mask material.

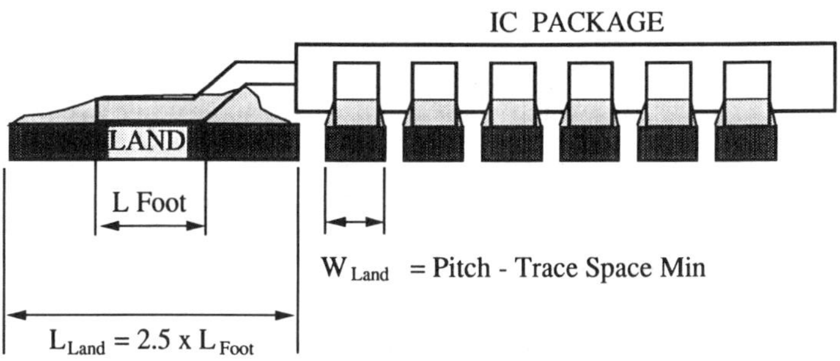

FIGURE 13.19. The land area equations for fine pitch applications. A land length of 1.6 × 2.4 × the foot length may yield equally well as the 2.5 × recommended, in all cases user tests are recommended.

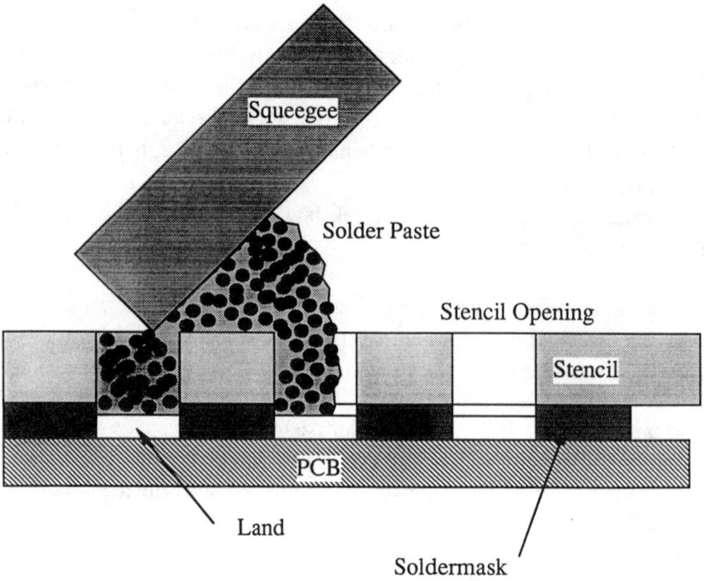

FIGURE 13.20. One benefit of solder mask is its ability to support the stencil during printing.

13.4 TRACE ROUTING AND PROTECTION

The traces used to interconnect the land areas and other components should be bare copper covered with an insulating soldermask. This board type is commonly called a SMOBC board, for solder mask over bare copper. The exposed land and via areas may be bare copper with a temporary protective mask or a fused solderable metal, such as a dipped and hot air leveled tin-lead solder coating.

Since the soldermask over the copper trace discourages solder wicking along the covered trace there are few restrictions on the trace routing. This is a change from the early surface mount design guidelines which recommended that traces intersect with lands only at 90° angles.

The prime consideration for routing to fine pitch lands is to try to keep the thermal mass as equal as possible. This provides relatively equal reflow time for each solder joint at both initial assembly and during rework. An example of equal and unequal thermal masses in shown in Figure 13.21. Often the vias that are tied to internal ground planes provide a high degree of heat sinking. The amount of heat sinking is enough to cause a thermal imbalance during reflow. To avoid this imbalance, it is helpful to use thermal relief pads, as shown in Figure 13.22.

Vias can be placed anywhere relative to the package as long as they are seperated from any land connections by a short trace. The purpose of this configuration, shown in Figure 13.23, is twofold. First, the via is separated from the land to prevent solder paste from wicking to and down the hole. Second, if during rework

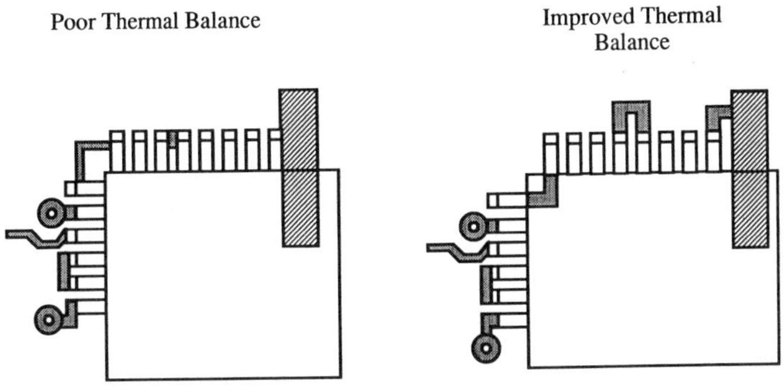

FIGURE 13.21. Unequal thermal sinks from the lands of fine pitch packages result in solder joint variations and rework problems. When possible, the thermal masses and resistance should be matched for each lead.

the land lifts from the board, it will break at the trace instead of inside the via, thus avoiding costly repair to the separated multilayers.

The length of the trace separator between the land and the via pad should be at least 0.4 mm (0.015″) from the edge of the land to the edge of the hole. The trace needs to be covered with soldermask.

Note the shape of the via pad in Figure 13.23. The pads are drawn with fillets on the trace entry to the pad. This addition prevents unacceptable breakout of the drill if there is misalignment to the small drill holes. The consequences of a pad with and without filleting is shown in Figure 13.24.

Covering the via with soldermask, called *tenting,* has received much debate in recent years as double-sided surface mount assemblies become more common. The benefits of tenting vias may out weigh the negatives that are shown in table 13.4 for many product applications. The tenting of vias is a technique to cover the via so it will not allow the passage of air when the board is being secured with a

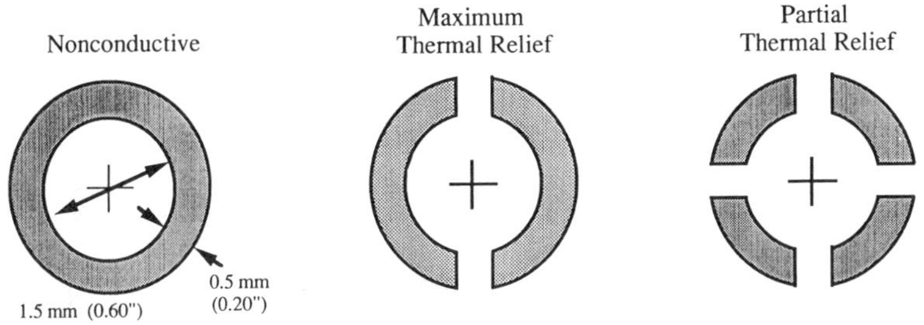

FIGURE 13.22. Via attachments to large copper planes can cause significant heat sinking and result in wide variations in solder melt times. The use of thermal relief designs increases the thermal resistance to the large copper areas and reduces the thermal sinking from solderjoint.

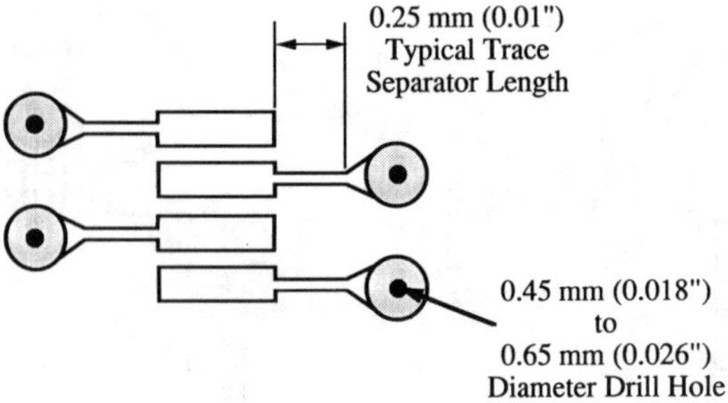

FIGURE 13.23. The preferred via to land connection, showing a fillet at the pad to trace connector. This fillet allows for a smaller pad without an increase in drill breakout.

vacuum hold down fixture. Vacuum hold down is common with in-circuit test systems and some stencil printer and placement systems. Previously, the vias would be filled with solder when the board was wave soldered, but double-sided SMT assemblies do not experience wave solder.

If the board will experience vacuum hold down and subsequent liquid cleaning, then the choice of soldermask material used to cover the via is critical. Some soldermask materials, especially certain photoimagable liquids, may only form a thin tent that can be punctured by the force of the vacuum. The concern with a

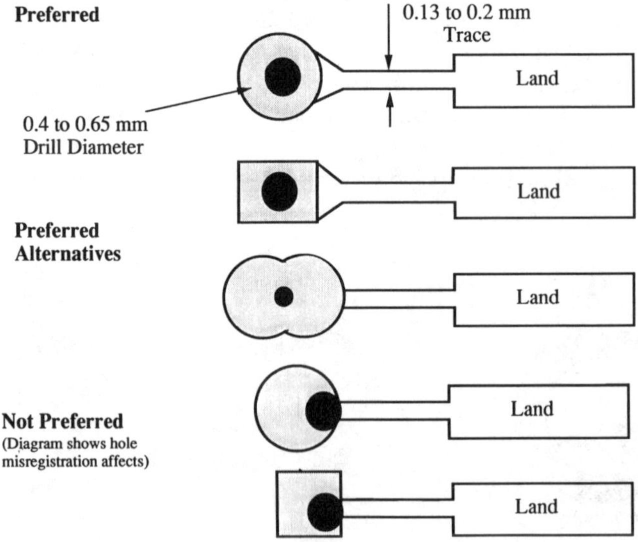

FIGURE 13.24. An Illustration of drill breakout due to excessively small and unfilleted pad sizes. (source: Electronic Packaging and Production, June, 1990)

TABLE 13.4 The Benefits and Negatives of Tented Via Holes.

Benefits of Tented Vias

- Minimizes loss of vacuum to hold the pcb during stencil printing and in-circuit electrical test.
- Minimizes solder wicking from lands down the via during reflow soldering.

Negatives of Tented Vias

- Shorter thermal cycle life compared to solder filled vias.
- Requires use of thick soldermask or a two step plug and mask process.
- Punctured or cracks in the soldermask tent may allow corrosive materials to attack the bare copper in the via hole.

punctured tent on a via is that it will trap liquids and corrosive residue. In actuality, untented and unfilled vias under low standoff large packages are just as prone to trap residues.

The final choice of tenting should be based on the product's end application, the flux and cleaning chemistries, and the electrical test method.

The recommended via for tenting is one with a drill diameter of about 0.5 mm (0.02″) and an aspect ratio of 4.0 or less for high reliability applications, and 6.0 or less for commercial and industrial applications. The aspect ratio is the via length divided by its diameter (Figure 13.25). If smaller diameter via holes are used then it is suggested that the designer request thicker copper plating, more ductile copper, or a laminate with a low z-axis coefficient of thermal expansion. High aspect ratio via holes are less tolerant of the z-axis expansion of the board and will break sooner than low aspect ratio vias.

13.5 THERMAL MANAGEMENT

For some severe situations, a heat sink and forced cool air is not sufficient to keep the integrated circuit or the solder joints cool enough to prevent component or solder joint failure. In these cases, the heat needs to be drawn away through the bottom of the package and the leads.

One simple method of drawing heat from the bottom of a package is to leave a large copper area under the package. This spreads the heat over the entire copper pad. It is also helpful to place several vias into the surface pad, connecting the pad

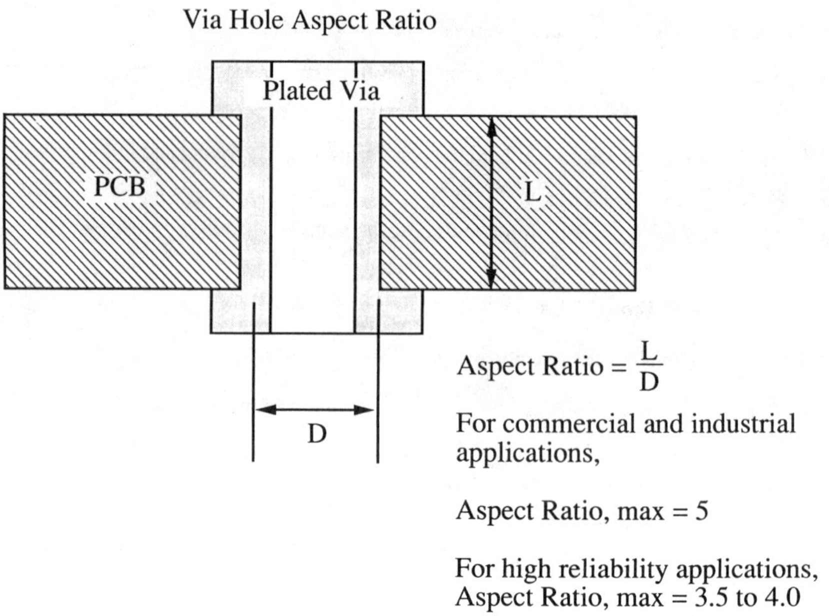

FIGURE 13.25. The aspect ratio of a via hole.

to the internal ground plane. The connecting vias from the heat sink pad to the ground plane should not have thermal relief cutouts. Do not connect the integrated circuit leads that require connection to ground, to the grounded heat sink pad unless a minimum width trace is utilized to provide some thermal resistance and reflow thermal balance with the other package interconnections. The heat sink pad is shown in Figure 13.26. Note that the pad should be free of soldermask to enable effective heat transfer. Since the area under the integrated circuit package is free of soldermask, it is necessary not to place any other vias under the package or, in the necessary cases, use tented vias. To maintain a high density design the traces will need to be routed under the package using blind and buried vias.

The use of thermal grease under the package is helpful, but, great care must be made in the selection and application of the thermal grease. If the grease gets on the lands it will all but destroy the solderability of those lands.

Thermal Vias

An additional method (also shown in Figure 13.26) that can be used to absorb the heat from the package is to place thermal vias under the package and in the lands. These vias should be completely plated with copper for optimum thermal conductivity and to prevent solder paste flow down the hole. The copper vias are very effective in conducting heat from the solder joints and the copper pad under the

FIGURE 13.26a. A photo of the conventional mentod of thermal management, that of gluing a heat sink onto a package. (Courtesy of the IPC.)

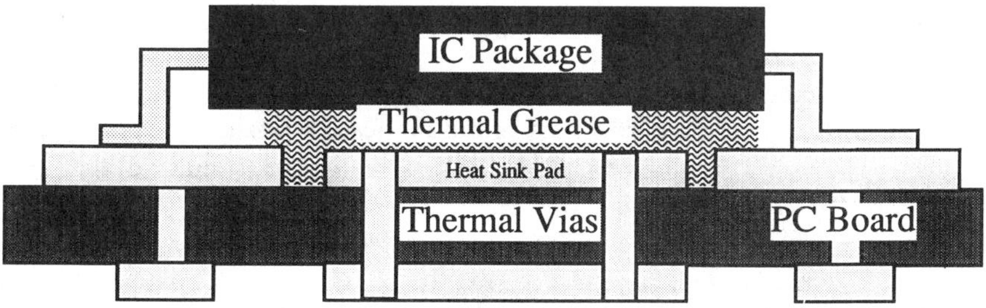

FIGURE 13.26b. An illustration of the use of a copper foil heat sink that is designed onto the board surface. The heat transfer from the package is improved if thermal grease is injected unlder the package.

package. A metal heat sink with selective insulation to prevent electrical shorting can be connected to the thermal vias to reduce the heat even further.

13.6 FABRICATION OF THE BOARD

The printed circuit board industry continues to adopt and change to meet the user's cost, quality, and technology demands. However, the output of all fabricators is only as good as the instruction provided. If the instructions to the fabricator are incorrect, in most cases so will the results (garbage in, garbage out). The fabrication of fine pitch circuit boards requires greater tolerance control of the etch artwork and the drill holes. On large boards, there is more of a need to be free of warpage, bow, and twist. And, the correct selection and application of a soldermask can make tremendous differences in solder yields.

Provide Correct Instructions

The first step a designer can take to improve the instructions to the fabricator is to provide correct and extensive fabrication notes with the fabrication package. These notes should include:

- The overall dimensions and tolerances allowable for the key features such as drill hole, trace, and soldermask sizes
- The copper plating thickness, and if necessary, the copper type and ductility
- The plating and leveling requirements on the exposed solderable areas, or the temporary coating if the board is to completely bare copper
- The laminate type and resin-to-laminate ratio if the dielectric coefficient is critical.
- The stacking (or layup) sequence of the various multilayers

Provide Electronic Data Artwork

Another important step is to provide the fabricator the etch artwork information in electronic form, i.e. the CAD data files, and not in photoplot form. Photoplots are acceptable elements of the fabricators build package, they are valuable as checkplots, to verify the correctness of the fabricators generated etch and drill artwork. The user's provided photoplots seldom match the accuracy of the fabricators, especially after exposure to different temperatures.

There are several electronic data transfer formats, with the Gerber format being the most common. Others include: VHSIC hardware description language (VHDL), the electronic design interchange format (EDIF), the initial graphics exchange definition (IGES), and the IPC-D-350 electronic data format.

These data transfer formats are capable of supplying data between some of the various levels of the design through assembly steps. These steps are lumped into the following three major categories; which are lumped together to form the CIM (computer integrated manufacturing) process:

1. CAE, computer aided engineering. This includes the tools for schematic creation, and simulation.
2. CAD, computer aided design. This tool performs the physical layout of the board, chassis and other mechanical features of the product.
3. CAM, computer aided manufacturing. This category includes the fabricator's phototool creation system, the test fixture system, the placement program generator and the visual inspection system.

It may also include the computer aided repair station (CAR) and the computer aided test system (CAT). Where these data formats fit into the CIM process is roughly illustrated in Figure 13.27.

Gerber format, also known as EIA-RS-274-D, gained it's acceptance as an electronic data format through its widespread use in photoplotters. The Gerber data format is adequate for photoplotting, but is very incomplete for other needs, such as placement and stenciling. The Gerber format (and its various derivatives such as PC Gerber) describes circuit geometry and basic features, but nothing more. The Gerber format is being uplifted to provide more information. This effort is known as Gerber plus or simply G1.

The G1 format effort attempts to bring the Gerber data transfer format closer to the capabilities of a format that is already structured to provide all of the information that a board fabricator, a stencil fabricator, and an assembler need. That format, the IPC-D-350, is designed to transmit, via modem or storage media (floppy-disk), all of the information required to define a circuit board, from plotting, through build. It also defines the bare board test vectors, the placement locations and some final test information. Figure 13.28 illustrates the various

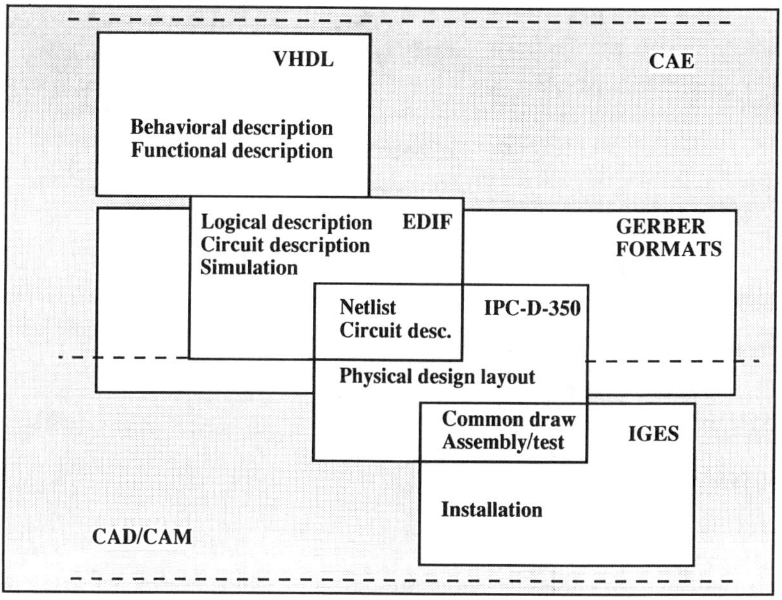

FIGURE 13.27. Several electronic data formats are used in the design, assembly, test and rework of assemblies. The approximate range of application of the major formats is illustrated here. (Source: IPC)

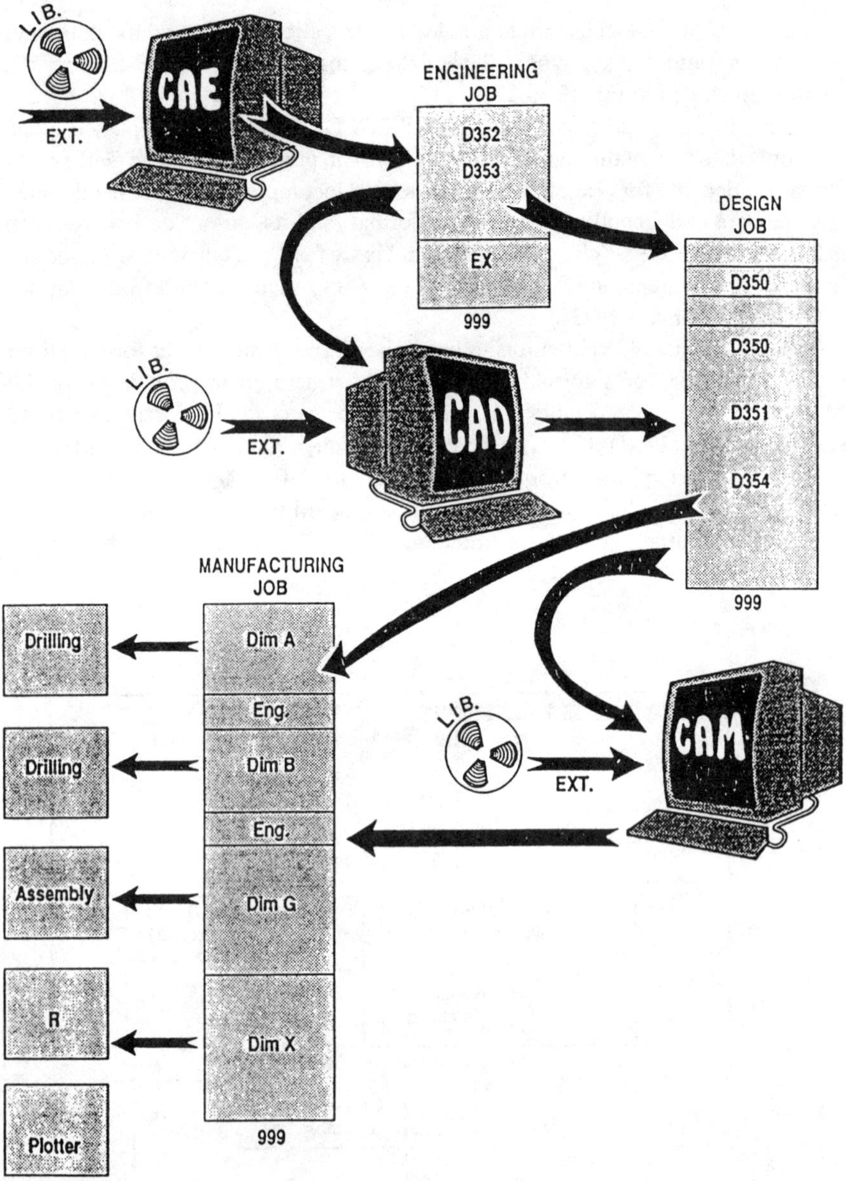

FIGURE 13.28. The various IPC-D-350 format levels and their application to the CAE, CAD, or CAM processes. (Source: IPC)

IPC-D-350 format levels, ranging from D350 through D354, and where they interface in the CAE to CAD to CAM process.

Configuring other systems to be compatible with IPC-D-350 systems is not without challenges. The task is much the same of organizing and using data files from personal computers working on different operating systems.

Figure 13.29 suggests how to set up a data transfer and archiving system. It suggests using the D350 file format as the core format file. Other formats, such as Gerber, IGES, and EDIF are converted to the D350 through man/machine manipulation. This means electronic data file translation of the data in the other formats through machine-to-machine connection, and additional information is manually inputted to build a complete D350 file from the incomplete other files.

Readers who desire knowing more about the D350 format are encouraged to request the IPC-DG-358 (to be renumbered as the IPC-DG-395 in late 1992)

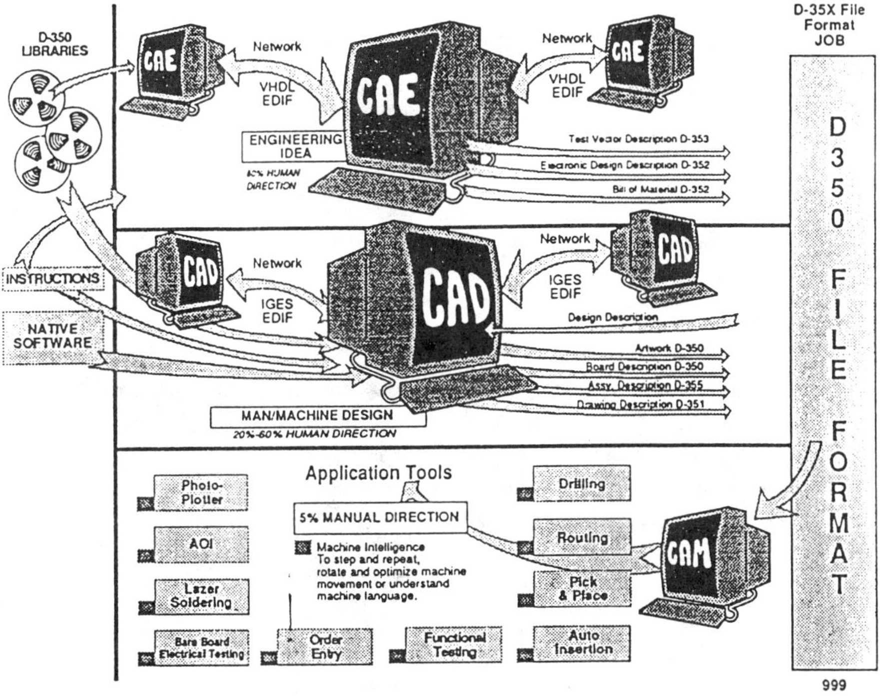

FIGURE 13.29. Data can be transferred and archieved into a D350 format via direct machine to machine transfer. However, missing data not normally supplied from the transferring data needs to be manually supplied.

document entitled a "Guide for Use of Digital Descriptions of Printed Boards, 'IPC-D-350 User's Guide' from the IPC, 7380 N. Lincoln Ave., Lincolnwood, IL 60646-1705.

Reducing Board Warpage

Board warpage, bow and twist is minimized by balancing the stacking of the various layers. An unbalanced and balanced multilayer stacking or layup is shown in Figure 13.30. To maintain layer balance, there should be an equal amount of copper thickness in both the top and bottom half of the board.

The required artwork layers for board fabrication and assembly are listed in Table 13.5. The outer silk screen layer is for board ID and component location markers. The adhesive layer is useful for programming adhesive dispensing machines. The required coordinate for adhesive dispensing and for part placement is the package centroid, which should be included in the CAD footprint as a consistant practice.

Specify the Correct Copper Weight

For trace widths and spaces less than 0.16 mm (0.007″) it is a good practice to specify 0.5 to 1.0 ounce copper weight. These copper weights correspond to a

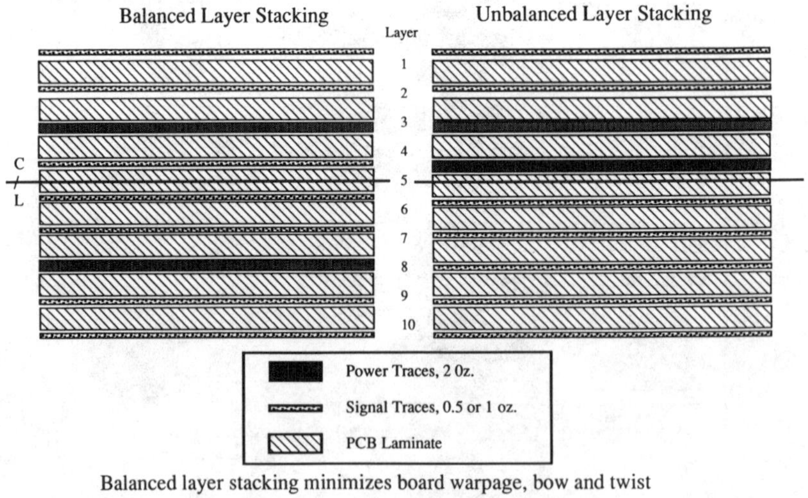

FIGURE 13.30. Board warpage is minimized by balancing the copper thicknesses equally on each side of the center of the board.

TABLE 13.5 The Artwork Layers Needed for Board Fabrication and Assembly.

	Layers Required
	Primary side traces or lands if a LEO board
	Primary side solder mask*
	Primary side silkscreen
X	Primary side paste master (all SMT/FPT lands)**
	All inner layer film
	Secondary side traces
	Secondary side solder Mask*
	Secondary side silkscreen
X	Secondary side paste master (all SMT/FPT lands if Reflow soldered)**
X	Secondary side adhesive master (if Wave soldering, size depends on adhesive used)
	* Solder mask is 1 : 1 if solder mask is photo-imageable or dry film process.
	Solder mask should be .010" oversize for wet film process.
	** Paste Master is 1 : 1 with the Land Area.
X	New layers necessitated by SMT

nominal thickness of 0.018 mm (0.0007″) and 0.035 mm (0.0014″), respectively. These thinner copper thicknesses are easier to etch for fine lines and spaces than the conventional 2 ounce copper.

If thicker copper is needed on the outer layers, e.g. due to current density, then specify the initial etch layer be the thinner copper, and after etch the final layer is to be plated up to 2 ounce copper. Generally, the fabricator will use a pattern plating process to provide the extra plating. This may result in unequal height traces and lands due to unequal plating rates. The unequal plating rates result when areas of low amounts of circuitry and small areas create higher plating

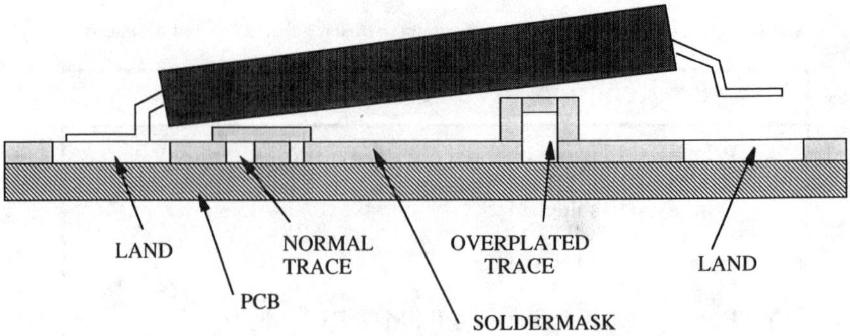

FIGURE 13.31. Pattern plating may result in unequal copper thicknesses as narrow features will plate thicker than wide features.

current densities than large copper areas. These plating differences can cause 2 or 3 times higher copper deposits on the small areas compared to the large areas. This effect is illustrated in Figure 13.31. It shows an overplated trace lifting an integrated circuit package off an under-plated land. The result is solder opens or weak solder joints.

The precautions to observe to avoid pattern plating nonuniformity are:

1. Discuss the problem with the fabricator and solicit design guidelines.
2. Equalize spacings between lines and traces.
3. Add internal “thieving” areas. These are nonfunctional copper areas left in low percentage copper and circuit areas, (see Figure 13.32).
4. Eliminate or minimize holes in ground plane areas which tend to overplate.

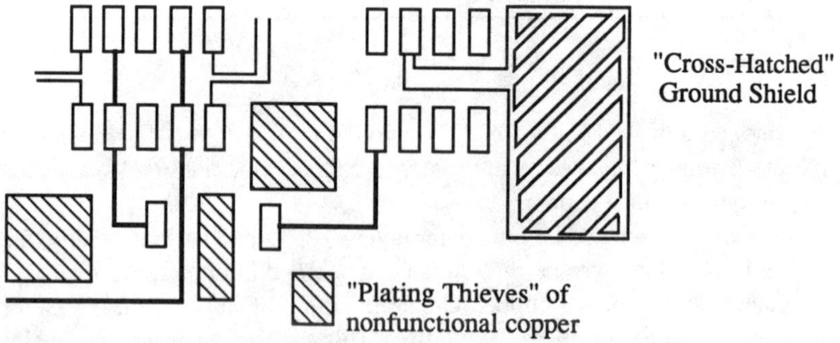

FIGURE 13.32. The unequal plating common to pattern plating is avoided by following some simple design rules.

13.7 STENCIL FABRICATION

As discussed in Chapter 5, the stenciling of solder paste into fine pitch land openings is a challenge. The artwork and fabrication instructions to the stencil maker need the same degree of concern as that needed for the printed circuit board fabricator. The stencil maker should receive etch artwork in electronic data form so that it can be resized to match the etch characteristics.

The stencils should be double-side etched and possibly laser trimmed to provide a straight, smooth was to assure easy and complete release of the solder paste from the narrow openings.

The paste opening is smaller than the land area to avoid excessive solder, shorts and bridging. Generally, the paste opening area is 50 to 60% of the land area for a stencil thickness is 0.20 mm (0.008″) to 0.15 mm (0.006″). Examples of stencil opening designs that are currently used by several assemblers are shown in Figure 13.33. The

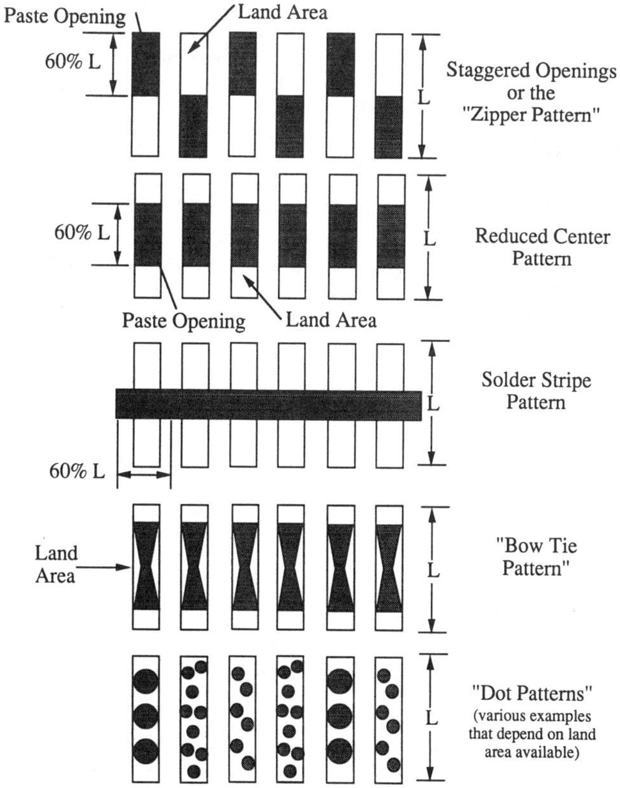

FIGURE 13.33. Examples of stencil openings that have been successfully used for printing solder paste on fine pitch land areas.

success of these designs is closely dependent on the solder paste type used, the squeegee shape and force, and the paste control during assembly.

References

1. Fjelstaed, J. 1991. Designing Manufacturable COB Circuits. In *Electronic Packaging and Production,* Feb. 1991.
2. Garrison, T. 1990. A Study of Mechanical Dimensional Commality for Fine Pitch Components. Proceedings of Nepcon West '90, Feb., 1990, Anaheim, CA.
3. Leppo, M. et. al. 1989. Substrates and Packages. Electronics Materials Handbook, Vol. 1. Materials. Park, OH: ASM International.
4. Marcoux, P., et al., 1989. An Introduction to TAB and Fine Pitch Technology SMC-TR-001. Lincolnwood, IL: Surface Mount Council.
5. Marcoux, P. 1989. SMT: Design for Manufacturability. Sunnyvale, CA: PPM Associates and Dearborn, MI: Society of Manufacturing Engineers.
6. Marcoux, P. 1992. PCA Design Guidelines Manual. Sunnyvale, CA: PPM Associates.
7. Prasad, R. 1989. SMT: Principles and Practices. New York: Van Nostrand Reinhold.
8. Scholten, L. 1991. Electronic Data Tooling Exchange. In *PC Fab,* Jan., 1991.
9. Solberg, V. 1990. Assembling Fine Pitch Devices on Fine Line PCBs Poses New Challenges. In *Electronic Packaging and Production,* Feb., 1990.
10. Vest, R. and Moore, R. 1990. Gearing Up for Fine Pitch Surface Mount Packages. In *Surface Mount Technology,* May, 1990.
11. —1991. Testability Guidelines, TP-101A. Edina, MN: Surface Mount Technology Association.
12. —1992. Electronic Design Data Exchange Guidelines, IPC-D-395 (formerly the IPC-D-350). Lincolnwood, IL: IPC.
13. —1992. Surface Mount Configurations and Land Patterns, IPC-SM-782A. Lincolnwood, IL: IPC.

Appendix A

Addresses of Standards Organizations

Following are the addresses of the IPC and EIA, as well as other sources for standards documents.

Electronic Industries Association (EIA)
2001 Pennsylvania Avenue, NW
Washington, DC 20006-1813

Institute for Interconnecting and Packaging Electronic Circuits (IPC)
7380 North Lincoln Avenue
Lincolnwood, IL 60646

Military documents are available from:
Standardization Documents
Order Desk, Building 4D
700 Robbins Avenue
Philadelphia, PA 19111-5094

Central office of the IEC:
International Electrotechnical Commission (IEC)
3 Rue de Varembe
1211 Geneva 20, Switzerland

IEC documents are available from:
American National Standards Institute (ANSI)
11 West 42nd Street
New York, NY 10036

Appendix B

Summary of SMT Semiconductor Outlines from JEDEC Publication 95

Generic Description	Registration Number
SOT-23	TO-236AA-AB
SOT-89	TO-243AA-AB
SOT-143	TO-253AA
DPAK	TO-252AA
SMT Header Family	TO-263AA-AB
MELF Diode	DO-213AA-AB
SOIC-3.75 mm Body	MS-012AA-AC (Standard)
SOIC-7.5 mm Body	MS-013AA-AF (Standard)
SOIC-11.2 mm (0.440") Body	MO-099AA-AB
SOP, Gullwing Leads	MO-117
SOJ-0.300" Body	MO-065AA-AB
SOJ-0.300" Body	MO-088AA-AF
SOJ-0.300" Body	MO-077AA-AC
SOJ-0.300" Body	
SOJ-0350" Body	MO-091AA-BA
SOJ-26/20-0.350" Body	MO-063
SOJ-32/28-0.400" Body	MO-061
SOJ-0.300" Body	MO-119
SOJ-0.350" Body	MO-120
SOJ-0.330" Body	MO-121
SOJ-12 mm Body	MO-123
SOJ-12.7 mm Body	MO-124
TSOJ-0.300" Body	MO-105
SSOP-0.300" Body	MO-118
PLCC-Square-0.050" Lead Spacing	MO-047AA-AH
PLCC-Rectangular-0.050" Lead Spacing	MO-016AA-AE (Standard)
PLCC-Rectangular-0.050" Lead Spacing	MO MO-052AA-AE
PLCC-Square, Ceramic-0.050" Lead Spacing, "J" Lead	MO-087
LCC-0.050" Lead Spacing	MS-002 thru MS-005 (Standard)
Leaded Socket 0.050" Lead Spacing	MS-009, MS-014 (Standard)
Plastic Quad Flat Pack-0.025" Lead Spacing	MS-069AA-AH
Plastic Quad Flat Pack-0.025" Lead Spacing, Bumpered, Thin Lead Family (Gullwing)	MO-069AA-AH
Plastic Quad Flat Pack-0.025" Lead Spacing, Bumpered, Thin Lead Family (Gullwing)	MO-071AA-BB

Generic Description	Registration Number
Plastic Quad Flat Pack-0.025" Lead Spacing, Bumpered, Low Profile (Gullwing)	MO-086AA-AH
Plastic Quad Flat Pack-0.050" Lead Spacing, Bumpered (Gullwing)	MO-089
TapePak/Molded Carrier Ring	MO-094AA-BD
TapePak/Molded Carrier Ring, Fine Pitch	MO-109
Tape Quad Flat Pack	MO-102AA-CD
Metric Quad Flat Pack (Body +3.2)	MO-108AA-FA
Metric Quad Flat Pack (Body + 3.9)	MO-112AA-FA
TAB - English Dimension	UO-017
TAB - Metric Dimension	UO-118
Ceramic Quad Flat Pack, 0.020" Lead Spacing, 256 Leads	MO-100AA
Ceramic Quad Flat Pack, 0.025" Lead Spacing	MO-082AA-AF
Ceramic Quad Flat Pack, 0.025" Lead Spacing, Guard Ring, 132 Leads	MO-104AA
Ceramic Quad Flat Pack, 0.015" Lead Spacing	MO-090AA-AF
Ceramic Quad Flat Pack, 0.050" Lead Spacing	MO-084AA-AF
Ceramic Leaded Chip Carrier	MO-107AA-AE
Ceramic Chip Carrier, "J" Lead. 0.050" Lead Spacing	MO-087AA-AE
Ceramic Round Lead, "J" Lead, 0.050" Lead Spacing	MO-110
Ceramic Round Lead, "Gull" Lead, 0.050" Lead Spacing	MO-111
Ceramic Quad Flat Pack, w/Tie Bar	MO-113
Cerquad Family w/Gullwing Leads	MO-114
Ceramic Quad Flat Pack, 132 Lead	MO-060
Ceramic Quad Flat Pack, 196 Lead	MO-125
6.35 mm Cerpak Leaded Flat Pack	MO-092AA-AD
Braze Lead Flat Pack	MO-098AA-AD
Top Brazed 48 Pin Flat Pack	MO-101AA-AB
Flat Pack Family. 0.535" Long, 0.303" Pitch	MO-106AA-AC
Flat Pack, 32 Lead	MO-115
Lateral (Leadless) Ceramic Chip Carrier, 0.025" Spacing	MO-056AA-HC
Lateral (Leadless) Ceramic Chip Carrier, 0.020" Spacing	MO-057AA-JC

Appendix C

Summary of Important Component, Material, Process and Design Standards

The following documents center on surface mount technology. These documents have been developed by standards organizations in the U.S. and internationally. The letters of each document indicate the organization that has responsibility for the document:

- EIA represents documents prepared by the Electronic Industries Association
- JEDEC represents documents of the Joint Electron Devices Engineering Council of the EIA
- IPC represents documents prepared by the Institute for Interconnecting and Packaging Electronic Circuits
- MIL represents documents prepared by the Military
- DoD represents documents prepared by the Department of Defense

Components, General

EIA-481-A	Taping of Surface Mount Components for Automatic Placement
EIA-481-1	8 mm and 2 mm Taping of Surface Mount Components for Automatic Handling
EIA-481-2	16 mm and 24 mm Embossed Carrier Taping of Surface Mount Components for Automated Handling
EIA-481-3	32 mm, 44 mm, and 56 mm Embossed Carrier Taping of Surface Components for Automated Handling
EIA/IS-47	Contact Termination Finish Standard for Surface Mount Devices
EIA-PDP-100	Registered and Standard Mechanical Outlines for Electronic Parts
EIA-JEP-95	JEDEC Registered and Standard Mechanical Outlines for Semiconductor Devices

Components, Passive
Capacitors

EIA-469-B	Standard Test Method for Destructive Physical Analysis of High Reliability Ceramic Monolithic Capacitors
EIA-CB-11	Guidelines for the Surface Mounting of Multilayer Ceramic Chip Capacitors
EIA/IS-28	Fixed Tantalum Chip Capacitor Style 1 Protected Standard Capacitance Range
EIA/IS-29	Fixed Tantalum Chip Capacitor Style 1 Protected Extended Capacitance Range
EIA/IS-36	Chip Capacitors, Multi-Layer (Ceramic Dielectric)
EIA/IS-37	Multiple Layer High Voltage Capacitors (Radial Lead Chip Capacitors)
IEC-384-3	Sectional Specification, Fixed Multilayer Ceramic Chip Capacitors
IEC-384-10	Sectional Specification, Fixed Multilayer Ceramic Chip Capacitors IECQ Draft Blank Detail Specification, Fixed Multilayer Ceramic Chip Capacitors
IECQ-PQC-31	Sectional Specification, Fixed Tantalum Chip Capacitors with Solid Electrolyte
IECQ-PQC-32	Blank Detail Specification, Fixed Tantalum Chip Capacitor

Resistors

EIA-575	Resistors, Rectangular, Surface Mount, General Purpose
EIA-576	Resistors, Rectangular, Surface Mount, Precision
EIA/IS-34	Leaded Surface Mount Resistor Networks Fixed Film

Components, Active

EIA-JEP-95	JEDEC Registered and Standard Outlines for Semiconductor Devices
EIA-JESD21-C	Configurations for Solid State Memories
EIA-JESD22-B	Test Methods and Procedures for Solid State Devices Used in Transportation/Automotive Applications
EIA-JESD-11	Chip Carrier Pinouts Standardized for CMOS 4000, HC, and HCT Series of Logic Circuits

Components, Electromechanical
Connectors

EIA-506	Dimensional and Functional Characteristics Defining Sockets for Leadless Type A Chip Carriers (.050 Spacing)
EIA-507	Dimensional Characteristics Defining Edge Clips for Use with Dec 1983 Hybrid and Chip Carriers
EIA/IS-47	Contact Termination Finish Standard for Surface Mount Devices

Switches

IECQ-PQC-41 -US00003	Detail Specification, Dual-in-Line Switch, Surface Mountable, Slide Actuated
EIA-448-23	Surface Mountable Switches, Qualification Test
EIA-520EAAAA	Detail Specification for Surface Mountable Dual In-Line Switches of Certified Quality

Printed Boards

IPC-FC-250	Performance Specification for Single and Double-sided Flexible Printed Boards
IPC-RB-276	Performance Specification for Rigid Printed Boards
IPC-SD-320	Performance Specification for Rigid Single and Double-sided Printed Boards
IPC-ML-950	Performance Specification for Multilayer Printed Boards
MIL-P-50884	Military Specification Printed Wiring, Flexible, and Rigid Flex
MIL-P-55110	Military Specification Printed Wiring Boards, General Specification For

Materials

IPC-L-108	Specification for Thin Laminate Metal Clad primarily for High Temperature Multilayer Printed Boards
IPC-L-109	Specification for Glass Cloth, Resin Preimpregnated (B Stage) for High Temperature Multilayer Printed Boards
IPC-L-115	Specification for Plastic Sheet Laminated Metal Clad for High Temperature Performance Printed Boards
IPC-CF-148	Resin Coated Metal for Multilayer Printed Boards
IPC-CF-150	Copper Foil for Printed Wiring Applications
IPC-CF-152	Metallic Foil Specification for Copper/Invar/Copper (CIC) for Printed Wiring and Other Related Applications
IPC-SM-817	General Requirements for SMT Adhesives
IPC-SF-818	General Requirements for Electronic Soldering Fluxes
IPC-SP-819	General Requirements for Electronic Grade Solder Paste
IPC-CC-830	Qualification and performance of Electrical Insulation Compounds for Printed Board Assemblies
IPC-SM-840	Qualification and Performance of Permanent Polymer Coating (Solder Mask) for Printed Boards
MIL-F-14256	Flux, Soldering, Liquid (Rosin Base)

Interconnecting Substrates

IPC-RF-245	Performance Specification for Rigid-Flex Multilayer Printed Boards
IPC-MC-324	Performance Specification for Metal Core Boards
IPC-HM-860	Performance Specification for Hybrid Multilayer

Design Activities

IPC-T-50	Terms and Definitions for Electronic Interconnections
IPC-CM-78	Surface Mount and Interconnecting Chip Carrier Guidelines
IPC-H-855	Hybrid Microcircuit Design Guide
IPC-D-249	Design Standard for Flexible Single and Double-sided Printed Boards
IPC-D-317	Design Standard for Electronic Packaging Utilizing High Speed Techniques
IPC-D-319	Design Standards for Rigid Single and Double-sided Printed Boards
IPC-SM-782	Surface Mount Land Patterns (Configuration and Design Rules)
IPC-D-859	Design Standard for Multilayer Hybrid Circuits
IPC-D-949	Design Standard for Rigid Multilayer Printed Boards
IPD-D-275	Design Standard for Rigid Printed Boards and Rigid Printed Board Assemblies
MIL-STD-275	Military Standard Printed Wiring for Electronic Equipment
MIL-STD-2118	Design Standard for Flexible Printed Wiring

Component Mounting

EIA-CB-11	Guidelines for the Surface Mounting of Multilayer Ceramic Chip Capacitors
IPC-CM-770	Guidelines for Printed Board Component Mounting
IPC-SM-784	Guidelines for Direct Chip Attachment
SMC-TR-001	An Introduction to Tape Automated Bonding and Fine pitch Technology

Attachment Techniques

IPC-SM-780	Electronic Component Packaging and Interconnection with Emphasis on Surface Mounting

Soldering and Solderability

EIA/IS-46	Test Procedure for Resistance to Soldering (Vapor Phase Technique) for Surface Mount Devices
EIA/IS-49-A	Solderability Test Method for Leads and Terminations
EIA-448-19	Method 19 Test Standard for Electromechanical Components Environmental Effects of Machine Soldering Using a Vapor Phase System
IPC-TR-462	Solderability Evaluation of Printed Boards with Protective Coatings Over Long-term Storage
IPC-TR-464	Accelerated Aging for Solderability Evaluations
IPC-S-804	Solderability Test Method for Printed Wiring Boards (new revision ANSI/J-STD-003)
IPC-S-805	Solderability Test for Component Leads and Terminations (new revision ANSI/J-STD-002)

IPC-S-815	General Requirements for Soldering Electronic Interconnections (new revision ANSI/J-STD-001)
IPC-S-816	Troubleshooting for Surface Mount Soldering
IPC-AJ-820	Assembly and Joining Handbook
MIL-STD-2000A	Standard Requirements for Soldered Electrical and Electronic Assemblies

Quality Assessment

EIA-469-B	Standard Test Method for Destructive Physical Analysis of High Reliability Ceramic Monolithic Capacitors
EIA-510	Standard Test Method for Destructive Physical Analysis of Industrial Grade Ceramic Monolithic Capacitors
IPC-A-600	Acceptability of Printed Boards
IPC-A-610	Acceptability of Printed Board Assemblies
MIL-STD-883	Methods and Procedures for Microelectronics

Reliability

IPC-A-38	Fine Line Round Robin Test Pattern
IPC-A-48	Surface Mount Artwork
IPC-SC-60	Post Solder Solvent Cleaning Handbook
IPC-AC-62	Post Solder Aqueous Cleaning Handbook
IPC-AI-640	User Requirements for Automatic Inspection of Unpopulated Thick Film Hybrid Substrates
IPC-AI-641	User Guidelines for Automated Solder Joint Inspection Systems
IPC-AI-642	User Guidelines for Automated Inspection of Artwork and Innerlayers
IPC-AI-643	User Guidelines for Automatic Optical Inspection of Populated packaging and Interconnection Assemblies
IPC-SM-785	Guidelines for Accelerated Surface Mount Attachment Reliability Testing

Test Methods

EIA-JEDEC	Method B 105-A, Lead Integrity Plastic Leaded Chip Carrier (PLCC) Packages
EIA-JEDEC	Method B 102, Surface Mount Solderability Test (JESD22-B)
EIA-JEDEC	Method B 108, Coplanarity (intended for inclusion into JESD 22-C)
IPC-TM-650	Test Methods Manual

Repair

IPC-R-700	Guidelines for Repair and Modification of Printed Board Assemblies

Numerical Control Standards

IPC-NC-349	Computer Numerical Formatting for Drilling and Routing Equipment
IPC-D-350	Printed Board Description in Digital FormIPC-D-351 Printed Wiring Documentation in Digital Form

IPC-D-352	Electronic Design Database Description for Printed Boards
IPC-D-354	Library Format Description for Printed Boards in Digital Form
EIA-224-B	Character Code for Numerical Machine Control Perforated Tape
EIA-227-A	One-Inch Perforated Tape
EIA-267-B	Axis and Motion Nomenclature for Numerically Controlled Machines
EIA-274-D	Interchangeable Variable Block Data Format for Positioning, Contouring, and Contouring/Positioning Numerically Controlled Machines
EIA-281-B	Electrical and Construction Standards for Numerical Machine Control
EIA-358-B	Subset of American National Standard Code for Information Interchange for Numerical Machine Control Perforated Tape
EIA-408	Interface Between Numerical Control Equipment on Data Terminal Equipment Employing Parallel Binary Data Interchange
EIA-431	Electrical Interface Between Numerical Control and Machine Tools
EIA-441	Operator Interface Function of Numerical Controls
EIA-474	Flexible Disk Format for Numerical Control Equipment Information Interchange
EIA-484	Electrical and Mechanical Interface Characteristics and Line Control Protocol Using Communication Control Characters for Serial Data Link Between a Direct Numerical Control System and Numerical Control Equipment Employing Asynchronous Full Duplex Transmission
EIA-494	32 BIT Binary CL Exchange (BCL) Input Format for Numerically Controlled Machines

Glossary

Adhesive Materials used to hold components in place during wave or reflow soldering which may become a permanent part of the assembly, or be subsequently removed.

Blow Hole A void in a solder connection caused by outgassing.

Castellation Metallized features that are recessed on the edges of a chip carrier which are used to interconnect conducting surfaces or planes within or on the chip carrier.

Chip Carrier A low-profile rectangular component package, usually square, whose semiconductor chip cavity or mounting area is a large fraction of the package size and whose external connections are usually on all four sides of the package.

Chip Component Besides ICs, the term includes diodes, inductors, resistors, and capacitors. A "1206" notation specifies the size of the device (0.120″ × 0.060″), with other standard notations being 0805, 1210, 1812, etc.

(CLCC) Ceramic Leaded Chip Carrier A ceramic chip carrier whose external connections consist of leads around and down the sides of the package.

Cold Solder Joint A solder connection exhibiting poor wetting and a grayish, porous appearance due to insufficient heat, inadequate cleaning prior to soldering, or to excessive impurities in the solder.

Component An individual part or combination of parts that, when together, perform a design function(s).

Component Lead The solid or stranded wire or formed conductor that extends from a component and serves asa mechanical or electrical connection or both that is readily formable to a desired configuration.

Component Mounting Site A location on a packaging and interconnection structure that consists of a land pattern and conductor fanout to additional lands/pads for testing or vias that are associated with the mounting of a single component.

Constraining Core A supporting plane that is internal to a packaging and interconnecting structure.

Contact Angle The angle enclosed within the solder fillet, between a plane tangent to the solder/base-metal surface and a plane tangent to the solder/air interface.

Coplanarity Term used to describe the relationship of component leads to each other across the horizontal plane

(CTE or TCE) Coefficient Of Thermal Expansion The linear thermal expansion of a material per unit change as a result of temperature change.

Dewetting A condition which results when molten solder has coated a surface and then receded leaving irregularly shaped mounds of solder separated by areas covered with a thin solder film: Basis metal may or may not be exposed.
Double-Sided Assembly A packaging and interconnecting structure with components mounted on both the primary and secondary sides.

Fiducial An "etched" pattern on the printed circuit board used as an optical point of reference for measurement or calculation.
Fillet The configuration of solder around a component lead and land. A blending or rounding of interconnecting conductors or leads which eliminates sharp corners.
Flat Pack A component with two straight rows of leads, normally on 0.050″ centers, which are parallel to the component body.
Flux A chemically/physically active formation that is capable of enabling and promoting the wetting of metals with solder.
Flux Solder Connection A solder joint characterized by entrapped flux that often causes high electrical resistance.
Footprint(See preferred term LAND PATTERN).
(FPT) Fine Pitch Technology The term used to describe the assembly technology for those IC packages having lead spacings of generally 0.65 mm (0.0256″)or less.
Fine Pitch IC Component Component packages with lead spacings of 0.65 mm (0.0256″) or less center to center.

(IC) Integrated Circuit An assembly of miniature electronic components simultaneously produced in batch processing, on or within a single substrate to perform an electronic circuit function.
Infrared Radiation The band of electromagnetic wavelengths existing between the extreme of the visible and the shortest microwaves. The strong absorption of infrared by many substances renders it a useful means of applying heat energy.
Infrared Reflow Soldering A reflow soldering furnace using infrared heating as the primary source of heat introduction in the furnace environment.
Ionizable/Ionic Contaminants Process residues such as flux activators, fingerprints, etching and plating salts, etc., that exist as ions or when dissolved, increase electrical conductivity.

Land A portion of a conductive pattern usually, but not exclusively, used for the connection, or attachment, or both of components.
Land Pattern A combination of lands intended for the mounting and interconnection of a particular component.
(LCC) Leadless Chip Carrier A chip carrier whose external connections consist of metalized terminations.

(MELF) Metal Electrode Face A tubular shaped component with metalized terminations for surface mounting.

Migration An undesirable phenomenon whereby metal ions, notably silver, are transmitted through another metal in the molten state, or across an insulated surface, in the presence of moisture and an electrical potential.

Mixed Mounting Technology A component mounting technology that uses both through-hole and surface mounting technologies on the same packaging and interconnecting structure.

Nonwetting A condition whereby a surface has contacted molten solder but the solder has not adhered to all of the surface: Base metal remains exposed.

(P/I Structure) Packaging And Interconnecting Structure The generic term for a completely processed combination of substrates, metal planes, or constraining cores and interconnection wiring used for the purpose of mounting components.

Paste Flux A flux formulated in the form of a paste for special application, not to be confused with a solder paste or solder-paste flux.

(PLCC or PCC) Plastic Leaded Chip Carrier A plastic chip carrier whose external connections consist of leads around and down the sides of the package.

Primary Side The side of the packaging and interconnecting structure equivalent to layer #1 (the same as the "component side" when using through-hole component mounting technology).

(PCA or PWA) Printed Circuit or Wire Assemblyi.e. a PCB containing attached components and interconnecting traces.

(PCB) Printed Circuit Board a board containing interconnecting traces.

Reworking The act of repeating one or more manufacturing operations for the purpose of improving the yield of acceptable parts

Secondary Side That side of the packaging and interconnecting structure furthest from layer#1 (the same as the "solder side" when using through-hole component mounting technology).

Shadowing Occurs during wave soldering when the trailing termination of a component receives a smaller amount of solder than the leading termination due to the component body preventing the solder from flowing properly to the trailing termination. The body of the component may also cause "shadowing" of the terminations of another component.

Silk Screen A screen of a closely woven silk mesh stretched over a frame and used to hold an emulsion outlining a circuit pattern. Silk screens are used in screen printing solder paste. The term is used generically to describe any screen (stainless steel or nylon) used for screen printing.

(SIP) Single In-Line Package A through-hole component which terminates in one straight row of pins and lead wires.

Single-Sided Assembly A packaging and interconnecting structure with components mounted only on the primary side.

(SMA) Surface Mount Assembly Trademark by AWI. A circuit assembly being primarily (60%) or wholly constructed using surface mount components.

(SMC)Surface Mounted Component A component designed to be mounted and soldered to lands on the surface of a packaging and interconnecting structure rather than inserted into through-holes in the structure.

(SMD)Surface Mount Device Trademark by Philips. The term is similar to SMC, surface mounted component.

(SMT) Surface Mount Technology The mounting process of components on planar surface.

(SO)Small Outline Similar to miniature dual in-line package, but with gull wing lead forms for surface mounting. Typical lead spacing:.050″.

(SOB) Small Outline Bridge A small outline full wave bridge similar in form to a SOT-143, but with greater body height.

Solder A metallic alloy which has a melting temperature below 427°C (800°F).

Solder Balls Small spheres of solder adhering to laminate, mask or conductor surfaces.

Solder Bridging Unwanted formation of a conductive path by solder between conductors.

Solder Connection An electrical/mechanical connection which employs solder for the joining of two or more metal parts.

Solder Connection, Disturbed A cold solder joint resulting from motion between the joined members when the solder was solidifying.

Solder Fillet A blended or meniscoid (rounded) configuration of solder around a component or wire lead and land.

Solder Flux A chemically active formulation capable of promoting the wetting of metals with solder.

Solder Paste (Cream) Mesh size controlled solder particles (spherical or nearly spherical) combined with a solder-paste-flux consisting of flux constituents to control viscosity, thickness, slumping, drying rate, etc.

Solder Projection An undesirable protrusion of solder from a solidified solder joint or coating.

Solder Webbing A continuous film or curtain of solder parallel to, but not necessarily adhering to, surface or between separate conductive patterns that should be free of solder.

Solderability The ability ofametal to be wetted by molten solder.

Soldering A process of joining metallic surfaces with solder, without the melting of the base material.

(SOT) Small Outline Transistor A surface mounted transistor package available in several forms. SOT-23, a three leaded device. SOT-89, a three leaded medium current device. SOT-143, a four leaded device for gated transistors and dual diodes.

Static Electricity An electric charge that has accumulated or built up on the surface of a material. The material may be conductive or nonconductive. In conductors, the charge can pass through the material. In nonconductors, the charge cannot pass and is in effect, locked in place, hence the term static electricity.

Stencil A thin sheet material (metal) with a circuit pattern cut into it. Used for printing solder paste on a screen printer in lieu of a silk screen.

Stress Relief The formed portion of a component lead or wire lead, providing sufficient lead length to minimize stress between terminations.

Supporting Plane A planar structure that is a part of a packaging and interconnecting structure to provide mechanical support, thermal-mechanical constraint, thermal conduction, and/or electrical characteristics. It may be either internal or external to the packaging and interconnecting structure.

Surface Mounting The electrical connection of components to the surface of a conductive pattern that does not utilize component lead holes.

Surface Mount IC Component Component packages having lead spacings from 0.8 mm (0.03″) to 1.3 mm (0.05″) center to center.

Terminal A metallic termination device used for making electrical connections.

Thermal Expansion Mismatch The absolute difference in thermal expansion of two materials.

Tinning A process for the application of solder coating on component leads, conductors and terminals to enhance solderability.

"Tombstones" The term describing component shift, by standing up vertically on one termination during the reflow process.

Vapor Phase Soldering Soldering accomplished by primarily conductive heat transfer during condensation of a high-temperature vapor of boiling fluorocarbon. The latent heat of vaporization, given up during condensation of vapor on the relatively cool workpiece is the major source of heat for conductive transfer.

Via A plated-through-hole used as a through connection between two or more layers of a packaging and interconnecting structure in which there is nointention to insert a component lead or other reinforcing material.

Blind Via A via that is connected to either the primary or secondary side of a multilayer interconnecting structure but not both sides.

Hidden Via A via located under a component.

Tented Via A blind or thru-via that has the exposed surface on the primary, or secondary, or both sides of a packaging and interconnecting structure fully covered by a masking material, such as a dry film polymer covering (solder mask), pre-impregnated glass cloth (prepreg), etc., in order to prevent hole access by process solutions, solder, or contamination.

Thru-Via A via that is connected between the primary and secondary side of a double sided or multilayer packaging and interconnecting structure.

Visual Examination The qualitative observation of physical characteristics, utilizing the unaided eye or within stipulated levels of magnification.

Void The absence of substances in a localized area.

Wetting The formation of a relatively uniform, smooth, unbroken and adherent film of solder to a basis metal.

Whiskers Slender acicular (needle-shaped) metallic growth between conductors and lands.

Index